Safety and Health in Industry

A Handbook

Third Edition

Safety and Health in Industry

A Handbook

Third Edition

A M Sarma
Former Member of the Faculty
Tata Institute of Social Sciences
Mumbai

BS Publications
An imprint of **BSP Books Pvt. Ltd.**
4-4-309/316, Giriraj Lane,
Sultan Bazar, Hyderabad - 500 095.

Safety and Health in Industry A Handbook, Third Edition ***by*** **A M Sarma**

Published by:

BSP BS Publications
An imprint of **BSP Books Pvt. Ltd.**
4-4-309/316, Giriraj Lane, Sultan Bazar,
Hyderabad - 500 095.
Phone: 040 - 23445688
e-mail: info@bspbooks.net
website: www.bspbooks.net

ISBN: 978-93-95038-79-9 (Hardback)

Preface to Third Edition

The third revised edition of the book is an on-going contribution to the available literature on safety and health in India. It is dedicated to all those who are engaged in the onerous task of health, safety and welfare of working class throughout the country.

All the chapters have been expanded and streamlined to make the text up-to-date and useful to the readers. For the most part, the new edition is designed to increase the utility of the book by adding new material in almost all the chapters.

The book contains altogether 15 chapters and 10 appendices. A new chapter Organisational Role on Safety and Health has been added. Appendix A and B are the two new additions.

I hope that this revised edition will continue to be of assistance to students, teachers and professionals.

I am indebted to the BS Publications for bringing out the revised edition of this book in time.

A M Sarma
arsola1939@hotmail.com

Preface to First Edition

Occupational accidents and diseases remain the most appalling human tragedy of modern industry and one of its most serious forms of economic and social waste. An accident is a phenomenon of diverse and multiple ethiology and often is the result of an occurrence, or series of occurrences, chronologically remote from the accident itself. These characteristics make the search for causative factors of accidents costly, difficult, and time consuming. In some industrialised countries, industrial accidents are responsible for the loss of four or five times as many working days as industrial disputes.

The economic burden of industrial accidents on the community cannot be expressed in terms of compensation alone. It also includes loss of productivity, disruption of production schedules, damage to machinery and equipment, and in the case of large scale accidents major social dislocations.

Originally, the main thrust of accident prevention was to improve the unhealthy working conditions. The first international standards on health and safety were designed either to do away with the more flagrant abuses impairing health, such as the employment of very young children, long hours of work, the absence of any type of maternity protection, prevalence of night work by women and children, and to combat the risks of occupational diseases. Higher health safety standards are a primary responsibility of enlightened social policy and efficient management.

There has been a steady development in safety and health legislation aimed at improving the lot of those who work in mills, factories, mines, plantations, and in offices. Developments in the field of safety and health is a characteristic feature throughout the world, indicating an increasing concern for the quality of working life.

A major thrust of this book has been to provide an authoritative, up-to-date guide in the areas of industrial safety and health. The book is intended for students, practitioners, and all others who are responsible in one way or other for employees' health and safety. I am very happy to place the book before the readers and I do hope that it will measure up to their expectation.

I owe grateful thanks to many authors on the same subject, professionals on the field, and to my family members and friends for the help and encouragement they have given me in writing this book.

I am also indebted to Mr. Nikhil Shah and Mr. Anil Shah of B.S Publications for bringing out this book in time.

Any criticism and suggestions from the readers for further improvement of the book will be highly appreciated.

A M Sarma
arsola1939@hotmail.com

Contents

Preface to Third Edition (v)

Preface to First Edition (vii)

1. Workplace Safety and Health – An Overview **1-20**

- Safety Management as a Profession 2
- Traditional Safety Management 3
- Accident Frequency and Severity Rates 4
- Safety Analysis 4
- Safety Culture 5
- Safety Policy Statements 6
- Systems Safety Engineering 7
- Occupational Hazards 8
- Systems Approach to Safety and Health 9
- Process Safety Management (PSM) 11
- Safety Whose Responsibility 12
- Emergency Preparedness 15
- Health Promotion 15
- Case 18
- Summary 19
- Discussion Questions 19

2. Accidents and their Prevention **21-66**

- Anatomy of an Accident 22
- Causes of Accidents 26
- Theories on Accident Causation 28
- Approaches to Reducing Accidents 29
- Classification of Accidents 32
- Costs of Accidents 34
- Accident Prevention 38
- Safe Work Environment 41
- Job Safety Analysis 47
- Investigation of Accidents 49
- Ergonomics 51

- Personal Protective Equipment (PPE) 53
- Risk Assessment 56
- Risk Perception 59
- Promotion of Health and Safety 59
- Basic Safety Programming 61
- Goals for the Future 62
- Case 63
- Summary 63
- Discussion Questions 65

3. DEVELOPMENT OF INDUSTRIAL HEALTH AND SAFETY 67-80

- Developments in Occupational Health 69
- Occupational Safety and Health in Britain 72
- Occupational Safety and Health in the USA 74
- Occupational Safety and Health in India 77
- Summary 80
- Discussion Questions 80

4. SAFETY AND HEALTH ORGANISATION 81-134

- Safety Organisation 81
- Conventional Safety Programming 82
- Safety Planning 83
- Safety Policies 84
- Safety Culture and Climate 89
- Safety Standards 90
- Safety Control 91
- Safety Promotion 91
- Safety Committee(s) 93
- Safety Specialists 95
- Safety Department 98
- Measuring Safety 101
- Safety Supervision 102
- Safety Responsibility 104
- Safety Inspection 106

- Safety Tours 108
- Safety Rules 109
- Safety Education and Training 110
- Safety Communication 114
- Safety Appraisal and Evaluation 116
- Safety Audit 120
- Health Programme 121
- Health Services 122
- Health Education 123
- Industrial Medical Department 124
- Mental Health in Industry 125
- Risk Management and Loss Control 128
- Record Keeping 130
- Research and Surveys 130
- TQM in Safety, Health and Environment 131
- Case 1 132
- Case 2 132
- Summary 133
- Discussion Questions 134

5. ORGANISATIONAL ROLE ON SAFETY AND HEALTH **135-148**

- Role of Management 135
- Role of Employees 137
- Role of Trade Unions and Employee Representatives 138
- Role of Safety Organisation 138
- Behavioural Model for Safety 139
- Strategic Issues 140
- Accident/Injury Statistics 142
- Safety Training and Competence 142
- Safety Audit 143
- Safe Working Procedures 144
- Contract System 144

- Health-Related Issues 145
- Summary 148
- Discussion Questions 148

6. Fire Prevention and Control 149-160

- Chemistry of Fire 149
- Classification of Fire 150
- Methods of Fire Extinguishment 150
- Fire Hazards 151
- Fire Explosion 153
- Fire Inspection 155
- Fire Safety 156
- First Aid and Emergency Procedures 158
- Summary 160
- Discussion Questions 160

7. Occupational Health and Safety 161-181

- Occupational Health 162
- Healthcare Services 163
- Occupational Physician 165
- Occupational Health in Developing Countries 167
- Occupational Safety 170
- Occupational Safety in Developing Countries 172
- Occupational Safety and Health in Small Undertakings 174
- Promoting Occupational Health 176
- ILO Strategy 176
- Case 178
- Summary 180
- Discussion Questions 181

8. Industrial Hygiene 183-198

- Genesis 184
- Industrial Hygienist 185
- Occupational Health Hazards 186
- Recognition of Hazards 190

- Occupational Diseases 191
- Work-Related Diseases 195
- Study Assignment 195
- Case 196
- Summary 196
- Discussion Questions 197

9. SAFETY IN CONSTRUCTION **199-224**

- Characteristics 200
- Construction Hazards 203
- Causes of Accidents 204
- Prevention of Accidents 206
- Safety Legislation 209
- Accident Reports 210
- Safety Unit 210
- Role of Parties 211
- Safety Management 213
- Duties of Safety Engineer 215
- First Aid 216
- Safety Control and Regulations 216
- Safety Programme 218
- Construction Team 220
- Safety Programme Audit 221
- Summary 224
- Discussion Questions 224

10. HEALTH AND SAFETY AT WORKPLACES **225-250**

- Factories 225
- Mines 231
- Ports and Docks 239
- Plantations 240
- Offices 243
- Buildings 245
- Transportation (Road) 247

- Study Assignment 249
- Case 249
- Summary 250
- Discussion Questions 250

11. ACCIDENT COMPENSATION STATUTES 251-260

- The Dangerous Machines (Regulations) Act, 1983 251
- The Public Liability Insurance Act, 1991 253
- The Employees' Compensation Act, 1923 254
- The Employees' State Insurance Act, 1948 257
- Summary 260
- Discussion Questions 260

12. SAFETY AUDIT 261-278

- Objectives 261
- Benefits 262
- Types 262
- Procedure 263
- Stages 263
- Functions 264
- Choice of the Auditor 265
- Coverage of Audit 265
- Safety Audit Questions 266
- Summary 276
- Discussion Questions 277

13. STRESS AT WORK 279-300

- Causes of Stress 280
- Costs of Stress 282
- Handling of Stress 282
- Sources of Stress and Control Measures 285
- Stress-Related Problems 289
- Stress Reduction 292
- Alcoholism 294
- Drug Abuse 296

- Smoking 296
- Case 1 297
- Case 2 298
- Summary 300
- Discussion Questions 300

14. ENVIRONMENTAL POLLUTION AND PROTECTION 301-326

- Air Pollution 302
- Water Pollution 303
- Noise Pollution 304
- Environmental Policies 308
- Environmental Programmes 312
- Pollution Control 314
- The Water (Prevention and Control of Pollution) Act, 1974 317
- The Air (Prevention and Control of Pollution) Act, 1981 318
- The Environment (Protection) Act, 1986 320
- Summary 324
- Discussion Questions 325

15. DISASTER MANAGEMENT 327-344

- Objectives of Disaster Planning 327
- Emergency Planning 328
- Disaster Preventive and Pre-emptive Measures 331
- Disaster Control Plan 334
- Development of a Contingency Plan 336
- Mutual Aid and Response Group (MARG) 337
- Bhopal Gas Disaster 338
- Summary 342
- Discussion Questions 344

APPENDICES

Appendix A ***Relevant Labour Laws*** **347**

Appendix B ***Health Safety and Environment*** **359**

Appendix C *ILO Conventions Concerning Safety and Health in Construction* 363
Appendix D *Model Guidelines on Construction Safety* 377
Appendix E *Workplace Health and Safety Survey* 399
Appendix F *Cases of Industrial Accidents* 405
Appendix G *Safety Checklist and Safety Programme Rating Form* 413
Appendix H *Environmental Legislation at a Glance* 419
Appendix I *Glossary* 429
Appendix J *Software on Industrial Safety* 441
BIBLIOGRAPHY 445
INDEX 449

1

Workplace Safety and Health - An Overview

Safety is that profession which is concerned with the scientific analysis of the causes of accidental deaths and injuries in a given environment and their elimination or reduction. In the industrialised areas of the world, accidents now cause more deaths than all infectious diseases and more than any single illness, except those related to cancer and heart disease.

The most path-breaking advances in safety during the last 40 or 50 years have been made in industry. The initial stimulus for the industrial safety movement came from the largest industries, in which severe injuries and loss of life were common occurrences. In the 19th century, the safety of an employee was generally considered to be his own responsibility. The liability of the employer for accidental injuries depended on certain common-law doctrines that operated against the employee. However, an important segment of progressive industrial management saw the fallacies in these concepts and felt that management should do more for the safety of the workers.

The social progress of the times was reflected in the development of safety movement which was given impetus by workmen's compensation laws enacted in most of the industrialised countries in the late 19th and early 20th centuries. A new concept was inherent in these laws, namely that the employer should accept primary responsibility for injuries at work, even though they could not be attributed to any one person or cause. Industry became legally liable to pay compensation for personal injuries to its workers, and survivors benefits to dependants.

The three specialised skills–safety engineering, industrial hygiene, and industrial medicine–were developed as methods of preventing injury to the worker and for reducing costs. The initial emphasis in safety engineering was on the installation of guards and protective devices, which led to a sudden, sharp decline in machine accidents. Industrial medicine has concentrated on controlling environmental hazards i.e., the inhalation of toxic gases, fumes, and such other harmful effluents. Industrial hygiene has focused on minimising the consequences of injury and disease and with restoring disabled workers to gainful employment.

From an emphasis on mechanical guarding and education of workers in safety, experience tended to shift the focus to an application of basic safety engineering principles in the design of machinery, plant layout, and working methods. More emphasis was placed on the role of supervisory personnel, and safety was recognised as an integral part of normal operating procedure. The role of personal errors and accident proneness had been dealt with by functional job analysis. Variables relating to the health of the worker had been stressed, and working methods were modified to ensure a safe and optimum working environment as far as employee health was concerned.

These developments were a part of the movement for setting up of various organisations in different countries to perform specialised safety functions. Safety principles have been incorporated into engineering courses, and departments of safety engineering have been established in engineering schools and colleges. The most important directing force for accident prevention, however, has remained within industry itself. Although many outside agencies have contributed to safety performance, the direct and immediate responsibility for the application of safety management principles, employee education, and on-the-job safety enforcement rested upon the individual employer. In most instances, laws represented only minimal requirements and are way behind the goal of ultimate objectives.

Safety Management as a Profession

The field of safety management has not developed fully as a unified, specific discipline, and its practitioners have operated under a wide variety of position titles, job descriptions, responsibilities, and reporting levels in industry and the loss-prevention activities of insurance companies. There has also been little uniformity in the development of educational programmes in technical institutes and universities for the specialised training needed for this field.

The most recent trends in occupational safety and health include increased emphasis on prevention by the anticipation of hazard potentials; changing legal concepts concerning product liability and negligent design or manufacture; developing emphasis on consumer protection; and the development of national and international legislation and controls.

Four general areas that have been identified as the major functions carried out by the safety engineer/safety professional are:

1. identification and appraisal of accident-producing conditions and practices;
2. development of accident and loss-control methods, procedures and programmes;
3. communication of accident and loss-control information to those directly involved; and
4. measurement and evaluation of the accident and loss-control systems, and the modifications needed to obtain optimum results.

The safety professional contributes to the prevention of accidents by using his specialised knowledge of accident causation and control by (a) seeking to eliminate hazards and causation factors; (b) designing machines and procedures to reduce the hazards; and (c) minimising their effects by prescribing specialised equipment to prevent injury or reduce the severity.

One of the most important functions of the safety professional is to identify hazards in the working area and to evaluate the loss-producing aspects of a given method or operational system. In the past, a great deal of attention has been devoted to analysing the existing hazards and products, while the newer methods involve not only learning from accidents that have occurred but also emphasising on the need for advance analysis of accident potential and the prevention of accidents prior to their occurrence.

Traditional Safety Management

The form of safety management followed by most of the industries, called as 'traditional safety management', has the following characteristics:

1. Top down communication
2. Minimal employee participation
3. Dependence on discipline to influence safety behaviour
4. Centred on technical requirements aiming at short-term results
5. Use of safety techniques after accident and injury
6. Lack of integration of safety activities with other functions in the organisation.
7. Responsibility of safety officials for safety, but does not have the authority to make changes.

Five Steps to Managing Safety

1. Know and understand your safety policy and procedures.
2. Plan ways to reduce risk and remove hazards.
3. Organise people and resources to create a safe working environment and safe systems of work.

4. Measure your safety record by statistics and discussion with your staff.
5. Investigate accidents fully.

Accident Frequency and Severity Rates

Most statistical analysis of accidents are based on either frequency or severity. The frequency rate indicates how many injuries are occurring in relation to the number of man-hours worked, and the severity rate indicates how serious those injuries are in terms of days lost in relation to hours of exposure. The former has been most widely used and can be most quickly and easily computed. It does not, however, indicate the seriousness of an accident, since it merely gives the rate at which accidents are occurring. The severity rate was developed to meet this need and to give an expression of the total time lost per million man-hours of exposure. Other composite indexes usually involve weighted averages of some type, obtained by multiplying the severity rate by ten and adding the frequency rate.

The safety professional is concerned with reducing both the frequency with which accidents occur and the frequency with which they threaten. The frequency and severity rates vary widely from one industry to another. Certain industries, such as communications, have low-frequency and low-severity rates; mining and lumbering are at the opposite extreme. These differences reflect not only the inherent dangers in an industry but also efficiency of the safety programme.

Safety Analysis

One widely used analysis of the circumstances of accidents include the following:

1. identification and location of the materials, machines, and tools most frequently involved in accidents and the jobs most likely to produce injuries;
2. location of the departments and occupations involved;
3. disclosure of the unsafe practices, which may necessitate retraining or shifting of employees; and
4. study of the role of supervisors, and provision of information about the principal hazards.

Other significant factors include the type of physical injury incurred; the source of the injury in relation to the object, The substance or motion that inflicted it; the specific part of the material or agency that was hazardous and finally, the unsafe act that took place, presumably in violation of commonly accepted safety practice. Other aspects include age, sex, occupation, and type of work being performed.

All possible faults in equipment and in the work areas in industry, as well as the capacities of the operator, should be subjected to an advance analysis designed to prevent accidents. If defects are present, it is only a matter of time before some operator "fails" and has an accident. Advance analysis assumes the following:

1. An operational job analysis that should include a survey of the nature of the task, the work surroundings, the location of controls and instruments, and the way the operator performs his duties.
2. A "functional concept of accidents" i.e. anticipation of the errors that may occur while the operator is working at the machine.
3. An assumption of human limitations.

One method of identifying unsafe conditions involves selecting a random sample of workers from various portions of the plant or population, exposed to various hazards. An interviewer questions the workers, asking them to describe near-miss safety errors they have made in the past, or unsafe conditions they have observed. The data are classified into hazard categories to ascertain factors which have contributed to an actual or potential loss-producing event.

Safety Culture

An organisational culture develops over many years and is normally well ingrained amongst the top managers in the organization, it is, therefore, extremely difficult to change. The organisational culture in general, and the existence or non-existence of a safety culture within the organisation in particular has, not surprisingly, a profound effect on the behaviour of people within the organisation. The statement "this is the way we do things around here should reflect an uncompromising commitment from the organisation to work safely.

Establishing and developing safety culture within an organisation and managing safety from a behavioural perspective, is the most powerful approach to the management of safety.

In almost all organisations with a very good safety record, two important principles involved in the establishment of a safety culture have been identified. Firstly, the organisation should clearly define the role, responsibility, and accountability of directors, senior managers, middle managers, and employees generally, for safety; and secondly, there should be a conviction that high standards are achievable through proper management.

In creating and maintaining the safety culture in an organisation, the following are important:

- A genuine and visible commitment from the top.

- An acceptance that improving safety performance is a long-term goal which requires a sustained effort and interest.
- A policy statement of high expectation which conveys a sense of optimism.
- Employee ownership, involvement, training and communication.
- Good codes of practice and safety standards.
- The provision of adequate resources for safety.
- Safety is seen as a line management responsibility.
- Setting challenging targets and performance measurement against those targets.
- Thorough investigation, documentation and dissemination of information relating to all incidents and deviations, irrespective of whether injury or damage occurred.
- Compliance with standards should be ensured through auditing.
- Good safety behaviour should be accepted as a condition of employment.
- All deficiencies in safety should be remedied promptly.
- Managers at all levels should regularly assess safety performance.
- Factors that influence the behaviour of managers, supervisors and employees should be properly managed.

Safety Policy Statements

Policy statements will reflect the culture of the company. They should be relatively short, concise documents strategically developed by the highest corporate or company authority. Statements should be clearly written to ensure full understanding by all concerned, and supported by clearly defined organisational arrangements by which the policy is implemented. They should be reviewed regularly to ensure they are up-to-date concerning organisational change and legal requirements.

Every policy statement together with supporting documents should cover all or most of the following elements or principles:

- Management's commitment to provide a safe and healthy work environment that encompasses techniques, methods, organisation, and facilities improvement. Management is responsible for ensuring that employees are provided with a working environment that is free of recognised hazards.
- Continuous improvement by working together with long-term views of safety and health, environment, economy, and social integration.

- Management systems should incorporate innovative and progressive systems that lead to minimal safety and health impacts.
- Education, training and communication help develop and promote mutual understanding of safety and health issues for all stakeholders.
- Research and workplace technologies, innovation, technical co-operation, and sound engineering are fundamental to facilitate improvements.

Government regulations must be recognised where applicable. Cooperation with the Government in a responsible manner fosters safety benefits and cost effective legislation.

Systems Safety Engineering

In recent years, a great deal of attention has been devoted to "systems safety engineering". In brief, the primary objective of systems safety engineering is the reduction or elimination of all hazardous consequences of equipment operation in an industrial plant or similar large man-machine complex. It characteristically involves a systematic application of special analytical techniques, derived criteria, evaluation methodology, and management skills. Since it emphasizes prevention rather than correction of problems, particular attention is to be placed upon early engineering design and procedural analysis. It has to take into consideration all aspects i.e. planning, design, development, fabrication, test, installation, maintenance, operation, and overall evaluation of man-machine systems.

Accidents result from deficiencies in human operation, defects in tools and mechanical equipment, and a variety of environmental influences. An accident may include mechanical or electrical failure, defective materials, and such environmental conditions as extreme heat or noise levels inhibiting inter-communication. Human failure may derive from such factors as distraction, fatigue, worry, improper attitude, lack of skill, and carelessness.

***Design failures and human limitations*:** In many instances failures in the designing of equipment are so subtle that those responsible for analysing and reporting accidents are not aware of them. It is so because the designers of equipment are unaware of human limitations.

In order to improve the ease, efficiency, and safety of operating all types of equipment in shops and hangars, consideration has to be given to layout of working areas. Additional factors influencing the efficiency and safety of workers relate to the physical variables in the environment. Some of these include temperature, humidity, ventilation, noise, and vibration. Extremes of any of the factors result in discomfort, lowered efficiency, and increased accident risk.

***Safeguarding machinery in working areas*:** The basic principles of safeguarding continue to be of great importance, however, in many industrial operations. The primary purpose of machine guarding is to prevent injury from direct contact with moving parts, and from mechanical or electrical failures in the equipment, and from predictable human errors or failures. The principles of guarding are also important in relation to plant layout and the movement of men and materials.

***Safety principles in materials handling*:** Prevention of unsafe work practices, such as improper lifting, carrying too heavy a load, incorrect gripping is still an important safety function. Many injuries can be avoided by proper design of the work station or of the physical activities involved. One of the most commonly reported injuries is to the back and spinal column. Manual handling of materials is the cause of a large percentage of all fatal injuries and of permanent and temporary disabilities.

Occupational Hazards

***Electrical hazards*:** Many accidents are caused by use of switches and items improperly selected for the specific purpose. In most countries, independent testing laboratories test and approve electrical equipment for use in industry and in the home. Such laboratories test for shock and fire hazards and may also evaluate radiation and heat hazards.

***Chemical hazards*:** Modern chemical plants are designed and constructed to provide protection against chemicals. Many common industrial chemicals involve toxic and fire hazards. Some of the chemical substances used may be poisonous or harmful, corrosive, or combustible and explosive. The safety engineer is concerned with the identification of exposures to chemical agents and how this exposure can be prevented. In industry, toxic materials are most commonly absorbed either through the respiratory system, or through the skin, or through the gastro-intestinal track. Recommended practices have been developed for the storage, transport, and handling of hazardous and flammable materials.

***Radiation hazards*:** Radiation is emitted from either the atom's unstable nucleus or its unstable orbital area. The problem of greatest concern to the safety specialist is the alteration of cells in the body induced by radioactive exposure. Radioactive sources increasingly are being used in production processes. The safety specialist in cooperation with the company medical department has to supervise the handling and storage of radioactive materials, the disposal of residue and wastes, and the proper shielding of new installations. Comprehensive regulations covering such activities have been developed in each country.

Miscellaneous hazards: Specific processes and activities in industries gives rise to various accidents and hazards, for instance:

- Many factors need to be considered in hazard prevention to ensure safe working in engineering, vehicle maintenance and other types of work activities.
- In order to eliminate or reduce shop floor accidents, floors should be sound and kept clean with adequate drainage where wet processes are undertaken.
- Various hazards associated with machinery and equipment should be operated only by workers who should be adequately trained and supervised in their use.
- Many accidents are caused by the misuse of hand tools or the use of defective hand tools. Hand tools should be examined on a regular basis and defective tools are to be rejected.
- The principal hazard in laboratories is fire. So, ideally all laboratories should be of single-storey construction and separate from other buildings. Good housekeeping is a key factor in safe laboratory practice. Employees should be trained in the elements of sound personal hygiene.
- Fire is the major hazard in offices. Also, equipment and materials used in offices present a wide range of hazards.
- A number of injuries take place in catering operations. Good standards of safety in catering originate from the design of the kitchen area. Kitchens also represent a considerable fire hazard, particularly where housekeeping standards are poor.
- There are certain hazards common to gas welding and electric arc welding, such as fire and explosion burns, toxic fumes and gases. Any leakage of fuel gas is potentially hazardous, as ignition may lead to rapid or explosive combustion, particularly in confined spaces or unventilated areas.
- Many hazards arise from poor standards of maintenance and/or repair of equipment, improper use, and use of unsuitable materials.

The main hazards associated with maintenance operations can be physical, mechanical, electrical, chemical, and structural. Hence, there should be safe access to the processing plants and buildings.

Systems Approach to Safety and Health

Effective safety management considers the type of safety problems, accidents, employees and technology in the organisational setting. Further, the systems approach to safety recognises the importance of the human element in safety.

Any comprehensive and systematic approach to safety begins with the organisational commitment. This effort should be co-ordinated from the top to involve all members of the organisation and be reflected in their actions and work. Once a commitment is made to organisational safety, planning efforts must be co-ordinated, with duties assigned to supervisors, managers, safety specialists, and personnel specialists. The focus of any systematic approach to safety is the continued diligence of workers, managers, and other personnel.

Basic principles of accident prevention:

1. An unsafe act, an unsafe condition, an accident – all these are symptoms of something wrong in the management system.
2. Certain circumstances of severe injuries can be identified and controlled.
3. Safety should be managed like any other company function.
4. Management should direct safety action by setting achievable goals by planning, organising and controlling.
5. The key to effective line safety performance is management procedures that fix accountability.
6. The function of safety is to locate and define the actual errors that allow accidents to occur.

Involvement of employees in safety is done through the use of safety committees. These committees are often composed of workers of different levels and from a variety of departments. At least one member of the committee is usually from the personnel unit.

A safety committee generally has a regular scheduled meeting, has specific responsibilities for conducting safety reviews, and makes recommendations for changes necessary to avoid future accidents. A safety programme includes a set of safety rules. Specific rules are made in order to protect employees from possible hazardous working conditions.

Six basic requirements while developing safety rules:

1. They should be in the written form.
2. They should be practical.
3. They should highlight an unsafe act or conditions.
4. They should be limited to safety matters only.
5. They should assist in implementing safety rules.
6. They should be adopted and strictly enforced.

Every accident at the workplace needs investigation. An accident investigation is a methodical effort to collect and interpret facts. It is a systematic look at the nature and extent of the accident and loss involved. It is an enquiry as to how and why the accident occurred; a consideration of what could be done to prevent similar accidents.

Five questions that can be applied to accident investigation are:

* Who was injured?
* Where did the accident happen?
* When did the accident occur?
* What were the contributing and immediate causes of the accident?
* How can a similar type of accident be prevented?

Safety education and training

Safety education for all levels of management and for employees is a vital ingredient for any successful safety programme. Top and middle management require education in the fundamentals of safety and the need for an effective accident prevention programme. The supervisors must understand their key role in the safety effort, namely, that they are primarily responsible for preventing accidents. They must conduct safety training programmes for their employees who are directly under their supervision. In addition to safety training, periodical safety meetings are to be organised. Among the topics that may be covered are how to prevent accidents, importance of good housekeeping, handling materials safely, first aid, machine hazards, fire prevention, and protecting oneself.

One way to encourage workplace safety is to involve all employees at various times in safety training sessions and at committee meetings. In addition to safety training, continuous communication programmes to develop safety consciousness are necessary. Posting safety policies and rules is part of this effort. Contests, incentives, posters, bulletin boards are all means through which employers can heighten safety awareness.

A short outline of information to be included in designing the training format:

- Purpose of the specific training and safety programme
- Company policy
- Responsibilities for the administration of the programme
- Hazards and their effects
- Safe behaviour
- Specific hazard controls – engineering, administrative, training and supervision controls
- Procedures for reporting uncontrolled hazards
- Emergency procedures and preparedness

Process Safety Management (PSM)

PSM is a management process that allows leaders involved in commercial manufacturing and chemical businesses to proactively recognise, understand, and control process hazards, such that process related injuries and incidents may be prevented and organisations can continue to progress towards its value maximisation goals. PSM when simplified is the application of a management system and supporting tool to eliminate, prevent, and minimise the impact of process-related incidents and events.

A PSM programme must be systematic and holistic in its approach to managing process hazards and must consider the following:

1. The process design
2. Process technology
3. Process changes
4. Operational and maintenance activities and processes
5. Non routine activities and procedures
6. Emergency preparedness plans and procedures
7. Training programmes
8. Other elements that affect the process

Several elements of PSM are focused on personnel management strategies for ensuring the health and safety of all workers. These include:

1. Employee training and competency
2. Contractor safety management
3. Incident investigation
4. Management of change (personnel)
5. Emergency preparedness, planning and management

Success in PSM requires the involvement of many supporting systems and processes. The safe and efficient operations of facilities and technology are also a core focus of PSM. These include process safety information, process hazard analysis, mechanical integrity of equipment, and operating procedures. The safe and efficient operations of facilities and technology are a core focus of PSM.

The key elements of PSM are:

1. Preserving the health and safety of all workers at all times on a process facility.
2. Ensuring that all processes and systems are conducted safely at all times.
3. Maintaining the integrity of the facilities and processes such that the health and safety of all workers are maintained.

PSM introduces a bold and complete new approach for managing the health and safety of workers and for reducing the impact of capital losses associated with catastrophic failures and disasters in the workplace.

Safety Whose Responsibility

Safety is primarily the responsibility of the management. This responsibility should rest on the shoulders of all cadres of management, such as plant manager, production manager, chief engineer, human resource manager, maintenance engineer, individual foreman, safety officer or director.

Every organisation should formulate and implement a safety policy. The procedure to be adopted naturally depends upon the size of a company, the number of plants it operates, the nature of the industry in which it is engaged, the production technology it uses, and the attitude of the top management. After it has spelt out its safety policy, a company should establish a safety programme, the primary goals of which should be to reduce the number of hazardous factors which are likely to cause accidents, and to develop safe working habits among its employees.

While management has the responsibility to provide safe working conditions for employees, supervisors must insist on safe work practices in the workplace. Safety is every one's business. However, the supervisor is in a better position to spot safety hazards and to make sure that the work is done in a safe manner by each and every person on the job.

In a nutshell, the management's responsibility towards health and safety will be to:

- contribute in eliminating the hazard at source through improving the system of work.
- isolate hazardous processes and substances so that workers do not come into contact with them.
- ensure that suitable items of protective equipment and clothing are used and worn.
- train workers on how to avoid risks to health.
- maintain plant and equipment to eliminate the possibility of harmful emissions and injuries to people at work.
- keep premises and machinery clean and free from toxic substances.

Safety audit: The safe operation of industrial plants is the responsibility of the top executives. To effectively discharge this responsibility, there is a need for laying down a broad safety policy as well as a comprehensive programme to ensure its implementation. Thereafter, it is important for the senior management to review the performance of the safety activities and programmes of the company at periodical intervals. Safety audit is a useful technique to undertake a systematic critical appraisal of the effectiveness of a company's safety programme. Such a periodic review provides an independent assessment of the correctness of the basic direction as well as identifies the specific areas for action to improve the programme.

Safety audit is a tool for ensuring that the plant operation and maintenance procedures match the design intent and standards. It keeps operating personnel alert to the process hazards; reviews operating procedures for necessary revisions; seeks to identify equipment or process changes that could have introduced new hazards; initiates application of new technology to existing hazards; and reviews the adequacy of maintenance and safety inspections.

Periodic safety audits should be conducted to determine what types and categories of accidents occur in the organisation. The following questions should be asked in a safety audit:

1. Was the incident fully investigated?
2. Was the data used to provide additional safety training to the injured employee, or to all employees where appropriate?
3. Did the investigation help eliminate hazards, provide personal protective equipment?
4. Is a safety awareness campaign planned for the coming year? What is the nature of the campaign?

***Safety rating system*:** A sector specific comprehensive rating system has to be developed to carry out qualitative, quantitative and objective assessment of an organisation's occupational safety and health (OSH) system as per the industry specific standards and adhering to both national and international best practices.

The safety rating system should cover certain major elements like

(a) OSH management
(b) Physical hazards
(c) Chemical hazards
(d) Fire and explosion hazards
(e) Industrial hygiene, and occupational health
(f) Accident/incident reporting, investigation and analysis
(g) Emergency preparedness
(h) Safety inspection.

***Accident reports and records*:** Reports and records of accidents should be maintained by an organisation in detail and should contain the following items.

- The total number of employees in the unit who are exposed to different types of accidents.
- The severity of the accident – whether it resulted in a broken bone, a deep cut – and the time that was lost as a result of it.

- The kind of work or occupation in which the employee was engaged.
- The date, time and day, and the shift during which the accident occurred.
- The total number of years during which the employee was engaged on that particular job when the accident occurred.
- Personal data, including the age and health of the injured employee.
- The immediate cause of the accident – whether it was the result of a malfunctioning of a machine, or whether the employee failed to use the safety devices provided for the purpose of preventing accidents, and so on.

Emergency Preparedness

The management has to prepare and implement an emergency plan to control major accidents and hazards. The basic and essential features of emergency preparedness should be to:

- Access the size and nature of the events foreseen and the probability of their occurrence.
- Work out up-to-date details of on-site and off-site emergency plans to be separately prepared and be consistent with each other.
- Liaison with the local authority including the emergency services.
- Lay down procedures for: (a) raising the alarm, (b) communication both within and outside the works, and (c) rescue/evacuation of personnel and containment of emergency.
- Appoint key personnel and specify their duties and responsibilities.
- Establish an emergency control centre equipped with adequate means of communication.
- Work-out appropriate details of actions and the decision to be taken on-site and off-site.
- Arrange for publicizing authoritative information to all concerned.
- Conduct training and rehearsal programmes.
- Review and update the effectiveness of emergency preparedness.
- Declare cessation of emergency in time and give a clear signal.

Health Promotion

Health has been defined by WHO as a state of complete physical, mental and social well-being and not merely the absence of disease or infirmity. Health promotion emphasises the positive dimensions of health. Health promotion is a mediating strategy between people and their environments, combining personal choice with social responsibility for health to create a healthier future.

The scope of health promotion in the workplace is potentially very large and intervention programmes can be designed to cover many subjects related to lifestyle such as exercises, healthy eating, and the management of stress. Health promotion among employees at the workplace should be an integral element of a company's business strategy. The reason is that a healthy worker is a productive worker. Moreover, improvements in employee health result in better work attitudes, higher morale and job satisfaction, lower employee turnover and reduced absenteeism.

Health promotion focuses on prevention rather than treatment or cure. Hence, the health programme in an organisation is normally planned around improvement and prevention of controllable risk factors such as smoking, obesity, high level of cholesterol, stress, hypertension, and low level of physical fitness which are responsible for many major diseases. The industrial health programme of many progressive organisations takes positive steps to maintain good employee health off the job as well as on the job. Some organisations have also established programmes in the field of mental health.

Examples of Occupational Health and Safety Policy Statements

***Example* 1**

Company X is committed to achieving the highest performance in occupational health and safety with the aim of creating and maintaining a safe and healthy working environment throughout its businesses. Consistent with this, the company will:

- Seek continuous improvement in its occupational health and safety performance taking into account the evolving community expectations, management practices, scientific knowledge and technology.
- Comply with all applicable laws, regulations and standards and where adequate laws do not exist, adopt and apply standards that reflect the company's commitment to health and safety.
- Involve employees and contractors in the improvement of occupational health and safety performance.
- Train and hold individual employees accountable for their area of responsibility.
- Manage risk by implementing management systems to identify, assess, monitor and control hazards and by reviewing performance.
- Ensure that employees, contractors and visitors are informed of and understand their obligations in respect of this policy.

- Communicate openly with employees, government and the community on occupational health and safety issues; and contribute to the development of relevant occupational health and safety policies, legislation and regulations.
- Support relevant occupational health and safety research.

***Example* 2**

At Company Y there is nothing more important than the health and safety of employees. The company is committed to:

- Integrating health and safety practices into all aspects of work and continuing to incorporate these practices into product and process design, manufacturing and business planning.
- Providing innovative and preventive health and safety programmes. The company will continually optimise the effectiveness and integrity of programmes through open communication, comprehensive training and education, audits and workplace assessments.
- Developing and understanding among those in leadership of their responsibilities and their accountability to provide a safe and healthy workplace.
- Developing and understanding among all employees of their responsibility to work safely, their accountability for individual performance, and the assignment of appropriate authority to implement these responsibilities.
- Meeting or exceeding the requirements of applicable legislation and regulations by superior performance in health and safety matters and thereby demonstrate leadership within the industry.

Safety philosophy of company A

- Safety is not a style statement; but an evolving science, to be meticulously practiced by every team member.
- Benchmark company's operations for meeting exacting global standards; this should be reflected in all its units.
- Create proactive safety philosophy with multiple safety levels at all operational units.
- Provide extensive training programmes ensuring information dissemination and 24 × 7 preventive safety measures.
- Prepare an action plan in consultation with E.I. DuPont India Private Limited to meet global safety standards and create a world-class safety culture in the company.
- Change the culture by providing safety training to employees and instil in them the importance of safety.

Reorganise the organisational structure by installing safety committees and sub-committees and involving the senior management in the structure to infuse the importance of health and safety.

2. Safety and Health Policy of Company B

The occupational safety and health policy of the company is based on and supported by the following eight principles:

- All injuries and occupational illnesses are preventable.
- All operationable exposures can be safeguarded.
- Safety evaluation of all business processes is vital.
- Working safely is a condition of employment.
- Training employees to work safely is essential.
- Management audits on safety are a must.
- Employee involvement is essential in all safety efforts.
- All deficiencies must be reported and corrected promptly.

CASE

The Hindustan Glass Works, a manufacturer of glassware with 900 employees, has had the following injury record for the past three years. The earliest year is shown first: injuries involving days away from work is 44, 53, 47. In the previous year, one worker suffered permanent disablement to burns; and another lost four fingers on his right hand.

The company employs a full-time safety manager and has a safety committee. There is a relatively strong union affiliated to a national organisation.

The union and management are in the midst of somewhat bitter collective bargaining. The union representatives in the third day of negotiation have introduced a demand that the union be given equal authority with management in all aspects of safety. They maintain that working conditions are not safe enough and so the union should be given authority to set minimum safety standards, to stop production when these are not complied with, and to co-opt as an equal partner with the present safety committee in inspecting and determining safety violation and proposing remedial action.

The hazards must be identified and systems established to reduce, or preferably eliminate the risks.

Question

1. What do you think the management representatives should do? Should this union demand be granted?

SUMMARY

- Accidents are the result of unsafe personal acts, unsafe conditions, or a combination of both, and it is generally agreed that unsafe personal acts cause the bulk of organisational accidents.
- Organisational health hazards include all toxic substances found in the work environments that may be hazardous to employee health.
- In controlling major accident hazards, it is the industries who have to take lead as they are the ones who create the hazards and it is they who have the means to prevent or control any major incidents leading to uncontrolled developments at its source and at the inception stage.
- Accident frequency and accident severity are the two most widely-accepted methods for measuring an organisation's safety record.
- Organisational interest in health and safety is influenced by social values and economic considerations and legal compulsions.
- Organisational efforts to improve safety and health usually begins with a company safety policy.
- Three general areas of activities in safety and health are investigating, reducing risks at the workplace, and monitoring the safety efforts through reporting and follow-up.
- Related safety techniques include careful selection of employees, safety research, accident investigation and analysis, safety committees, protective equipment, and safety training and communication.
- Managing health and safety through legislation, administrative reorganisation, and most importantly the education and reorientation of the attitudes of industrial community and general public, is a great challenge to be faced by all concerned.

DISCUSSION QUESTIONS

1. Why is safety management regarded as a profession?
2. Accidents result from deficiencies in human operation, defects in tools and mechanical equipment, and a variety of environmental influences. Elucidate.

2

Accidents and their Prevention

All safety work revolves around and stems from those ubiquitous events called an accident. But what is an accident? The dictionary defines an accident as an undesirable or unfortunate happening; casualty; mishap and anything that happens unexpectedly without plan, or by chance. Thus, it may be defined as an unexpected, unplanned, uncontrolled occurrence. It is doubtful that any single definition can be devised that would cover all types of events that could be called “accidents”. The psychologists would focus on the so-called behavioural accidents – those unintentional acts such as forgetting an appointment, losing things, or making a wrong turn. The industrial hygienist would likely view occupational disease, dermatitis, silicosis, and so on as the consequence of a series of recurring events or accidents.

Accidents disrupt the normal functions of a person or persons and causes injury or near-injury. During an accident, a person’s body comes into contact with or is exposed to some object, other person, or substance, which is injurious; or the movement of a person causes injury or creates the probability of injury.

Accidents and injuries are not the same. An accident is any occurrence that interrupts or interferes with the orderly progress of the activity in question. For instance, a hoisting chain may break, dropping a load of steel bars to the floor. This is an accident, but not an injury. An “injury” may be defined as a harmful condition sustained by the body as the result of an accident. It can take the form of an abrasion, a bruise, a laceration, a fracture, a foreign object in the body, a puncture wound, a burn, or an electrical shock. Injury is the consequence of an accident, but many “no-injury” accidents also occur. Although psychologists are concerned with accidents that may or may not result in injury, they generally pay more attention to those accidents involving injuries.

There is a possibility of accident in every sphere of human life, at home, whilst travelling, at play and at work. The victims of an accident may not be directly involved in the activity which gives rise to it – they may be nearby workers, bystanders or those living in the vicinity.

Many countries collect data on accidents at work, during travel or in the home. It is often obligatory to notify certain categories of accidents (e.g., boiler explosions or crane failures that cause no injury, or accidents that cause injury of particular severity (e.g., prevent a person from doing his normal job for more than three days). The number of notified accidents causing injury can often be compared with the number of workers engaged in a particular activity to produce a frequency rate (e.g., number of notified accidents where 1,00,000 workers or working shifts) and so enable more hazardous occupations and activities to be identified and priorities for preventive action accessed. Due to differences in compiling accident statistics in various countries, international comparisons are not generally valid.

Occupational accident refers to any organic or functional injury or damage to the body, limbs or health, due to external, sudden and violent cause occurring during work or total, or partial, permanent or temporary incapacity for work.

The ILO's Employment Injury Benefits Recommendations, 1964 (No. 121) sums up the generally accepted definition of an industrial accident as follows:

(a) accidents, regardless of their cause, sustained during working hours at or near the place of work or at any other place of employment;
(b) accidents sustained within reasonable periods before and after working hours in connection with transporting, cleaning, preparing, securing, conserving, storing and packing work tools or clothes;
(c) accidents sustained while on the direct way between the place of work and
(i) the employee's main or secondary residence; or
(ii) the place where the employee usually takes his meals; or
(iii) the place where he usually receives his remuneration."

Anatomy of an Accident

The common agencies of industrial accidents are:

1. Machines, pumps, prime movers
2. Elevators and hoisting apparatus

3. Conveyors
4. Boilers, pipe and pressure apparatus
5. Vehicles
6. Transmission apparatus
7. Electrical apparatus and tools
8. Hand tools
9. Chemicals
10. Other materials
11. Other working surfaces.

There is an accident potential in almost every situation. However, before any accident can actually happen, two sets of factors–those in the environment and those in the individual–must be brought together. The identification and evaluation of each of the many components in the anatomy of an accident is of paramount importance. A detailed outline of these four components is as follows:

1. The accident
   A. Fall
   B. Slip
   C. Slide
   D. Collision
   E. Being caught in or between
   F. Eruption or explosion
   G. Burn.
2. Results of the Accident
   A. Annoyance
   B. Production delays
   C. Reduced quality
   D. Spoilage
   E. Property damage
   F. Minor injuries
   G. Disabling injuries
   H. Fatality.
3. Immediate causes of accidents
   A. Unsafe actions
      (i) Protective equipment or guard provided but not used
      (ii) Hazardous method of handling

Fig. 2.1 Protection Aganist Fire.

  - (iii) Improper tools or equipment used despite availability of proper tools
  - (iv) Haphazard movement
  - (v) Horseplay.

B. Unsafe conditions
  - (i) Ineffective safety device
  - (ii) No safety device although one is needed
  - (iii) Hazardous housekeeping
  - (iv) Defective equipment, tools, or machines
  - (v) Improper dress or apparel for job
  - (vi) Improper illumination, ventilation, and so on.

4. Contributing causes of accidents:
   1. Inadequate codes or standards.
   2. Failure by management to enforce safety rules and regulations.
   3. Faulty design or lack of maintenance.
   4. Inadequate personal protective equipment.
   5. Lack of proper inspection and maintenance.
   6. Failure on the part of the supervisor to improve safety standards.
   7. Employees' lack of safety awareness and training.
   8. Lack of emotional stability and temperament.
   9. Poor physical and mental conditions of worker like fatigue, deafness, poor eyesight, improper attitude, and so on.

For a complete understanding of the anatomy of an accident, a full knowledge of the contributing and immediate causes of accidents is essential. When an accident does occur, it is an indication that something has gone wrong. Somewhere along the line someone has not done a good job of accident prevention, and unsafe acts and conditions have been tolerated. The lack of safety awareness is a significant factor in accident causation.

The accident control steps are the following:

1. Supervisor safety performance
   - A. Job-hazard analysis
   - B. Enforcement of safety rules
   - C. Adequate safety knowledge
   - D. Promotion of employee participation in safety
   - E. Proper job placement
   - F. Development of safe working conditions.
2. Mental condition of person
   - A. Regular safety contacts by supervisor
   - B. Adequate safety indoctrination and on-the-job safety training
   - C. Safety promotions and publicity

D. Employee participation in safety programme
E. Regularly scheduled safety meetings
F. Adequate supervisor-employee communication on all matters concerning safety of the employee.

3. Physical condition of person
   A. Preplacement physical examinations
   B. Periodic re-examinations
   C. Proper job placement
   D. Adequate medical facilities
   E. Careful examination of physical condition of worker on all transfers and changes in job
   F. Recognition of the physical limitations of workers new to a job, especially if heavy work is involved.

Causes of Accidents

In every sphere of human activity there is the possibility of an accident, and work is no exception. Industrial accidents are the end-products of unsafe acts and unsafe conditions of work. However, accidents are preventable – they don't just happen. They usually occur as a result of the combination of a number of factors of which the three main ones are technical equipment, the working environment, and the worker. For e.g., safety equipment may be lacking in the factory or machinery may have been poorly designed with inadequate safety devices. The working environment may be so noisy that it is impossible to hear safety signals. Also, the workers themselves may be a contributory factor in that they may not have received adequate training or may have little experience of the task.

Environmental Causes

- improper guarding
- substances or equipment defective through use or abuse
- substances or equipment defective through design or construction
- unsafe procedure
- unsafe housekeeping
- improper illumination
- improper ventilation
- improper dress or apparel.

Behavioural Causes

- improper attitude
- lack of knowledge or skill
- physical or mental defect.

People who might be considered as accident prone include those who face difficulties, permanent or short-lived, such as:

1. A lack of aptitude for the work.
2. A lack of certain skills and coordination.
3. A possible literacy problem.
4. An attitude or personality problem.
5. Alcohol or drug problems.
6. Personal life stresses.

There are unlimited number of hazards that can be found in almost any workplace. There are obvious unsafe working conditions, such as unguarded machinery, slippery floors or inadequate fire precautions, but there are also a number of categories of insidious hazards including:

- chemical hazards, arising from liquids, solids, dusts, fumes, vapours and gases;
- physical hazards, such as noise, vibration, unsatisfactory lighting, radiation and extreme temperatures;
- biological hazards, such as bacteria, viruses, infectious waste and infestations;
- psychological hazards, resulting from stress and strain;
- hazards associated with the non-application of ergonomic principles, for example badly designed machinery, mechanical devices and tools used by workers, improper seating and workstation design, or poorly designed work practices.

Workers do not create hazards – in many cases the hazards are built into the workplace. This means that the solution is to remove the hazards, not to try to get workers to adapt to unsafe conditions. The most effective accident and disease prevention begins when work processes are still in the design stage and when safe conditions can be built into the work process.

All workplace hazards can be controlled by a variety of methods. The goal of controlling hazards is to prevent workers from being exposed to occupational hazards. Some methods of hazard control are more efficient than others, but a combination of methods usually provides a safer workplace than relying on only one method.

Hazard control programme includes the following components: 1. Hazard identification, 2. Ranking hazards by risks, 3. Establishing preventive and control measures, 4. Monitoring, and 5. Evaluating programme effectiveness and feedback.

Hazard control is not the responsibility of just one individual or one department. It is the collaborative effort of all the departments associated with safety management. The activity of hazard control is a continuing and ongoing one.

Researchers in different disciplines address the problem of industrial accidents from many perspectives, but all concerned with reducing frequency and severity. Industrial psychologists focus on individual characteristics associated with accidents and investigate traditional approaches (personnel selection, work design, and training) in accident reduction. Industrial hygienists view occupational diseases as consequences of recurring events or accidents. Safety engineers see accidents as results of a sequence of acts or events with undesirable consequences, such as personal injury, property damage, or work interruption.

Theories on Accident Causation

Domino Theory: One of the most interesting theories of accident causation was formulated by H.W. Heinrich in 1959, and is known as the domino theory. This theory explains the accident process in terms of five factors:

- ancestry and social environment
- personal fault
- the unsafe act and/or mechanical or physical hazard
- the accident
- the injury.

These factors are of a fixed and logical order. Each one is dependent on the one immediately preceding it, so that if one is absent, no injury can occur. This theory can be visualised as five standing dominoes and the behaviour of these dominoes is studied when they are subjected to a disturbing force. When the first social environment fails, the other four automatically follow, unless one of the factors has been connected, that is, removed, thereby creating a gap in the required sequence for producing an accident.

Cooper (1998) described the Domino Theory of accident causation to demonstrate the influence of management in reducing accidents. The Domino Theory of accident causation takes the view that poor management control creates either poor personal factors or poor job factors. In combination, these two factors lead to either unsafe acts or unsafe conditions.

Multiple Causation Theory: There is certainly nothing wrong with the domino theory. It is obviously a good approach to safety, and it obviously is just what it was intended to be - a very practical system for removing the things that are causing accidents. Today, we know that behind every accident there lie many contributing factors, causes, and subcauses. The theory of multiple causation states that these factors combine together in random fashion, causing accidents. Under the multiple causation theory, some of the contributing factors listed were:

1. Why was the defective ladder not found during normal inspections?
2. Why did the supervisor allow its use?
3. Didn't the injured employee know it should not be used?
4. Was the employee properly trained?
5. Was the employee reminded not to use the ladder?
6. Did the supervisor examine the job first?

Exhibit 2.1 What to do in the Event of an Accident

1. Call for help if necessary.
2. Eliminate risk of further accident (e.g. switch off machine, electricity, prop up insecure materials).
3. Summon first aid.
4. Make sure someone stays at site of accident to warn others and comfort casualty.
5. Put warnings round accident area and keep people away.
6. Co-operate as requested when a first aider supervisor or other competent person takes over the accident site.
7. Report details of accident to the appropriate authorities and record all details in the accident book.

Approaches to Reducing Accidents

There are three approaches to achieving workplace safety. The first is based on individual differences and involves selecting people who have the characteristics thought necessary or desirable for specified work demands. This can be termed as personal selection approach. The second approach is to train people in the necessary skills, knowledge, or attitudes. This can be termed as personal training approach. The third approach is modifying the workplace to provide a better match with abilities and characteristics. This is the ergonomic or work design approach. These approaches are all used to reduce industrial accidents, and to create a safer work environment.

Engineering Approach: It entails modifying the tools, equipment, and machinery to eliminate unsafe conditions. Machines must be designed to complement human skills and limitations. The training approach is based on the belief that safety skills, like all skills can be improved through training. This approach stresses the importance of creating an environment conducive to safety and of acquiring the knowledge and skills needed to reduce accidents. Each method approaches safety from a different perspective, but all have proved effective in reducing accidents.

Behavioural Approach: Safety engineers, supervisors, and managers must realise the effectiveness of psychological factors in accident prevention. Human errors of omission or commission contribute to a disproportionately high percentage of all accidents. How and why such errors are committed must be understood, particularly by supervisors and managers.

In safety programming the goal is to make employees more receptive to being trained in safe practices, more willing to accept new safety devices and more favourably inclined to adopt safety as a way of life both on and off the job. To achieve this important goal, supervisors and managers must deal effectively with the human side of safety.

Any discussion that influences human behaviour and worker efficiency would not be complete with atleast some mention of the biorhythm theory. Proponents claim the theory has been used successfully in accident prevention programme. Essentially, the biorhythm theory is based on the subjective interpretation of observations of human behaviour collected by two researchers working independently during the late nineteenth century. The theory suggests that human behaviour is characterised by three inherent cycles called biological rhythms or biorhythms. The physical cycle influences tasks of a physical nature, the emotional cycle is a factor in situations of high emotional content, and the intellectual cycle impacts pursuits requiring awareness and judgement.

Total Safety Approach: "Total safety" includes not only personal injuries on account of mechanical, electrical, chemical and related hazards, on account of fires and explosions, but also impairment to health due to occupational health hazards. In a sense, comfort and welfare of employees should also be included under this big umbrella of total safety. Thus the entire gamut of "working conditions" would form this subject.

Efficiency and safety in industrial operations can be greatly increased by careful planning of the location, design and layout of a new plant or of an existing one in which major alterations are to be made. Numerous accidents, occupational diseases, fires and explosions are preventable, if suitable measures are taken right from the earliest planning stage. Factors determining the size, shape and types of buildings, their general layout. Can vary greatly. Some of the more important of these factors are nature of the manufacturing process, nature of the materials to be stored, handled, processed, besides maintenance problems, design of mechanical handling equipment and the working conditions necessary for the process. The goal of production has to be accomplished in such a manner that no one is hurt and losses are minimised.

Design of industrial plant buildings with respect to fire safety, would include provision of adequate number of fire escapes, staircases, ladders, and ramps. In respect of working at dangerous heights also, these are important. In this context, Rule 70 of Maharashtra Factories Rules, 1963, has been amended in August, 1989. These are very comprehensive provisions and give ample guidance to design architects and also entrepreneurs, in the matter of safe design of industrial plant buildings. The provisions include means of fire protection, access for fire-fighting, means of protection against general ignition, spontaneous ignition, lightning; with respect to storage of flammable liquids in work rooms, storage of compressed gas cylinders, prevention of accumulation of flammable dust, gas, fume or vapour in air or flammable waste material on the floor. Some more important parameters in the building fire codes with respect to layout of fire exits, corridors, passageways, stairways, staircases, doors and ramps and also covering their design, construction, width, treads, risers, preferred angles of inclination have since been brought in this statutory rule since August 1989. These must be complied with, in advance so as to save costly subsequent alterations.

Accident Proneness: An employee with comparatively short service with the company has drawn to himself through having sustained a series of injuries, or one with several years' service on a single job suffers a like series of injuries, and the general cry is "accident-prone" in reference to each case.

The accident-prone individual is a person who will have an abnormally high rate of injuries, regardless of his job, work environment, age, or length of service. It is a matter of record that the accident-prone concept originated following the discovery that a small percentage of the population had a high percentage of injuries. The currently accepted theory that most accidents are sustained by a small fixed group of "accident prone" individuals is open to question. Most people move in and out of the so-called accident prone group depending on age, mental and physical state, environmental factors, and other conditions that vary with the passage of time rather than remaining fixed with the individual.

Ultimately, all industrial accidents are - either directly or indirectly - attributable to human failings. People are not machines; their performance is not fully predictable and mistakes are committed. Accident causation is a complex subject and various theories have been put forward to explain how accidents happen and how they can be avoided in the future. For example, the "pure chance theory" implies that accidents are "acts of God", that there is no discernible pattern in the chain of events leading upto the accident. The "accident prone theory" suggests that some workers are more likely than others to have an accident due to innate personal characteristics. In other words, these workers are always likely to have an accident and very little can be done to prevent it.

Organisations need not limit themselves to one approach to reducing accidents. In fact, the prudent organisation should adopt a comprehensive programme using multiple approaches together. Each approach has its own merits; when used in combination, they improve the likelihood of reducing accidents. Research indicates that some variables are useful for predicting which people are most likely to have accidents. To the extent that these differences can be reliably predicted, organisations can be staffed with people who exhibit safe behaviour.

Classification of Accidents

There are many different methods of classifying industrial accidents. Some classify accidents according to where the fault lies, others classify them according to the cause. The First International Conference of Labour Statisticians, organised by the ILO in 1923, which recommended a simple classification system of accidents by cause containing the main headings: machinery; transport equipment; explosions and fire; poisonous, hot or corrosive substances; electricity; fall of persons; stepping on or striking against objects; falling objects; handling without machinery; hand tools; animals; and other causes.

In October 1962, the Tenth International Conference of Labour Statisticians, convened by the ILO, adopted a standard multiple classification system to replace the 1923 cause classification. According to this system, industrial accidents are to be classified under each of the four headings: (a) the type of accident; (b) the agency; (c) the nature of the injury; and (d) the bodily location of the injury.

Whatever form of classification is adopted, it appears that the most common causes of accidents are to be found, not in the most dangerous machines or the most dangerous substances but in quite ordinary actions like stumbling, falling, the faulty handling and lifting of goods or use of handtools, and being struck by falling objects.

Very rarely does an accident arise from a single cause: More frequently there is a combination of factors which must all be simultaneously present. The crane that fails through overload is not simply the result of physical stress but the consequences of a system of work which failed to determine the weight to be lifted and plan the operation, failed to check the crane and its safety devices to ensure that it was suitable for the purpose and failed to see that the crane driver know what was expected of him and was trained to do it. Above all, failure to maintain good housekeeping standards is responsible for a high proportion of falling, tripping and impact injuries. It may also contribute to many machinery accidents, as when a slipping floor causes a fall on to dangerous moving machinery. Inadequate or unsuitable lighting is another contributory part in many accidents of every kind.

Usually, an accident is required by legislation to be notified to the competent authority if it is fatal or if the injured workman is incapacitated from carrying all his normal work for a period of time. There is no universal rule as to how many days of work incapacity make the accident notifiable. In the USSR and New York State it is one day; in France and in India it is two days; in Germany and U.K. it is three days; and in Malaysia it is four days.

Unsafe mechanical or physical conditions are those factors that are present due to defects in condition, errors in design, faulty planning, or omission of essential safety requirements for maintaining a relatively hazard-free physical environment. The following are the seven categories in which unsafe physical conditions may be grouped:

1. Inadequate mechanical guarding.
2. Defective condition of equipment.
3. Unsafe design or construction.
4. Hazardous process, operation, or arrangement.
5. Inadequate or improper illumination.
6. Inadequate or incorrect ventilation.
7. Unsafe dress or apparel.

Unsafe personal acts are those types of behaviour that lead to injuries. The following are some of the personal acts that result in injuries.

1. Working unsafely such as improper lifting, hazardous placement, incorrect mixing, and so on.
2. Performing operations for which supervisory permission has not been granted.
3. Removing safety devices or altering their operation so that they are ineffective.
4. Operating at unsafe speed.
5. Use of unsafe or improper equipment.
6. Using the equipment unsafely.
7. Failure to use safety attire or personal protective devices.

The work of analysing causes of accidents can be outlined in ten steps:

1. Secure the supervisor's report of the accident.
2. Secure the injured worker's report.
3. Secure the reports of witnesses.
4. Secure the report of the nurse or doctor.
5. Investigate the accident.
6. Record all the facts.
7. Tabulate the facts of the accident together with those of other accidents.

8. Study all the facts.
9. Determine what action should be taken.
10. Assign responsibilities for carrying out the action decided.

Costs of Accidents

H.W. Heinrich listed eleven elements of indirect accident cost. They are:

1. Cost of lost time of injured employee.
2. Cost of time lost by other employees who stop work;
   (a) out of curiosity
   (b) out of sympathy
   (c) to assist injured employee
   (d) for other reasons.
3. Cost of time lost by foremen, supervisors, or other executives as follows:
   (a) assisting injured employee
   (b) investigating the cause of the accident
   (c) arranging for the injured employee's work to be continued by some other worker
   (d) selecting, training, or breaking-in a new worker to replace the injured man
   (e) preparing state accident reports, or attending hearings before state officials.
4. Cost of time spent on the case by first-aid attendant and hospital department staff, when not paid for by the insurance carrier.
5. Cost due to damage to the machine, tools, or other property, or to the spoilage of material.
6. Incidental costs due to interference with production, failure to fill orders on time, loss of bonuses, payment of forfeits, and other similar causes.
7. Cost to employer under employee welfare and benefit systems.
8. Cost to employer in continuing the wages of the injured worker in full, after his return - even though the services of worker (who is not yet fully recovered) may, for a time, be worth only after half of their normal value.
9. Cost due to the loss of profit on the injured worker's productivity and on idle machines.
10. Cost of subsequent injuries that occur in consequnce of the excitement or weakened morale due to the original accident.
11. Overhead cost per injured worker – the expense of light, heat, rent and other such items, which continue while the injured employee is a non-producer.

As Heinrich puts it, "this list does not include all the points that might well receive consideration, although it does clearly outline the vicious and seemingly endless cycle of events that follow in the train of accidents".

Work-related accidents or diseases are very costly and can have many serious direct and indirect effects on the lives of workers and their families. For workers some of the direct costs of an injury or illness are:

- the pain and suffering of the injury or illness;
- the loss of income;
- the possible loss of a job; and
- health care costs.

It has been estimated that the indirect costs of an accident or illness can be four to ten times greater than the direct costs, or even more. An occupational illness or accident can have so many indirect costs to workers that it is often difficult to measure them. One of the most obvious indirect costs is the human suffering caused to workers families, which cannot be compensated with money.

The costs to employers of occupational accidents or illnesses are also estimated to be enormous. For a small business, the cost of even one accident can be a financial disaster.

For employers, some of the direct costs are:

- payment for work not performed;
- medical and compensation payments;
- repair or replacement of damaged machinery, equipment, and products;
- reduction or a temporary halt in production;
- increased training expenses and administration costs;
- possible reduction on the quality of work;
- claims made in the court;
- fines imposed in criminal courts; and
- insurance premiums.

Some of the indirect costs for employers are:

- the injured/ill worker has to be replaced;
- a new worker has to be trained and given time to adjust;
- it takes time before the new worker is producing at the rate of original worker;
- time must be devoted to obligatory investigations, to the writing of reports and filling of forms;

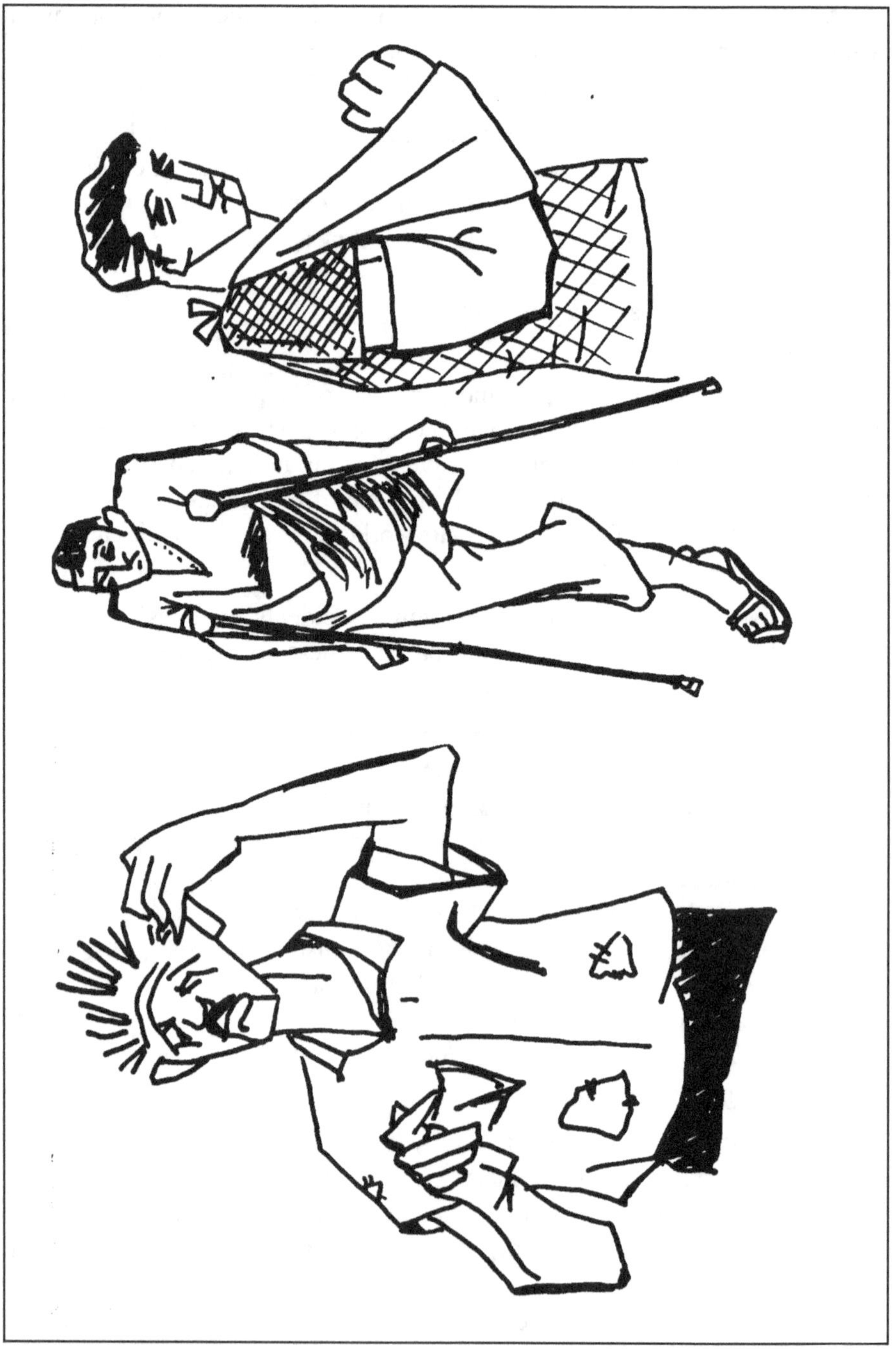

Fig. 2.2 Accidents Cost a Lot.

- accidents often arouse the concern of fellow workers and influence labour relations in a negative way;
- poor health and safety conditions in the workplace can also result poor public relations;
- medical and first aid treatment;
- transport to hospital;
- lost time of the injured employee;
- loss of an injured person's personal skills;
- lost time of other persons - managers, supervisors and employees;
- replacement of labour, including training, retraining and extra supervision;
- health and safety administration and investigation;
- loss of employee morale resulting in reduced productivity;
- counter productive trade union activity.

Cost Benefit Analysis

Quantifiable and non-quantifiable benefits

Quantifiable Benefits

- Fewer incidents and injuries
- Less severe injuries
- Fewer work-related fatalities and diseases
- Reduced absenteeism
- Reduced less time away from work
- Reduced healthcare treatment cost
- Fewer employee turnovers
- Reduction in insurance cost
- Lower operating and production cost

Non-Quantifiable Benefits

- A more motivated workforce
- Ability to attract and retain the best and brightest
- Greater stakeholder commitment
- More community support and engagement
- Higher trust and credibility among all stockholders

Accident Prevention

Accident prevention is both a science and an art. It represents, above all other things, control – control of man performance, machine performance, and physical environment. The word 'control' is used widely because it connotes prevention as well as correction of unsafe conditions and circumstances (Heinrich).

By finding out the causes of accidents, necessary steps can be taken to prevent them. The basic safety principle is relatively simple and should be easily understood. Basically, what is required is an acceptance of and firm belief in three fundamental safety concepts: 1. accidents are caused; 2. steps must be taken to prevent accidents; and 3. the same type of accident will recur if corrective action is not taken.

There are six principles of accident prevention. These are:

1. Accident prevention is an essential part of good management and of good workmanship.
2. Management and workers must co-operate wholeheartedly in securing freedom from accidents.
3. Top management must take the lead in organising safety in the workplaces.
4. There must be a definite and known safety policy in each workplace.
5. The organisation and resources necessary to carry out the policy must exist.
6. The best available knowledge and methods must be applied.

Typical accident prevention activities should include:

- safety inspections
- hazard surveys
- accident investigations
- monthly accident returns
- machinery acceptance certificates
- safety device checks
- joint consultation
- publicity
- competitions
- suggestion schemes
- personal protective equipment
- safety training
- fire precautions
- safe systems of work.

Safe systems of work are fundamental to accident prevention. Where safe systems of work are used, following considerations should be given in their preparation and implementation:

1. Safe design
2. Safe installation
3. Safe premises and plant
4. Safe tools and equipment
5. Correct use of plant, tools and equipment
6. Effective planned maintenance of plant and equipment
7. Proper working environment ensuring adequate lighting, heating and ventilation
8. Trained and competent employees
9. Adequate and competent supervision
10. Enforcement of safety policy and rules
11. Additional protection for vulnerable employees
12. Formal issue and proper utilisation of all necessary protective equipment and clothing
13. Continuous emphasis on adherence to the agreed safe method of work by all employees at all levels.
14. Regular reviews of all written systems of work to ensure:
    (a) compliance with current legislation,
    (b) systems are still workable in practice,
    (c) plant modifications are taken into account of,
    (d) substitute` materials are allowed for,
    (e) new work methods are incorporated into the system,
    (f) advances in technology are implemented,
    (g) proper precautions are taken in the light of accident experience, and
    (h) continued involvement in, and awareness of the importance of, written safe systems of work.
15. Regular feedback to all concerned–possibly by safety committees and job contact training sessions–following any changes in existing safe systems of work.

Exhibit 2.2 Ten Principles of the DuPont Corporation in the Field of Health and Safety at Work

1. All injuries and occupational illnesses can be prevented.
2. Management is directly responsible for preventing injuries and illnesses, with each level accountable to the one above and responsible for the level below.
3. Safety is a condition of employment, that is, each employee must assume responsibility for working safely. Safety is as important as production, quality and cost control.

4. Training is an essential element for safe workplaces. Safety awareness does not come naturally – management must teach, motivate and sustain employees' safety knowledge to eliminate injuries.
5. Safety audits must be conducted. Management must audit performance in the workplace.
6. All deficiencies must be corrected promptly, either by modifying facilities, changing procedures, providing better employee training or disciplining constructively and consistently. Follow-up audits are to be used to verify effectiveness.
7. It is essential to investigate all unsafe practices and incidents with injury potential, as well as injuries.
8. Safety off the job is as important as safety on the job.
9. It is good business to prevent illnesses and injuries. They involve tremendous cost – direct and indirect. The highest cost is human suffering.
10. People are the most critical element in the success of a safety and health programme. Management responsibility must be complemented by employees' suggestions and their active involvement.

Source: Stranks, J., Human Factors and Safety, P. 9-10

The prime objective in accident prevention is to control hazards at work, so as to reduce or eliminate accidents. It can be said that accidents result from inadequate hazard management.

The term 'hazard' may be defined as the result of a departure from the normal situation, which has the potential to cause injury, damage or loss. Essentially, there are three steps in the management of hazards: (i) identification, (ii) evaluation or assessment and (iii) control (elimination or reduction). Within an organisation, there are several ways by which hazards may be identified. These include: (i) workplace inspections, (ii) management/worker discussions, (iii) independent audits, (iv) job safety analysis, and (v) hazard and operability studies.

Once a list of hazards within a company has been identified, each hazard on the list should be evaluated. This assessment should take account of legal, humanitarian and economic considerations in addition to the frequency. From these assessments a list of priorities for hazard control can be established and used as the basis of control action.

The control of hazards within an organisation requires careful planning and its achievement will involve both short-term (temporary) and long-term (permanent) measures. These measures can be: (i) eliminate hazard at source; (ii) reduce hazard at source; (iii) remove employee from hazard; (iv) contain hazard by enclosure; (v) reduce employee's exposure to hazard and (vi) utilise protective equipment. The long-term aim must always be to eliminate the hazard at source, but, while attempting to achieve this aim, other short-term actions will be necessary.

System safety techniques may be applied in order to eliminate any machinery malfunctions or mistakes in design that could have serious consequences. According to Bird and Loftus, the stages associated with system safety are as follows:

1. The pre-accident identification of potential hazards.
2. The timely incorporation of effective safety-related design and operational specification, provisions, and criteria.
3. The early evaluation of design and procedures for compliance with applicable safety requirements and criteria.
4. The continued survelliance over all safety aspects throughout the total life-span-including disposal of the sytem.

It is well recognised today that accident prevention is skilled and highly technical work of great importance to modern industry. Accident prevention is a managerial activity in that it must treat both people and things. Thus accident prevention management is a significant profession drawing upon established and systematic principles. It is a combination of specialised body of engineering techniques and humanities.

Safe Work Environment

A number of factors in the work environment may lead to accidents. These factors could be insufficient illumination, ventilation and temperature, noise, heat, dust, and so on.

***Lighting*:** Lighting is important as a safety factor in the physical environment of a worker. Artificial lighting has become the major source of illumination because natural light is undependable source, specially in the winter months when working schedules do not coincide with daylight. Several investigations into the relation between production and lighting have shown that proper lighting may result in maximum production and a minimum inefficiency. Lighting factors that contribute to accidents include direct glare, reflected glare from the work, and dark shadows. Adequate lighting is particularly important for accident prevention in places where there is a risk of stumbling or falling. Where there are large number of persons in workrooms, it is essential for gangways, staircases and exits to be kept lighted under all circumstances.

Although human beings have extraordinary ability to adapt to their environment, their well-being, morale and fatigue are affected by light and colour. Cases of visual disorders at workplaces are frequent, and their causes are manifold. They should always be taken seriously and lighting engineers should try to provide for optimum visibility conditions. According to the present state of the art, good lighting should meet the following conditions:

(a) adequate and proper illumination
(b) uniform lighting
(c) avoidance of glare
(d) appropriate contrast
(e) correct colour

***Ventilation and Temperature*:** Industrial ventilation is regarded as an integral part of air conditioning when it is used in combination with heating, cooling and humidifying appliances to bring the interior of a workroom to a satisfactory condition of the product or for the thermal comfort of the occupants. When used alone, the purpose of ventilation is usually either to keep the occupants cool or to reduce the concentration of a contaminant in the air inhaled by them. Ventilation is normally accomplished by the simultaneous removal of air from a workplace and replacement with fresh air. Ventilation helps to control heat, gases, or dust.

Ventilation, whether general ventilation or local exhaust ventilation, falls mainly within the province of industrial hygiene, but is of some importance from the safety stand point. Ventilation systems, however, require careful designing, particularly exhaust ventilation. If it is badly designed, it can be worse than no ventilation at all. Good ventilation seems to provide air which is pure and clean and an atmosphere which is stimulating and refreshing.

***Heat*:** People must keep the temperature of their vital organs within narrow limits if they are to survive exposure to intemperate environments. As heat impinges upon man, his first response is a sensation of discomfort, job inefficiency, and so on.

A classification of disorders caused by exposures to high levels of environmental heat is as follows:

(a) systematic disorders – heat stroke, heat exhaustion, water defficiency, salt deficiency.
(b) skin disorders – prickley heat, sweating deficiency, cancer of the skin.
(c) psychoneurotic disorders – mild chronic heat fatigue, acute loss of emotional control.

***Dust and Fumes*:** Dust may be defined as a disperse system of heterogenous solid particles in a gas (air) whose broad size distribution is predominantly that of a colloid. Dust generally originates from the mechanical comminution of a coarser material.

Dust is defined by the ILO as 'an aerosol compound of solid inanimate particles'. The term *aerosol* implies that air borne particles are carried in or contained in air, which may be inhaled. An aerosol can embrace liquid droplets as well as solid particles.

Dust particles are carried with the air stream into the lung during inhalation. A small number of these particles may be deposited in the lung, depending on their size. Specific illnesses can be linked with certain types of dust. This is the case of quartz dust (silicosis), asbestos dust (asbestosis), allergenic dust (various allergies), and so on. The extent to which any type of dust represents a health risk depends on exposure which includes the nature of the dust, its concentration and the duration of exposure, as well as upon individual factors such as the general constitution and state of health of the person concerned.

Workers in a dusty occupation should undergo a pre-employment medical examination followed by periodical medical check-ups. Exposure to dust imposes a heavy load on the different functions of the lung and in particular on the self-clearance mechanism. The basic techniques of dust control are:

(a) elimination
(b) the substitution of a less hazardous material
(c) segregation and enclosure of dusty processes
(d) ventilation and filtration

***Noise*:** For hundreds of years, many people have been paying a high price for industrial noise, resulting in permanent loss of hearing, and other disorders. High level of noise can lead to hearing loss, cause annoyance, loss of sleep, and interfere with concentration and communication in the workplace.

Not all sound is noise – noise is sound that people do not like. Noise can be annoying and it can interfere with one's ability to work by causing stress and disturbing concentration. Noise can cause accidents by interfering with communication and warning signals. Noise can cause chronic health problems including loss of hearing, temporary or permanent.

The sounds of industry, growing in volume over the years, have heralded not only technical and economic progress, but also the threat of an ever-increasing incidence of hearing loss and other noise-related hazards to expose employees. Noise is not a new hazard. Indeed, noise-induced hearing loss was observed centuries ago. The advent of steam power during the Industrial Revolution first brought general attention to noise as an occupational hazard. The increasing mechanisation that has occurred in all industries and in most trades has since aggravated the noise problem.

Excessive noise makes communication between workers very difficult, makes hearing of warning signals impossible, causes misunderstandings and leads to permanent loss of hearing. For these reasons, noise must not be overlooked in safety planning. In addition, noise can be extremely tiring, and, in this respect, it has the same ill-effects as other types of fatigue.

Industrial noise levels are increasing. Effective protection of workers against the harmful results of noise is therefore a problem of great actuality. Ear protectors prevent excessive sound energy from entering the external ear canal. Earplugs are inserted in the external ear canal and remain in position without any special fixing device.

***Guarding the Machines*:** The 'Model Code' of Safety Regulations for Industrial Establishments for the Guidance of Governments and Industry (Rule 82) deals with guarding machinery. It reads as follows: Guards should be so designed, constructed and used that they will:

(a) provide positive protection;
(b) prevent all access to the danger zone;
(c) not cause the operator discomfort or inconvenience;
(d) not interfere unnecessarily with production;
(e) operate automatically or with minimum effort;
(f) preferably constitute a built-in feature;
(g) be suitable for the job and the machine;
(h) provide for machine oiling, inspection, adjustment and repair;
(i) withstand long use with minimum maintenance;
(j) resist normal wear and shock;
(k) be durable and fire and erosion resistant;
(l) not constitute a hazard by themselves; and
(m) protect against unforeseen operational contingencies.

***Housekeeping*:** Housekeeping signifies not only cleanliness, but a place for everything and everything in its place. A clean and orderly place makes employees respect the company, plant, and working area. It assists in improving the quality of the products, the efficiency and safety of the worker, his morale and pride. A customer or visitor has more confidence in an organisation when he finds that he is taken care of.

Housekeeping should be ranked at the top of list of things to be done to prevent accidents and increase production. It can be said that housekeeping is the one activity which, more than any other, is the spearhead of a safety programme.

Management must have a sincere desire for a clean and orderly plant and must be able to express this desire clearly by actions, by direct order, or by both. Perhaps it may adopt and publicise a housekeeping policy, which should simply state that the company wants a clean and orderly plan and should be translated into specific orders from executives to department heads and superintendents, to line supervisors, to employees.

Housekeeping involves the inspection of the following items:

1. Loose material and objects under foot
2. Loose material and objects overhead
3. Piling
4. Projecting nails
5. Disposal of scrap and waste
6. Grease, water, or oil spillage
7. Tool housekeeping
8. Marked aisle lines
9. Window cleanliness
10. Painting
11. General cleanliness
12. Orderliness
13. Fire hazards

Steps which management must take in organising and maintaining a good housekeeping programme can be summarised as follows:

1. Make clear to all members of the organisation what is required of them.
2. Enumerate the advantages of good housekeeping for the employees and for the operating managers.
3. Make managers and supervisors responsible and accountable for the good housekeeping of their departments, shops, and work areas.
4. Inform managers what should be done and how; then enforce the rules.
5. Provide the equipment and personnel necessary to get the job done.
6. Reward when a job is being done well and penalise when a job is being done poorly.

The advantages of a well-kept plant are many. The most important factors can be summarised as follows:

1. A reduction in injuries sustained on the job.
2. Increased efficiency, brought about by an easier flow of materials and improved performance.

Fig. 2.3 Poor Housekeeping.

3. Improvement in employees' morale; as employees want to work in an orderly environment.
4. Better labour relations; reduction in complaints by employees.
5. Better community relations; winning the esteem of the community.
6. Minimisation of losses from fire hazards.

Benefits accruing from a safe work environment include reduction in lost employee hour, decrease in hospital cost, reduced sick leave, down-trend in pollution costs, low property damages, and significant reduction in insurance premiums which ultimately result in higher productivity.

Job Safety Analysis

Job safety analysis (or job hazard analysis) is an accident prevention technique that should be used in conjunction with the development of job safety instructions, safety systems of work, and job safety training. The technique of job safety analysis has evolved from the work study techniques known as method study and work measurement. The method study engineers' aim is to improve methods of production. In this, they use a technique known as SREDIM principle:

Select (work to be studied)
Record (how work is done)
Examine (the total situation)
Develop (best method for doing work)
Install (this method into the company's operations)
Maintain (this defined and measured method)

Work measurement is utilised to break the job down into its component parts and, by measuring the quantity of work in each of the component parts, make human effort more effective. Job safety analysis uses the SREDIM principle but measures the risk (rather than the work content) in each of the component parts of the job under review.

The following information is required for effective job safety analysis:

(a) job title
(b) department or section
(c) job operations
(d) machinery and equipment used
(e) materials used i.e. raw materials and finished products
(f) protection needed e.g. machinery guarding, PPE
(g) hazards that may be encountered
(h) degree of risk involved
(i) work organisation
(j) specific tasks

Job safety analysis is based upon four principles:

1. Any job or task can be separated into a series of relative simple steps.

2. The potential risks associated with each step can be identified.
3. The procedures can be developed to control or eliminate potential risks.
4. The procedures can be continuously improved.

The basic procedure for job safety analysis is as follows:

1. Select the job to be analysed. (SELECT)
2. Break the job down into its component parts in an orderly and chronological sequence of job steps. (RECORD)
3. Critically observe and examine each component part of the job to determine the risk of accident. (EXAMINE)
4. Develop control measures to eliminate or reduce the risk of accident. (DEVELOP)
5. Formulate written and safe systems of work and job safety instructions for the job. (INSTALL)
6. Review safe systems of work and job practices at regular intervals to ensure their utilisation. (MAINTAIN)

The benefits of job safety analysis are multitudinous and affect production as well as safe operation. From a safety viewpoint, the advantages are:

1. Discovery of existent physical hazards.
2. Discovery and elimination or safeguarding of motions, positions, or actions that are hazardous.
3. Determination of the qualifications required for the safe performance of the work, such as physical fitness, motor skills, special abilities, and so on.
4. Determination of equipment and tools needed for safety.
5. Establishment of standards needed for safety, including instruction and training of workmen.
6. Proper organisation of methods consistent with accepted efficient and safe practices.
7. Preplanning, preparedness, and proper performance to execute properly the requirements of the operations.

A job safety analysis is a procedure to make jobs safe. The safety analysis identifies the hazards or potential accidents associated with each step of the job, determines the severity of the hazard, and tries to establish a probability of occurrence factor for an accident. The safety analysis also develops a solution that will either eliminate or control each hazard.

When selecting a job to be analysed one should first look into past accident frequency. Certainly, the jobs with the highest frequency should be the first to be analysed. The safety professional should work with the supervisors responsible for the areas where the accidents occur. In planning and setting up the production job, the following provisions should be made for safety:

1. Comfortable seat with proper height in relation to table.
2. All moving transmission parts fully enclosed.
3. Machine controls conveniently located and protected against accidental contact.
4. Provide lighting of adequate intensity free from shadows or glare.

The hazards usually present in the maintenance job are:

1. Interference with adjacent operations.
2. Contact with adjacent machines or equipment.
3. Falls in setting up and removing hoisting rig or in working on scaffold or machine frame, or from tripping over loose material or from slippery footing due to oil spillage.
4. Flying particles.
5. Hand tool hazards.
6. Operations connected with the lifting, hoisting, moving and placing of mill parts.
7. Explosions of babbitt due to failure to guard adequately against the presence of moisture.
8. Burns.
9. Electric shock or burns if electric powered tools or extension lights are used.

General safe practices in maintenance work should include:

1. Use of proper tools for specific tasks.
2. Planning for proper headroom while working.
3. Placing tools conveniently for reaching.
4. Keeping tools in A-1 condition.
5. Prevention of reaching over moving equipment by establishing a rule to stop all machinery before starting to work, and locking out or tagging the switch.
6. Correct positions for lifting.
7. Consideration of hazards to other workmen.
8. Reporting of noted hazards.

Investigation of Accidents

Accident investigation is an important aspect of any effective safety and health programme. Accident investigation is necessay to reduce the likelihood of their being repeated. Accidents should be investigated as soon as possible after their occurrence.

There are four main methods of gathering accident information:

1. Interviewing the accident victims.
2. Interviewing accident witnesses.

3. Investigation of the accident scene.
4. Reenactment of the accident.

The following factors are to be considered in investigation of accidents:

1. Analyse the case carefully and impartially.
2. Investigation has to be done by right people.
3. Do not dismiss it by saying the man was careless.
4. Take some action to prevent a recurrence.
5. Avoid looking for excuses and get the causes.
6. Avoid trying to convict someone for negligence, and remember that the worker, the supervisor, and the company are usually at fault though the degree may vary.
7. A knowledge of all circumstances surrounding an accident is essential.
8. Make use of the information obtained.

The purpose of an accident investigation is: (a) not to fix the blame on somebody but to find out the causes of the accident; and (b) to take appropriate preventive measures. Basically, whatever type of investigation is undertaken, it must answer the following questions:

- Who was injured?
- When did the accident occur?
- Where did the accident occur?
- Why did the accident occur?
- What happended and what were the contributing factors?
- How to prevent such an accident again?

Effective accident investigation is essential to the success of an accident prevention programme and should not be regarded as simply a necessary evil. Every accident has a cause and result. Accident investigation is an important means of developing preventive techniques. Investigations should reveal causes, agencies, and locations of predominant types of accidents, and it is essential that preventive measures and recommendations based on the investigations be put into effect as soon as possible.

In accident investigation certain key facts about each injury or property damage and the accident that produced it should be identified and recorded.

The following are some of the key facts:

1. Nature of injury or loss
2 Source of the injury or loss
3. Type of accident
4. Hazardous condition
5. Personal data
6. Location of accident
7. Cause of accident

An accident investigation should always be made on the spot. It will be much easier if the investigator finds the situation at the scene of the accident exactly as it was when the accident took place. Consequently, after an accident, the site should be left undisturbed unless changes have to be made to ensure the safety of persons or to prevent further damage. The investigator should carefully inspect the site, seek technical assistance if any, and then examine the witnesses. When an accident is due to unforeseen chemical reactions, laboratory tests are also necessary in order to ascertain what exactly happened.

Accidents, whether they result in injury, damage, disease or loss, need to be controlled. To enable an accident control system to be developed, it is necessary that all accidents are reported, recorded, investigated and analysed, so that after remedial action has been decided, plans can be drawn up to prevent a recurrence. The most important question to be asked in any accident investigation is: 'What action has been taken to prevent a recurrence?'

The following are some principles of accident investigation:

1. Common sense and clear thinking are must.
2. Investigator should be familiar with the equipment, operations and processes.
3. Investigator should not be working under a foreman or supervisor of the department when the accident had taken place.
4. Each clue should be investigated thoroughly.
5. Definite recommendations for corrective actions should be made.
6. Investigation should be done as soon as possible after the occurrence.

Table 2.1 Accident Investigation Report.

1. Name and designation of the injured person
2. Date and time of the accident
3. Exact location in which the accident took place
4. What was the injured doing at the time of accident
5. Describe briefly how the accident occured (add sketches and additional sheets, if necessary)
6. What was the unsafe act/condition which caused the accident?
7. Estimated lost time from accident or illness or seriousness of the accident
8. What are the precautions taken to prevent similar accidents?
9. Any other information

Ergonomics

The term "ergonomics" is derived from two Greek words, 'ergon' meaning work and 'namos' meaning law or rule. One of the simplest definitions of ergonomics is 'the science of making the job fit the worker'. Another definition is 'the study of man in his working environment'. It deals with the variety of ways in which people interact with their work environment, including design and function of controls, displays, safety devices, lighting, temperature, workplace layout, tools and work organisation. Ergonomics is the scientific study of the inter-relationships between people and their work. Ergonomics, or 'human factors engineering' or 'the scientific study of work', seeks to create working environments in which people receive prime consideration. Ergonomics is also sometimes narrowly referred to as the study of the 'man-machine interface'. This interface is significant in the design of working layouts and safe systems of work and in setting work rates.

Ergonomics is a multi-disciplinary study which incorporates the expertise of engineers, occupational physicians, occupational health nurses, occupational hygienists, health and safety specialists, organisation and method study people and research scientists. It is a team approach to the examination of numerous aspects of the working environment.

Ergonomics is an inter-disciplinary field that draws from anatomy, psychology, engineering, medicine, anthropology, and other related disciplines. One of the greatest values of ergonomics is that it causes people to think and promote innovation. For ergonomic issues, a team approach often helps make for an effective programme. Those involved should represent safety/medical, engineering, operations management, human resource management, and production employees and/or union representatives.

By applying the basic principles of this field, employers can reduce costs related to issues like workers' compensation, turnover and absenteeism. Plant operations can be made more efficient by workplace designs. Ergonomics provides a set of conceptual guideposts for adapting workplaces, products and services to fit human needs. The field provides a strategy for engineering design and a philosophy for good management.

The basic philosophy behind the ergonomic approach to safety is that of designing to take into account the limitations of the human operator. The goals of the occupation of ergonomics are two-pronged: first, to design equipment and environment which fit the individual's capacities and needs and by so doing allow him to perform effectively, and second, to ensure that design of work systems and environments does not violate requirements for physical and mental well-being, including the acceptability of the system to the operator and his level of general comfort whilst working.

The basic approach of ergonomics is that of 'fitting the task to the man' as contrasted with the complementary approach of 'fitting the man to the task' which emphasises the requirements for selection and training. The ergonomic approach to health and safety examines, in particular, the physical and mental capacities and limitations of operators, taking into account, at the same time, psychological factors, such as learning, individual skills, perception, attitudes, and physical factors, such as strength, stamina and body dimensions. It is also concerned with the working environment and the potential for environmental stress associated with, for instance, extremes of temperature, inadequate lighting, noise and vibration.

Ergonomics, therefore, is a very broad area of study largely concerned with maximising human performance and, at the same time, eliminating as far as possible, the potential for human error.

Ergonomics, a relatively young discipline, is basically concerned with maximising human efficiency, comfort and skill. It explores and optimises the interactions between man, machine and his working environment, and thus directly contributes to his health and safety in the work system.

Ergonomics has a direct and important part to play in the prevention of occupational ill health and accidental injury at work. Application of ergonomics can make a major contribution in ensuring a safer and healthier workplace.

Ergonomics is concerned with the provision of efficient work stations, devices and products that match human capabilities and needs. It has made, and will continue to make, many important contributions to industrial safety. Ergonomic principles, as applied to the design of the office system as a whole, can both improve work efficiency and safety, and also reduce the occurrence of associated occupational ill health.

Personal Protective Equipment (PPE)

Personal Protective Equipment commonly referred to as "PPE " is equipment worn to minimise exposure to hazards that causes serious workplace injuries and illnesses. These injuries and illnesses may result from contact with chemical, radiological, electrical, mechanical, or other workplace hazards.

Protective equipment exists for almost every type of hazard and for all situations. The best way of preventing accidents is undoubtedly by eliminating the hazard, or controlling it as close to the source as possible. Personal protection should be considered as the last line of defence, because often the equipment is cumbersome to wear and restricts movement. Therefore, it is sometimes discarded by workers. In many countries, standards for personal protective equipment have been drawn up by national standard institutes. Whatever the nature of the hazard, the equipment or clothing must provide adequate protection against the specific hazard and such equipment or clothing should cause the minimum of discomfort.

A wide variety of devices have been developed for protection of exposed or vulnerable parts of the body against many specific hazards. Examples of protective devices include safety helmets, face shields, safety belts, safety goggles, safety gloves, ear plugs and safety shoes. If many injuries involve the eyes, safety goggles may be required, or if they are already provided, their use may be enforced when working at certain jobs. Safety gloves are to be used to protect fingers and hands from injuries. Safety shoes have been developed for protection in a wide variety of situations.

When purchasing safety equipment, management should select only the best equipment available for the intended hazard protection. Some equipment standards, especially those relating to hazardous operations and rescue equipment, require that the equipment be certified, approved, listed, and tested for its intended use.

Once safety equipment is provided, the employer must insist that workers use it and use it properly. Both supervisors and workers must be instructed how to use and to recognise the capabilities and limitations of the equipment. Safety equipment must be inspected before and after use.

***Head Protection*:** There are many occupations that require the use of head protection to protect workers from falling and flying objects. Particularly hazardous operations requiring the use of head protection include construction work, mining, ship building, logging, general maintenance, and the foundry and steel industries. Workers who are liable to be struck by falling or flying objects, or otherwise exposed to head injuries, should wear hard hats or helmets, which are strong enough to protect them, but not too heavy.

***Eye and Face Protection*:** Injuries to the eyes and face can not only disable, but also disfigure workers. Flying objects cause most eye injuries. Eye protection instruments must be properly selected and used. Many types of personal protective equipment shield the face against light impact, splashes of chemicals or hot metals, heat radiation, and other hazards of similar nature.

It is important to ensure not only that protective equipment is worn, but that it is worn correctly. For example, workers not accustomed to wearing glasses sometimes reject the various types of eye protective equipment because it is a nuisance and causes discomfort.

***Hand Protection*:** The hands and fingers are exposed to all kinds of cuts, scratches, bruises, and burns. Although fingers are hard to protect, they can be shielded from many common injuries by proper protective equipment.

Gloves should not only protect workers from hazards but should also allow the fingers and hands to move freely. The kind of glove required will vary according to the injury to be prevented. It should be remembered that it is dangerous to wear gloves when working at drilling machines, power presses, and other machines in which a glove might be caught.

***Skin Protection*:** Protective creams may be used to supplement protective devices and clothing; however, the creams are to be used carefully.

***Foot and Leg Protection*:** In most instances of foot injury, the severity of the injury could have been reduced significantly if the injured person had been wearing appropriate foot equipment. Foot protection is needed in almost all industrial operations, and employers should see to it that employees wear the proper protective equipment wherever hazards exist. For leg protection proper guards are to be used by persons who work around molten metal in foundaries and steel mills.

Safety shoes should protect workers against accidents caused by heavy objects dropping on the field, protruding nails, molten metal and acids. Sometimes, special kinds of footwear are necessary.

***Lung Protection*:** In the control of occupational diseases caused by breathing contaminated air, the primary objective is to prevent atmospheric contamination. Suitable respirators must be provided by the employer when they are necessary for the protection of the health of the employee. Respiratory protective devices can be classified as (i) air purifying respirators; (ii) atmosphere (air) supplied respirators; and (iii) self-contained breathing devices.

The following are the guidelines for effective use of PPE:

- It should provide adequate protection against the particular hazard for which they are designed.
- It should be reasonably comfortable when worn under the designated conditions.
- It should fit properly without unduly interfering with the movements of the user.
- It should be durable.
- It should be so designed that it can be disinfected.
- It should be easily cleanable.
- It should be kept clean and in good condition to minimise the degree of risk.
- It should be correctly selected and tested.

- It should always be applied and used whenever and wherever needed.

Points to Remember About PPE

- PPE is the least effective method for controlling hazards in the workplace and should be used only when hazards cannot be controlled sufficiently by other methods.
- PPE can be uncomfortable, decrease work performance and can create new health and safety hazards. Workers in PPE should take regular breaks.
- Hot or humid working conditions decrease the effectiveness of PPE. Under these conditions, workers should take frequent breaks and drink plenty of fluids.
- The type of PPE required depends on the hazard, the way exposure affects the body and the exposure time.
- All workers using PPE should be trained in the proper use, maintenance and limitations of PPE.

PPE information and training is an employer's prerogative and responsibility. Employers are required to train each worker as to:

- When it is necessary
- What type is necessary
- How to properly put it on, adjust, wear, and take it off
- The limitations of the equipment
- Proper care, maintenance, and disposal of the equipment

Risk Assessment

Risk assessment means identification of hazards and their elimination. Risks are inherent in:

- Materials
- New plant and equipment
- Modifications
- Systems of work
- Products.

Why risk assessment:

1. It improves safety performance by reducing accidents.
2. It helps to recognise, manage and control risk.
3. It is good for employees, for business and public.

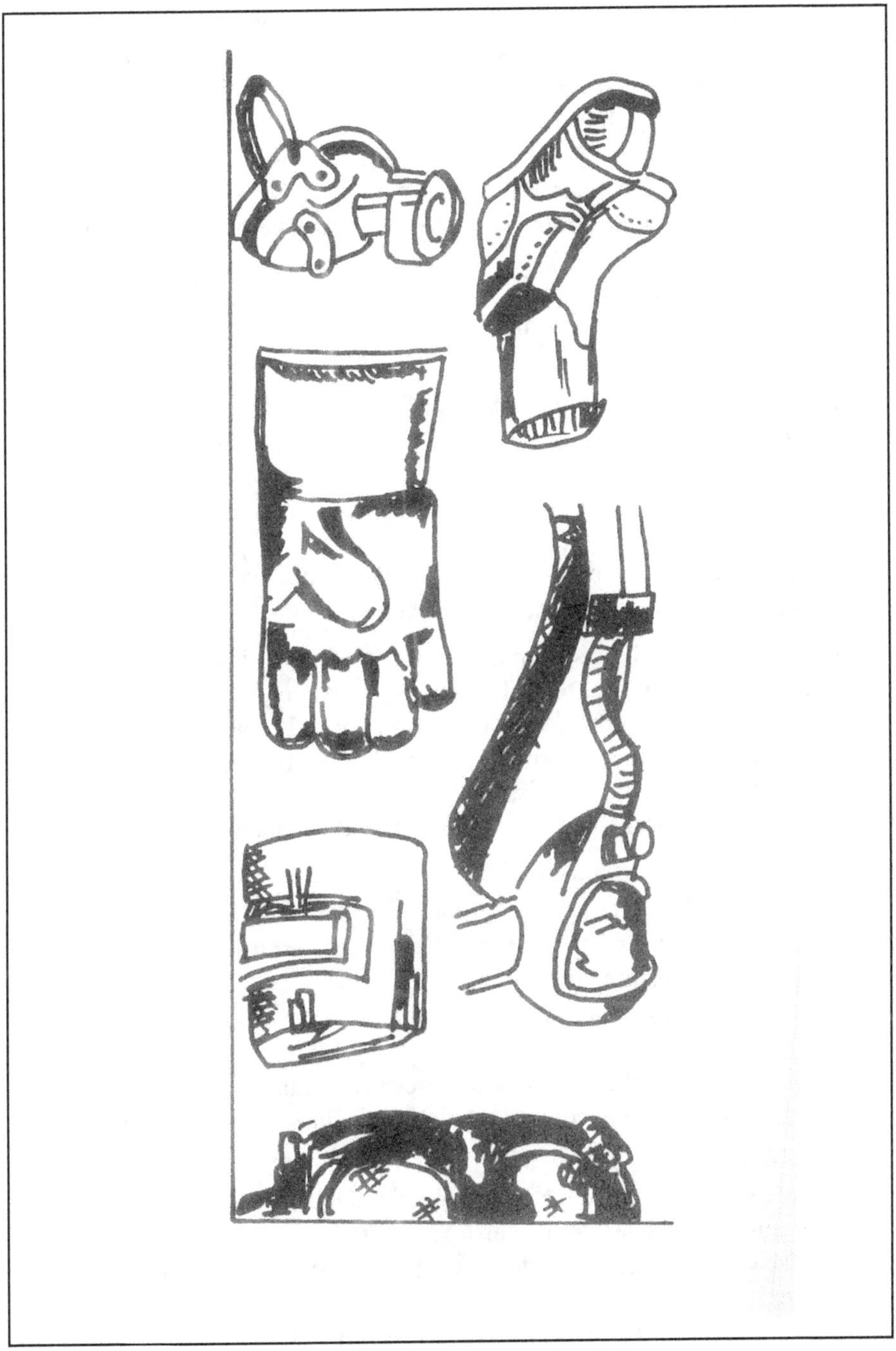

Fig. 2.4 Personal Protective Eqipment.

Risk assessment is a safety system where people most at risk of an accident are assessed, reasons are explored, and risks are reduced. There are five steps in risk assessment:

1. Look for the hazards.
2. Decide who is likely to be harmed and how.
3. Evaluate the risks arising from the hazards and decide whether existing precautions are adequate or whether more should be done.
4. Record the findings.
5. Review the assessment from time to time and revise it, if necessary.

It is necessary to quantify risk in order to make decision making easier and to make products more cost effective. Risk is expressed in terms of frequency i.e. financial risk, risk to life and environment. We can therefore assess the level of risk by looking at the combination of frequency and consequence. Event happens less often as a result of protective measures. Risk assessment of employees who handle or are exposed to hazardous substances must be made as early as possible.

Risk assessment is a systematic process of assessing causes of health risks related to the use of a hazardous substance. It helps to make a judgement about the risks and to decide on control measures that are to be taken. It has to be done whenever a new job is created, a new process is initiated, and a new situation is faced. It needs to be kept up-to-date and to be reviewed whenever there is a major accident or incident.

The risk assessment must include:

- the identification of the hazardous substances;
- the assessment of the risk created by the hazardous substance, the working environment, and work processes;
- a decision whether workers may be exposed to the hazardous substance;
- a decision on the control measures (including health surveillance and monitoring) needed in relation to the hazardous substance.

Steps to reduce the risk:

1. Keep the workplace clean and free from hazards.
2. Remove obstacles, and clear-up spillages.
3. Use your judgement, do your job correctly, and use equipment properly.
4. Identify hazards, remove them if you can, and put up warning signs.
5. Know who the first-aid personnel are; how to contact them; and where the emergency equipment, first-aid boxes and alarms are situated.
6. Know what to do in case of fire, accident or emergency.

7. Lift objects correctly and do not overstrain yourself.
8. Learn about the hazards in your occupation and environment, and take necessary precautions.
9. Make sure that wiring is secured and safely insulated. If wires temporarily cross paths, put up warning signs.
10. Use equipment in which you have been trained.

Risk Perception

Risk perception refers to risk taking by individuals or by an individual in his/her capacity to accept a certain amount of risk. In short, ability to discern risk is tied up to risk tolerance. Known hazards and risk factors are generally taken care of by using standard safe design, engineering, installation and selection of safe operational processes and practices. However, these risk factors do not discourage individuals taking risks. The reasons are diverse human factors, mostly psychological. An organisation has to deal with additional psychological risk factors such as risk tolerance and workplace emotions.

Safety perception of employees is built on internalisation and integration of self and organisational safety beliefs, values and behavioural regulations. In measuring and managing of employee safety and risk perceptions, organisations need the leading indicators of perception change that can be developed by answering the following questions:

1. Does the organisation know about the current level of employee safety/risk perception?
2. Do the seniors care to know the psychological profile of their reports?
3. Do the perceptions lead to a positive safety attitude?
4. Do the decisions taken by superiors shape perceptions, attitudes and behaviours of their reports?
5. Are the employees competent to take decisions on safety matters related to their activities?
6. Do the incidents and work experiences influence employee safety perceptions, decisions, behaviours and stories?
7. What are the current behaviours and stories? Are these negative, positive or mixed?

Promotion of Health and Safety

In the promotion of health and safety among industrial wage earners, four general lines of procedure are possible:

1. The provision of a satisfactory working environment.
2. Intelligent employment policies and methods.
3. A constructive programme of betterment activities.
4. A well-organised and efficient medical and safety service.

A satisfactory working environment includes a safe, helpful place of employment. Machinery, tools, equipment, and processes are to be safeguarded so that hazards to body and health are reduced to a minimum. Protection is given against communicable diseases. Hours of work are reasonable and rest periods are frequent and well-timed.

The various means generally used at present to promote industrial safety may be classified as:

(a) *regulations*, i.e. mandatory prescriptions concerning such matters as general working conditions, the design, construction, maintenance, inspection, testing and operation of industrial equipment, the duties of employers and workers, training, medical supervision, first-aid, and medical examinations;

(b) *standardisation*, i.e. the laying down of official, semi-official or unofficial standards concerning, for e.g. the safe construction of certain types of industrial equipment, safe and hygienic practices, or personal protective devices;

(c) *inspection*, i.e. the enforcement of mandatory regulations;

(d) *technical research*, including such matters as investigation of the properties and characteristics of harmful materials, the study of machine guards, the testing of respiratory masks, the investigation of methods of preventing gas and dust explosions, or the search for the most suitable materials and designs for hoisting ropes and other hoisting equipment;

(e) *psychological research*, i.e. investigation of the psychological patterns prone to accidents;

(f) *medical research*, including, in particular, investigation of the physiological and pathological effects of environmental and technological factors;

(g) *statistical research* to ascertain what kinds of accidents occur, in what numbers, to what types of people, in what operations, and from what causes;

(h) *education*, involving the teaching of safety as a subject in engineering colleges, trade schools or apprenticeship courses;

(i) *training*, i.e. the practical instruction of workers, and especially new workers, in safety matters;

(j) *persuasion*, i.e. the employment of various methods of publicity and appeal to develop safety awareness;

(k) *safety measures* within the individual undertaking.

Basic Safety Programming

There are certain logical steps in safety programmes which are usually carried out in the following order:

1. ***Secure principal management's involvement*:** Obtaining a highly visible commitment to safety from management is regarded generally as the first essential.
2. ***Organise for achievement*:** The safety specialist is expected to marshall facts and resources, forming a coordinated effort.
3. ***Detail the operating plan*:** The company's safety objective, policies, rules and regulations, and the method chosen for their implementation should be communicated upon the programme's initiation.
4. ***Inspect operations*:** Knowledge about the conditions to be corrected and an ongoing evaluation of the progress being made is provided by plant inspections.
5. ***Consider engineering revisions*:** Corrections are expected to begin with consideration of means of removing physical hazards.
6. ***Use guards and protective devices as a last resort*:** If engineering revisions are not possible or will not complete the safety objective, use supplementary means to safeguard the exposure.
7. ***Provide education and training*:** Provide safety training and education on an on-going basis.

Seven factors have to be thought of, planned for, if a proposed safety programme is to have its fair chance of success.

1. ***Top management leadership*:** A successful safety programme can get along without good many things, but not without the support of top management. Top management's support shows itself in several ways. One of them is a clear-cut, simply stated and easily understood expression of policy. Management's support of the safety programme also shows itself in the chief executive's personal interest in who runs it and how it is run.
2. ***Somebody must be responsible*:** Stray dogs are always underfed. The safety programme that is passed like the buck, from one department to another, the programme which is half under personnel department, half under two or three other departments will likewise suffer from malnutrition.
3. ***The supervisory staff takes hold*:** The supervisor who is in direct daily contact with the production force has the final answer in any plant on questions of safety, especially in the control of unsafe practices, in training procedures, and in the building of proper attitudes on the part of employees. Another reason for a supervisor's participation in the safety programme is the kind of responsibility he holds for the lives and well-being of people under his direction.

4. ***The element of training*:** Since human beings are capable of doing almost anything, both unexpected and predictable, the problem of controlling human behaviour is a major one in the safety programme. Safety training of employees will depend upon the kind of work the employee is to do and the type of hazard most prevalent in that work. While training in safe practices is not substitute for supervision, adequate training in safe practices will do a great deal to minimise the supervisory role.
5. ***Employee belief in safety*:** One of the best conditioners of employee attitudes is a good system of job training. One way to arouse interest and a sense of personal responsibility in employees is to give them a share in working out the safety rules.
6. ***Inspection of buildings and operations*:** Inspection of buildings and machines and operations on a regular schedule serves several useful purposes. The primary purpose of inspection is to discover places, equipment and conditions which, through wear and tear and oversight, have created hazards.
7. ***Accident records and analysis*:** Like all book keeping, a record of a departments' accident history serves a particular purpose.

   Many industrial organisations today state that their first three concerns are safety, quality, and productivity respectively. While designing the processes, machines, jigs and fixtures, none should be considered perfect and ready for use until it is clear that the operation will be safe to the worker involved directly and will not constitute a hazard to other employees.

Goals for the Future

The ultimate goal of any safety programme should be to eliminate all accidents. No unplanned or uncontrolled activity with the potential to cause injury should happen. An accident-free working environment may be an idealistic goal, but is certainly achievable. The following principles are a valuable guide:

1. All injuries and occupational diseases can be prevented.
2. Management is directly responsible for preventing injuries and illnesses as at every level.
3. Each employee should assume responsibility for working safely.
4. Professionalism in safety is as important as professionalism in production, quality and cost control.
5. Training is an essential element for safe workplaces and for developing safety awareness.
6. Safety audits should be conducted at the workplace.
7. All deficiencies should be corrected promptly through modifying facilities and changing procedures.

8. It is essential to investigate all unsafe practices and incidents with injury potential, as well as injuries.
9. Safety off-the-job can be an important influence on job safety.
10. It is good business to prevent illnesses and injuries as they involve significant costs - direct and indirect; the highest cost is human suffering.
11. People are the most critical element in the success of a safety programme.
12. Management responsibility should be complemented by employees' suggestions and their active involvement.

Test your Knowledge

1. List three hazards with potential to cause harm in your workplace.
2. List three common causes of accidents in your workplace.
3. List three differences between a hazard and a risk?
4. List three key factors for good safety awareness in the workplace.
5. List three things that employers are legally bound to do.

CASE

An oil company had known for years that the captain of one of its oil tankers had suffered from a health problem of serious drinking, but nonetheless left him in command of oil tankers. On one of the tanker journeys, the captain was drunk on duty, and left an inexperienced officer in charge of navigating the tanker through hazardous icy conditions. The tanker ran aground a reef, and spilled over 40 million litres of crude oil, which coated about 2,400 km of coastline and killed thousands of birds, marine mammals and other aquatic creatures. How could such an accident could have been avoided?

SUMMARY

- The dictionary defines an accident as 'an unfortunate event' or a 'misfortune or mishap', especially causing injury or death.
- The terms *accidents* and *injuries* are mistakenly used interchangeably; but are different.
- One accident many involve several injuries or no injuries; and so the number of accidents and injuries experienced by a given organisation over a period of time are unlikely to be equal.
- Industrial accidents are the end products of unsafe acts and unsafe conditions of work.

- Accidents are preventable, and the hazards at the workplace can be removed by a variety of methods.
- Workers are to be prevented from being exposed to occupational hazards.
- Having safety means that employees: (a) understand and follow procedure; (b) report unsafe conditions; (c) encourage others to work safely; (d) find, solve and fix the problems in their areas of

Three essentials that must be provided for in the workplace are:

- (a) responsibility; (b) suggest improvements in procedures and equipment; and (c) avoid shortcuts of a risky nature.
- Safety activities usually include: (a) developing and administering the company's safety programme; (b) inspection to locate unsafe conditions or unsafe practices; (c) investigating injuries, particularly the more serious ones and ensuring timely corrective action; (d) maintaining work injury and illness records; (e) publicizing safety materials; (f) checking statutory compliance and rules and regulations; and (g) checking on or aiding in safety aspects of training.
- Safe systems of work are fundamental to accident prevention.
- Common experience reminds us that injurious occurrences are repeated despite knowledge of their causes or the availability of recommended controls.
- Job safety analysis is a technique which identifies all accident prevention measures appropriate to a particular job or area of work activity and the behavioural factors which most significantly influence whether or not those measures are to be taken.
- For reducing accidents, certain general control measures are to be observed: (a) use of national standards; (b) observance of safety rules and regulations; (c) testing and inspection; (d) marking and warnings; (e) work permit system; and (f) use of personal protective equipment.
- Indeed, implementation difficulties have been the critical problem for hazard control programmes.
- Ergonomics and human factors is a multidisciplinary activity striving to assemble information on people's capacities and capabilities in designing equipment, jobs, products, and workplaces.
- Human factors are emerging as a significant factor in safety and accident prevention.
- Safety philosophy helps to create a culture of accountability and responsibility apart from respect to human life and improved morale.

- It is important to regularly monitor the effectiveness of all technological initiatives.
- Technology has a significant impact in the reduction of injuries and deaths in the workplace.

DISCUSSION QUESTIONS

1. What are the causes of accidents?
2. What is meant by the term "unsafe mechanical or physical conditions"?
3. How to prevent industrial accidents?
4. What is the place of personal protective equipment in general, in preventing accidents?
5. "Much of good safety is really just good common sense", i.e., think safety; act safely; be safe. Enumerate
6. What specific things the management must do to prevent accidents?
7. Is safety more of a psychological problem or an engineering problem?
8. Keeping in view cost reduction as an objective of any safety programme, work out a plan for preventing injuries and deaths in your organisation.
9. Prepare a housekeeping inspection checklist suitable for the plant in which you work.

3

Development of Industrial Health and Safety

The concept of safety is probably as old as the history of mankind itself. Efforts towards protecting the safety and health of the industrial worker were initiated by philanthropists with the idea of mitigating the suffering of the weaker or vulnerable sections of the society. In the absence of any concrete information, it is difficult to say when, how or where it began. But as early as 1800s, there has been social concern generated about the hazards of occupation and working conditions in factories in many parts of the world. Between 1850 and 1900 most of the industrialised countries had introduced factory legislation covering employment of women and children, working hours, exposure to occupational diseases and compensation for injuries.

The first popular book on industrial safety was written by H.W. Heinrich of the Travellers Insurance Company Ltd., USA, in the year 1931. This book dealt at length with the technical aspects of accident prevention such as machine guarding, industrial lighting, fire prevention, ionizing radiation, and laid down guidelines for practical safety programmes in industry.

Since then there have been continuing efforts from different quarters for the betterment of working conditions and ensuring safety of the workmen. There have been a number of factors which have directly or indirectly contributed to this growth during the past few years:

1. Over these years, the trade union movement has grown and come to be recognised as an integral part of the social control system. The trade unions have contributed to the safety movement in many ways.
2. The safety legislation has undergone revisions during the past few decades specifying better conditions of work and increased

responsibility for safety on the part of employer. The enforcement of the laws have also improved.

3. In the field of management itself, new theories have emerged advocating a changed approach to the human tasks and giving more importance to the dignity of human beings.
4. The all-round consciousness generated and the enlightened management and the general society have led to the formation of many voluminous bodies to promote health and safety in industry, which have made substantial contribution to awareness, training and research.
5. One more factor which contributed to the growth of safety movement is the improvement of the standard of education of workers which made them conscious of the hazards around and their right for protection.

Firstly, the main motivating factors for management's efforts directed at accident prevention are the statutory provisions to be complied with.

Secondly, there are considerations relating to the prestige or image of the company. A poor accident record and frequent entanglement with the law, lowers the prestige of the organisation in the eyes of the public. More and more units are making efforts towards achieving better safety records.

The material considerations are not least important. Now-a-days, an accident can be very costly to the management in terms of cost of medical aid, medical facilities, compensation costs, loss of production and so on. Accidents also sometimes trigger industrial relations problems.

The current thinking is that the employer has not only to provide safe plant and equipment, but also create a climate where safe operations are possible.

During recent years the dimensions of "safety" have increased and a number of other areas of technology and management have found their application in the management of safety in industry. They include industrial hygiene, ergonomics and work physiology, industrial medicine, illumination, ventilation, industrial psychology, operational research, and so on.

One of the many achievements of the safety movement is that it has generated voluntary efforts towards prevention of accidents in industry. Industries in both public and private sectors have been encouraged and motivated to launch accident prevention programmes, with many of them achieving creditable performance. Creation of safety awareness at different levels is yet another achievement.

If the changes that have taken place in the past are any indication, the demands on the management as regards safety and health at works will substantially increase in the future. The standards employed need to be revised in tune with the demands of the time. On the whole, the safety and health task is going to be more demanding and challenging in the future.

Occupational health is a field which is growing rapidly all over the world. Its objectives have gradually broadened from dealing mainly with occupational hazards causing accidents and occupational disease to include all kinds of factors at work or related to working conditions. The activities of occupational health services have changed and their scope is now much wider than before, not only in preventing occupational hazards but also promoting the general health of the worker and the adjustment of work to man and of man to work.

Developments in Occupational Health

Indifference to work health and safety, has been a feature of both ancient and modern societies until relatively recent times. Rapid and extensive developments in occupational health began in the early 1940s when the Second World War made an impact on human beings. In both the developed and developing countries there has been a growing awareness of its importance.

Mining is one of the oldest industries and has always been a hazardous occupation. Conditions in the gold, silver and lead mines of ancient Greece and Egypt reveal an almost complete disregard for miners' health and safety. During the middle ages, the status of the miner had changed. Mining in Central Europe had become a skilled occupation, which led to the emancipation of the miner.

Eighteenth century brought great technological inventions and laid the foundations in Europe and North America, a modern society with its factory system. It exposed workers of all grades to the pressures of increasing production and associated physical and psychosocial hazards of work.

The more serious effects of health which followed the Industrial Revolution were not directly occupational in origin. Family life was disrupted when men moved into new industrial areas leaving their families behind, a situation that encouraged alcoholism and prostitution. Epidemics followed as a result of overcrowding in insanitary conditions. Work moved people from the countryside to the new industrial towns. Poor housing, overcrowding and lack of sanitation caused by concentration of an expanding population around the new factories led to the development of public health services which were designed to control disease and improve the health of these communities.

The health problems arising from industrial progress in developing countries today are, in many aspects, similar to those during industrialisation in the 19th century; these countries also have to face major threats from endemic disease and generalised poverty.

Inside the factories and mines of the 19th century the workers were exposed to the hazards of occupational disease and injury and the adverse effects of excessively long hours of work. As manufacturing techniques improve, machines became speedier and more dangerous. Little attention was paid to safety devices and workers were often simple people untrained to handle the new machinery. Toxic hazards increase due to prolonged exposure to a wider range of new chemicals which were introduced without considering their possible effect on workers. In the 19th century, manufacturers generally believed it was economically important to keep their new machines running continuously with cheap labour.

After the period of rapid expansion which followed the Second World War, occupational health service has continued to develop and consolidate. Group occupational health service have been set up to meet the needs of small plants. Most of the heavy industries and public organisations have established occupational health services. However, in developing countries, through a combination of failure to reorganise needs, and inadequate resources to meet them, services have often been either insufficient or non-existent.

During the 20th century the trade unions in many countries began to exert influence on occupational health by pressing for improvements in legislation and for the extension of compensation laws to cover occupational injuries and diseases. They have since become more directly involved in health and safety.

In 1973, WHO listed three major tasks in occupational health, namely, identifying and controlling known or suspected work factors that contribute to ill-health; educating management and workers to fulfil their responsibilities for health and safety; and promoting health programmes not primarily concerned with work-related injury and disease.

Without doubt, people at each stage of their development were distressed by injuries to themselves, damage to an important tool, or any other unexpected personal loss. Probably they were less concerned, if bothered at all in the beginning, when an injury occurred to someone else. As a sense of right and wrong developed, people reasoned that whoever caused injury should suffer equal loss and pain, but they did not focus on observation and analysis on the injury causation process.

Perhaps one clue to why personal safety does not command consistent and urgent attention lies in the original application of the idea of right and wrong. Penalties were used early to control injurious occurrences. No attention was given to how and why the injuries happened. It was simply believed that injuries would be controlled by introducing punishments as countermeasures.

More than 2,000 years before the Christian period, ancient Babylon concerned itself with the accidents of the time and prescribed a method for indemnifying the injured. Hammurabi (2100 B.C.) during the 30th year of his reign, regained independence for Babylon and ordered the compilation of a body of laws. These were carved on a diorite column in 3,600 lines of cuneiforms which now is in Paris. The Code of Hammurabi survived to influence Syro-Roman and later Mohammedan law.

Since shipping by sea was common place, the Code required the ship builder to make good any defect of construction and the damage it caused within a year of delivery. The captain was responsible for the freight and the ship. He was required to replace all loss at sea. Even if he refloated the ship, he had to pay a fine of half its value if it sank. The details were clearly worked out so that in the case of collision, the boat underway was made responsible for damages to the boat at anchor.

According to the Code, carelessness and neglect were severely punished. A veterinary surgeon who caused the death of an ox or ass paid a quarter of its value. Depending on the circumstances, a builder whose careless workmanship caused a fatal injury or goods damaged, he had to rebuild the house, make good any damages due to the defective building and repair the defect as well. Obviously, the Babylonians were concerned primarily with providing redress for damages.

The Code of Hammurabi provides significant evidence that there was a decided awareness that atleast 4,000 years ago of the need for ajusting and controlling unwanted losses. The existence of tribunals to adjudge awards is indicated in the Code. Unfortunately, the Code espoused the ancient eye-for-an-eye principle, which attempts to maintain control by the explicit threat of punishment equal to the severity of the offence. Thus it may have set a pattern for safety that largely became conventional. The Code contributed to the familiar notion that authorative regulations and the threat of discipline for violations are the principle means of furnishing safety.

In the great civilisations of yore, labour was not predominantly done by slaves. A popular assumption, therefore, that safety was of no concern because the worker was merely a slave is probably wrong. As a matter of fact, there may have been more concern for the safety of slaves than for free men. The slave was regarded as a valuable capital asset by its owner. There is good evidence of concern over harmful working conditions at the start of the Christian era, although the reason is obscure as to whether this interest was humanitarian or merely a result of the desire to protect one's investment.

Ramazzini, an Italian physician, published his classical book, *De morbis Artificium* in the early eighteenth century. With this, a proper study of the health of the worker can be said to have taken place. Subsequently, there grew up the practice of industrial medicine dealing in the earlier years largely with treatment of injury and sickness, but tending gradually to become more interested in preventive work. Now managements and workers find industrial medicine as essentially preventive in character rather than curative. Prevention is now the keynote, especially in more hazardous industries.

The principle of preventive medicine as applied to industry are put into practice by submitting all new entrants to a careful placement, and their periodical medical examination. By means of such examinations, latent disease may often be recognised and treated, with a greater chance of success than in the case when it has become fully established. Constant education of employees is required to make sure that personal hygiene is not neglected.

Good primary treatment of accident and illness is essential and, indeed, is itself a form of preventive medicine. Full, complete and technically accurate records are important to the industrial medical department. A good case-industry is often of immense value when a decision is being made as to the appropriate treatment of accident or illness. Good records have been described as "the seeing eyes of industrial medicine", and by means of them the nature of the illness occurring in the industry becomes clear.

The medical record and history of the individual worker, except in so far as they are required for legal purposes, must, of course, be regarded as confidential documents, the contents of which must not be divulged to a third party without the consent of the workman. The medical department would be guilty of an ethical offence if information gained in a professional capacity were so divulge; indeed, the confidential doctor-patient relationship is the only foundation on which a successful industrial medical service can be built.

Occupational Safety and Health in Britain

From the early days of Christianity until the end of the 15th century, information about the industrial work situations in England is scanty. Then, in England, an almost monotonous succession of statutes governing working conditions through the 18th century is noted. These laws apparently set the stage for dealing with labour as a public matter of concern to the whole state.

A special objective of charitable and philanthropic endeavour in the 17th and 18th centuries was found in the houses of industry in which small children, even under five years of age, would be trained for apprenticeship with employers. The evils and excesses which lay within this apprenticeship system gave the first impulse to a new venture in labour legislation which swept rapidly through the 19th century. In many ways both the employer and the worker were affected. One of the developments was an awakening to the need for controlling work hazards. It became the basis for organised safety programmes which were to follow. At this time injuries were no longer dealt with simply as an economic problem. A movement had started to determine the cause of injuries and effectuate their elimination.

The rapid development of steam power and its application to manufacturing led to the growth of employment of children in city factories rather than in apprenticeship systems. Before long this brought the general question of regulation and protection of child labour in English textile factories to the fore.

In 1795, the Manchester Board of Health was formed. It advised for legislation on the regulation of hours and conditions of labour in the factories. In 1802, the Health and Morals of Apprentices Act was passed, which in effect formed the first step toward regulated prevention of injury and protection of labour in English factories. It was aimed only at evils of the apprenticeship system. This Act did not apply to places employing fewer than 20 persons or 3 apprentices. Thus, the injuries to children in the English factory system prompted legislative intervention on behalf of safety for the first time.

The Mines Act, 1842 provided for punitive compensation for preventable injuries caused by unguarded mining machinery. It made provision for appointment of mine inspectors, excluded women and children from underground mining, and prohibited boys under the age of 10 years from working. However, the injury toll in the mines continued, and an organised governmental mine safety inspection programme was begun in 1850. This was followed by the Act of 1855 which specified seven general safety needs for the inspectors to investigate: ventilation, guarding of unused shafts, the proper means for signaling, correct gauges and valves for steam boilers, and so on.

A series of disastrous iron mine accidents and explosions resulted in an extension of the English law (Mines Act of 1860). A demand was made to compel the employment of only certificated managers for coal mines. As a consequence, another new regulation was decreed (Coal Mines Act of 1872). The Act of 1872 extended general safety rules, improved the method of formulating special safety rules, provided for certificated and competent management, and increased inspection. The use of safety lamps and the securing of roofs and sites was made compulsory, and the handling of explosives was regulated. Willful neglect of safety provisions became punishable in the case of employers. Factory and workshop laws and regulations also continued to grow.

Factory regulations were applied only to textile plants. English bleaching and dyeing works were included in the Acts of 1860 and 1862, and lace factories in the Act of 1861. The Workshop Regulation Act of 1867 (amended in 1870) practically completed the application of the principle of the factory legislation to all places in Great Britain.

The Factory and Workshop Act of 1883 provided that white led factories should not be operated without a certificate of confirmance with certain requirements. In 1889, certain procedures were specified for cotton cloth factories.

The increasing requirements of the safety and health regulations and the expanded number of industrial classifications which they covered did not stop the flow of work injuries. As a result of the inadequacies implicit in the common law approach to indemnifying injured workers, workers' compensation laws were enacted in Germany in 1885, in Great Britain in 1897, and in the United States in 1902. The compensation law in Britain required the employer to compensate the injured employee whether or not negligence could be proved. The compensation law in Britain overcame many of the defects of the earlier common law approach. Under the common law approach the employer had three powerful defenses: (a) the defense of contributory negligence; (b) the defense of assumption of risk; and (c) the fellow servant rule. Added to the legal defenses, the employer had another advantage – the injured employee often did not wish to jeopardize his job by suing the employer.

In the UK, Lord Robens Committee which went into the working of the then Factories Act 1961 proposed that the approach to industrial safety should be more adapted to modern technology and so recommended the principles of self regulation in 1972. The Flixborough accident led to the setting up of an Advisory Committee on Major Hazards (ACMH) in the UK in 1974. In the UK, the Flixborough and in Italy the Seveso accidents made the European Community Countries to issue directives to their member countries to formulate a suitable legislation to control the major hazard industries and installations.

Occupational Safety and Health in the USA

In the United States, the move toward safety regulation followed the British pattern. Textile factories were firmly established in America during the period from 1820 to 1840. Massachusetts is considered the first state to have recognised the necessity of pursuing the course taken by English factory legislation. In 1876, Massachusetts reconstructed its laws relating to the employment of children. Although the concepts for controlling accidents in the United States generally followed the regulatory model initiated in England, there has been one notable exception. American industry, particularly the larger companies initiated work safety programmes before the statutory safety regulations came into force.

One of the intentions of the proponents of workers' compensation was to advance occupational safety programmes. It was believed that the indemnification costs imposed on the employer would motivate that employer to institute safety programmes as a logical defence against such increasing expense. Arising out of the workers' compensation laws grew the opportunity for a new type of insurance coverage and service in the United States.

The role of the American government in the safety movement is exemplified clearly in the number and variety of regulations, laws, and court decisions which in one way or another have shaped the course of safety concepts and interests. Emerging from safety legislation are many governmental agencies established to initiate, direct, enforce, advise and in some cases, research programmes that bear upon the nation's injury control performance.

The creation of the U.S. department of labour had a significant bearing on the pursuit of occupational safety in the United States. Until the enactment of Williams-Steiger Act, 1970, more popularly known as Occupational Safety and Health Act (OSHA), the primary responsibility for governmental action concerning the prevention of occupational injuries and disease in the United States rested with the states. Early state safety laws were limited in scope and number.

OSHA (1970) became what many consider the most pervasive safety law ever passed. It authorises the federal government to set and enforce safety and health standards for all places of employment and to enforce the standards with criminal and civil penalties for violations. A new agency, the Occupational Safety and Health Administration was created. Technically, the law is administered and enforced by the secretary of labour. However, it provides for an assistant secretary of labour for occupational safety and health to whom the functions of the secretary are delegated. Briefly, it is the duty of the secretary of labour to establish occupational safety and health standards through inspections of workplaces. The Act establishes the National Institute for Occupational Safety and Health (NIOSH) in the department of Health and Human Services (HHS). The NIOSH is authorised to develop occupational safety and health standards, and fulfill the research and training functions of the secretary of HHS.

Under the Act, each employer is required to furnish to each of his employees a place of employment which is free from recognised hazards that are causing or are likely to cause death or physical harm to his employees. The employer is required to comply specifically the occupational safety and health standards promulgated under the Act.

Employees are provided certain rights under OSHA. Among the most important are:

1. the right to request the secretary of labour in writing for safety and health inspections in the plant, work establishment, or job site;
2. the right to maintain accurate records of employee exposures to potentially toxic materials or harmful physical agents and to have access to such records;
3. the right to identify dangerous substances by labelling or posting in the plant or at the job site;
4. the right to be informed promptly of any exposure to toxic materials or harmful physical agents that are present in excess of those prescribed under safety standards;
5. the right to know about the safety violations by the employer as noticed by the compliance officer.

OSHA requires all workers and contractors to comply with all safety and health standards that apply to their actions on the job. Employees should:

* Read the OSHA poster.
* Follow the employer's safety and health rules and wear or use all required gear and equipment.
* Follow safe work practices as directed by the employer.
* Report hazardous conditions to a supervisor or safety committee.
* Report hazardous conditions to OSHA, if employers do not fix them.
* Co-operate with OSHA inspectors.

The law also provides specific rights for the employer. These include:

1. the right to be advised by OSHA personnel of the reason for an inspection;
2. the right to participate in the walk-around inspection;
3. the right to file a notice of contest by the OSHA in case of a penalty;
4. the right to apply to OSHA for a temporary variance from a standard if the employer is unable to comply because of the unavailability of materials, equipment, or personnel to make changes within the required time.
5. the right to apply to OSHA for permanent variance from a standard if the employer can prove the facilities or methods of operation and provide protection that is atleast as effective as that required by the standard.

The Act provides for penalties for non-compliance with the safety and health standards promulgated under the Act. It also stipulates on the part of the employers to maintain safety records, to despatch safety reports to the appropriate authorities and display information and statistics. The employers' responsibilities include position of a safe and healthful workplace free of recognised hazards and to follow OSHA standards. Employers' responsibilities also include training, medical examinations, and record keeping. The OSHA has laid down certain standards or rules to protect workers against many hazards on the job. These standards limit the amount of hazardous chemical workers can be exposed to, require the use of certain safety practices and equipment, require employers to monitor hazards, and maintain records of workplace injuries and illnesses.

Occupational Safety and Health in India

The Constitution of India contains specific provisions on occupational safety and health of workers. Article 24 states that no child below the age of 14 years shall be employed to work in any factory or mine or engaged in any other hazardous employment. According to Article 39, the State shall, in particular, direct its policy towards securing:

(i) that the health and strength of workers, men and women, and the tender age of children are not abused and that citizens are not forced by economic necessity to enter vocations unsuited to their age or strength;

(ii) that children are given opportunities and facilities to develop in a healthy manner and in conditions of freedom and dignity and that childhood and youth are protected against exploitation and against moral and material abandonment;

(iii) the state shall make provision for securing just and human conditions of work and for maternity relief.

The Directorate General of Factory Advice Service and Labour Institutes (DGFASLI), and the Directorate General of Mines Safety (DGMS), the two field institutes of Labour Ministry strive to achieve the principles enshrined in the Constitution of India in the area of occupational safety and health in factories, mines, and ports.

Directorate General of Factory Advice Service and Labour Institutes (DGFASLI)

The Directorate General, Factory Advice Service and Labour Institutes, Mumbai which is an attached office of the Ministry of Labour functions as a technical arm of the Ministry in matters concerned with safety, health and welfare of workers in factories and docks. It assists the Central Government in formulation and review of policy and legislation on occupational safety and health in factories and docks; maintains a liaison with factory inspectorates of States and Union Territories in regard to implementation and enforcement of provisions of the Factories Act 1948 and renders advice on technical matters; enforces the Dock Workers' (Safety, Health and Welfare) Act, 1986; undertakes research in industrial safety, occupational health, industrial hygiene and industrial psychology; and provides training mainly in the field of industrial safety and health including one year diploma course in industrial safety and three months certificate course known as associate fellow of industrial health.

The structure of DGFASLI organisation comprising of the headquarters; four labour institutes and inspectorate dock safety offices in ten major ports. The headquarters at Mumbai comprises of three divisions, namely, factory advice service, dock safety and construction safety.

The Central Labour Institute at Mumbai started working from 1959. The Institute has assumed the status of a major national resource centre with the following divisions: 1. industrial safety, 2. industrial hygiene, 3. industrial medicine, 4. industrial physiology, 5. industrial psychology, 6. staff training, 7. productivity, 8. major accident hazard control, 9. management information services, and 10. safety and health communication.

The different divisions of the Institute undertake activities such as carrying out studies and surveys, organising training programmes, seminars and workshops, rendering services such as technical advice, safety audits, testing and issuance of performance reports for personal protective equipment, delivering talks, and so on.

The Regional Labour Institutes (RLIs) located at Calcutta, Kanpur and Madras are serving the respective regions of the country. Each of these Institutes are having following divisions/sections: industrial safety, industrial hygiene, industrial medicine, staff training and productivity, art and layout, major accident hazard control, and computer centre.

The inspectorates of dock safety are established at all the major ports of India. Under the Dock Workers (Safety, Health and Welfare) Act 1986, the director general is the chief inspector of dock safety for all the eleven major ports. They carry out inspection of ships and oil tankers at the major ports.

Directorate General of Mines Safety (DGMS)

The Directorate General of Mines Safety which is a subordinate office of the Ministry of Labour and has its headquarters at Dhanbad with its zonal, regional and sub-regional offices spread all over mining areas. It is entrusted with the responsibility of enforcing the provisions of the Mines Act, 1952, and the Rules and Regulations framed thereunder in coal, metalliferrous and oil mines. Apart from inspection of mines, the DGMS also undertakes investigation into all fatal accidents and certain serious accidents and dangerous occurrences, and makes recommendations for remedial measures to prevent recurrence of similar mishaps.

National Safety Council

The National Safety Council was set up on March 4, 1966, by the Ministry of Labour, Government of India, as an apex body to generate, develop, and sustain voluntary movement on safety and health at the national level.

The main objective of the Council which is an independent and self supporting national level institution, has been to generate, develop and sustain a movement of safety awareness at the national level. To achieve this objective, the Council conducts a variety of educational, training and promotional activities. These activities include specialised public courses, seminars and conferences on industrial safety and health, consultancy services, safety audit and information services. It also brings out a number of publications on industrial safety.

The Government of India secured the help of ILO in 1985 for advising it on the control of major hazards. In addition to the identification of major hazard installations in the country, the Government has brought out statutory/regulatory measures by way of comprehensive amendments in 1987 to the provisions of the Factories Act, 1948.

To sum up, in the words of planning commission, "with the adoption of advanced technology and increase in the use of various kinds of chemical substances in different sectors of economic activity, an increasing proportion of the workforce, as well as, the population in general, are exposed to work-hazards and environmental pollution. Modernisation of the industry has also brought in its train, problems of occupational hazards arising out of work-posture and man-machine environment. Greater attention than before will, therefore, have to be paid to the assessment and control of hazards to workers and the general population and to the development of safety devices, protective gears, appropriate design of machines and tools, plant layout and workplace layout. Among the programmes envisaged in the labour sector are application of ergonomics for improvement of working conditions in factories and docks, establishment of a system of chemicals, strengthening of the system for monitoring improvement of occupational health status and certification of personal protective equipment. In the field of mines safety, it is proposed to deal with problems relating to humidity, mine fires, ground control, stability of illumination, etc. It is also proposed to develop computer programmes for health monitoring of miners" (8th Five Year Plan 1992-1997, Vol. II, P. 157).

SUMMARY

- The concept of safety is probably as old as the history of mankind itself.
- Mining is one of the oldest industries and has always been a hazardous occupation.
- Inside the factories and mines of the 19th century, the workers were exposed to the hazards of occupational disease and injury.
- During the 20th century the trade unions in many countries created an awareness of occupational health hazards among workers and advocated for payment of compensation for injuries and diseases.
- For the first time, in the year 1844 a law was passed in England regulating the working hours of adult women and certain provisions were made for their health and safety.
- The Occupational Safety and Health Act (OSHA) is the basic piece of legislation governing the occupational safety and health programme in the United States.
- The OSHA provides for mandatory safety and health standards of three types: (a) interim standards; (b) emergency temporary standards; and (c) permanent standards.
- A health standard (OSHA) is a rule that requires "conditions, or the adoption or use of one or more practices, means, methods, operations, or processes, reasonably necessary or appropriate to provide safe or healthful employment and places of employment."
- The Constitution of India contains specific provisions on occupational safety and health of workers.
- The Government of India has set-up the Directorate General of Factory Advise Service, and different enforcement agencies, namely, Directorate General of Mines Safety, Ports and Docks; National Safety Council for promotion of all-round health and safety in their respective spheres.

DISCUSSION QUESTIONS

1. What are the functions of DGFASLi and National Safety Council?
2. What is the impact of industrial revolution on safety?
3. What are the various developments in the field of occupational health?
4. What are the various developments in the sphere of occupational safety?

4

Safety and Health Organisation

An organisation consists of two or more people working together co-operatively within identifiable boundaries to accomplish a common goal or objective. It is the process of a systematic determination of tasks and responsibilities and inter-relationships in the light of human and material resources with a view to achieving common goals. It is a goal-oriented relationship between people, tasks, resources, and managerial activities. The process of organising involves dividing all the work that has to be accomplished and assigning it to individuals, groups and departments.

Safety Organisation

Safety organisation may be defined as a definite, planned and organised set-up whose purpose is to enlist and maintain a combined effort of the entire personnel of an establishment or undertaking in the work of accident prevention. It constitutes a vehicle of systematic procedures by means of which interest is created and maintained and all safety activities are correlated and directed. The safety organisation is a "clearing house" of information on safety matters. The expression has a broader significance than merely the inclusion of safety committees, safety inspectors, safety meetings, and other similar details. It includes the interest, support, direction, and participation of the higher executives. As such, its nature and scope are governed by the principles of accident prevention.

Safety has been described as everyone's responsibility. As a generalisation this is true. Most functions in modern society are fulfilled through an organisational hierarchy. Many industrial companies today state that their first three concerns are safety, quality, and production, respectively.

Causative fact-finding investigations often occupy a number of key personnel for several days when serious events occur. Also, the injured person is lost to the workforce for a period of time. In some industries, traditionally the entire crew may quit for the day when a fatality occurs. Obviously, the organisation's mission cannot be accomplished when such unwanted events occur. Profit, production, and other considerations are negatively affected. An organisation should be aware that injuries are expensive. Not only do the injuries incur medical costs and indemnities, but valuable time is lost from productive work, and often property damage is incurred.

The elements of safety organisation:

- Safety policy
- Safe work policies
- Safety training
- Safety meeting
- Safety investigation and analysis
- Safety rules and regulations
- Safety promotion
- Safety inspection
- Safety records
- Emergency preparedness

Conventional Safety Programming

There are four basic steps in conventional safety programming:

1. Case analyses by categorising injurious events, identifying their causes, determining trends, and performing related evaluations of the events.
2. Communication of the knowledge derived from the case analyses by codifying the acquired information into standards and making the knowledge available in instructional formats such as training programmes, posters, visual reminders, films, and so on.
3. Inspections with a dual purpose of appraising the level of compliance as well as to detect unsafe conditions and practices before they produce injurious events.
4. A fourth step may be added to the industrial safety programme. Supervisor safety training is intended to orient the supervisor to safety achievement and responsibilities.

What is safety programme and its goals? Simply put "safety" is a management tool to:

- identify, eliminate and manage risks
- develop safety standards
- respond to accidents and injuries
- train employees in expected behaviours and standards

- monitor compliance, conditions and behaviours
- document compliance
- set goals for action and improvement.

In the past fifty years, industrial technology has developed more new equipment and processes than were developed in the preceding 1,000 years. Technology has created expanded requirements and need for accident prevention at the workplace. Many tasks have become increasingly complex and demanding, and their potential for serious injury has increased substantially. As a result, management has broadened its functions to include the application of safety programme management in its organisational systems. It further means that safety programming has become a function of the managerial class, right from the line supervisor upto the chief executive officer of the organisation.

Safety programme management is the control of the working environment, the equipment and processes, and the workers for the purpose of reducing accident injury and losses in the workplace.

Safety Planning

Good planning is as essential in safety as it is in production. If a new factory is to be built, or an existing factory reconstructed, there are many things affecting both production and safety that should be taken into account in the planning stage, such as the site, facilities for handling and storing materials and equipment, floors, lighting, heating, ventilation, lifts, pressure vessels, boilers, electrical installations, facilities for machinery maintenance and repair, and fire protection. It is essential that safety considerations be borne in mind at the time of the actual planning of the building of the factory. Once a factory is in operation, planning is still essential in a number of fields to ensure the highest possible standards of safety as well as efficiency.

Safety Planning: Key Questions

- What is your plan towards health and safety?
- How do you consider safety issues?
- Have you identified hazards and assessed risks?
- Have you set safety targets and standards?
- Do you have contingency plans for major disasters?

The manager of a factory can generally follow a number of principles in planning for safe and efficient production. Here are some examples:

1. Keep the handling of materials and articles to the minimum.
2. Provide adequate space for machinery and equipment.
3. Provide safe working surfaces on floors, stairs, platforms, gangways, and so on.

4. Provide for safe access to every place to which workers have to go.
5. Provide for the safety of maintenance and repair personnel, such as window-cleaners and of men working on overhead equipment.
6. Provide safe transport facilities.
7. Provide adequate means of escape in case of fire.
8. Isolate dangerous processes, such as spray-painting and processes with high-fire or explosion risks.
9. If possible, only buy machines with built-in safety devices.

There are a number of safety measures which are related to good order and good housekeeping all over the plant. If there is a place for everything, and if everything is in its place, a considerable number of accidents are unlikely to occur. Good odour and good housekeeping not only reduce accident risks by eliminating physical risks but also contribute to safety by their psychological effect. It is obvious that good housekeeping and good odour can be achieved more easily if workers support the idea and obey all instructions designed to promote it. Good housekeeping not only helps to prevent accidents but also makes work easier.

Safety Policies

Policies are those principles and rules of action which guide an organisation to achieve its objectives with due regard for its business ideals. When accident prevention is the objective, the setting down of the guiding principles and rules will provide a foundation upon which the safety organisation will be created.

Policies should not be confused with procedures. Policies are designed to guide us in what to do; procedures inform us how to do it. Policies are therefore guides to action in decision making; procedures or routines to follow which will enable achievement of the objectives.

Sound policies are necessary to:

- Promote intelligent co-operation at all levels.
- Prevent deviation from accepted courses of action.
- Provide a network for the intelligent exercise of initiative.
- Produce a foundation for co-ordination at various levels.
- Present a clear expression of management's objectives and intentions.

As a minimum, the following areas ought to be touched on in a safety policy:

1. Management's intent – what does management want?
2. The scope of activities covered – does the policy pertain only to on-the-job safety? Does it cover off – the – job safety also?
3. Responsibilities – who is responsible for what?

4. Accountability – where and how is it fixed?
5. Safety committees – will there be committees? What will they do? Why do they exist?
6. Authority – Who has it, and how much?
7. Standards – what rules will the company abide by?

Setting Safety Policy

Take a moment to answer the following key questions:

1. Do you have a clear policy for health and safety – is it a written document?
2. Does it specify who is responsible, and the arrangements for identifying hazards, assessing risks and controlling them?
3. Do your staff know about the policy and understand it? Are they involved in making it work?
4. Is it up-to-date?
5. Does it prevent injuries, reduce losses and really affects the way you work?

The safety policy statement is the foundation of the safety programme. The safety policy statement can be a driving force or a destructive force for the entire safety programme. The safety policy statement must be perceived as company policy, not the safety person's policy, not the human relation's department policy, nor any individual's policy.

Exhibit 4.1: The Philosophy and Objectives behind Safety Commitment of Company A

- The safety and health of all employees is our first priority.
- The only acceptable level of safety and health performance is one that prevents injury and accidents.
- Safety and health are integral part of production and all other business functions that cannot be separated or by-passed.
- Safety and health are a responsibility that must be shared equally and without exception by everyone within the organisation.
- Supervisors and management will be held accountable for the safety and health of the personnel for whom they are responsible.

The Health and Safety Policy Commitment of Company B

- Prevent injuries and occupational illnesses by continuous training and education of employees including contractors, apprentices, transport workers and suppliers.
- Implement effectively the occupational safety and health management practices.

- Select plant equipment, machinery, material, and personnel by meeting the requirements of health and safety standards.
- Conduct periodic risk assessment for health, safety and environment.
- Incorporate information on health and safety performance in its annual reports.
- Review the policy and level of implementation periodically to ensure that it remains relevant and up-to-date.
- Communicate the safety policy and environment guidelines to all concerned.

To be effective, safety policies should be stable, flexible, compatible, sincere, realistic, understandable and written. Organisational safety policies are institutional guidelines on all matters concerning safety administration. Policies specify the corporate aims and designate the responsibilities and authority for their achievement.

Developing a safety policy will include the following steps:

1. Write and announce the policy regarding the control of hazards for which the organisation has a responsibility. Designate accountability and delegate authority for implementing the policy.
2. Appoint a safety director.
3. Analyse the operational record of injuries, property damage, and work illnesses.
4. Appraise the scope and seriousness of operational hazards.
5. Select, organise, and schedule communication methods for employee safety training, safety maintenance, and informing organisation's safety progress and needs.
6. Establish a periodic review scheduled for auditing the programme and facilities.
7. Determine long-range objectives and short-term goals for the programme.

Management's acceptance of its responsibilities for safety can best be expressed by setting a policy to provide a safe and helpful workplace for its employees. This is a commitment, and hence it should be designed to lay down step by step company's plan for evolving a safe workplace. A total safety programme must always start at the top management level. Each management level must reflect considerable interest in the company's safety objectives and set good examples of compliance with safety rules.

Today, safety professionals realise that safety is a state of mind and what is really needed is "built-in-safety" and "integrated safety" as a part of company's policy. Such a safety policy aims to build safety into various procedures, and continuously audit the procedures to ensure that the controls are adequate.

A company's occupational safety and health policy should:

- Create worksite free from occupational health and safety hazards for its employees associated with it and those living in its neighbourhood.
- Pursue the safety efforts by adhering the occupational health and safety management system based on the requirements of internationally recognised standards.
- Demand accountability for safety performance and provide the resource to make safety programme work.
- Involve all employees for continual improvements in health and safety.
- Create and nurture a safety culture that supports flexibility, learning, and is proactive to change.
- Comply with the laws and rules that are applicable; and also other requirements of health and safety.

Exhibit 4.2 Safety Policy of Tata Steel

Tata Steel believes that a healthy worker is the surest basis for its continued success.

Tata Steel, therefore, is committed to the task of ensuring the safety and safeguarding the health of all its employees.

Importance will be given to continuous training for promoting safety consciousness among all employees.

Joint committees of executive and employees' representatives will supervise the company's safety measures.

Within their areas of responsibility, everyone will be accountable for:

- Establishing a safe and healthy work environment.
- Ensuring compliance with mandatory safety and health requirements.
- Proper maintenance and orderly housekeeping in order to control the risk of damage to plant and equipment.
- Insisting on safe work procedures being followed by employees, contractors and visitors.

Exhibit 4.3 Safety Policy of Rashtriya Chemicals and Fertilizers (RCF)

RCF firmly believes that all accidents are preventable, and the safety of the employees, industrial operations, products and customers, and the neighbouring communities are of paramount importance. To achieve this the company will effectively take the following steps:

- Every attempt will be made to reduce the possibility of accident occurrence and to improve and maintain the environment safe.
- Work environment will be monitored and safe and healthy working conditions will be maintained.
- Operating practice that will safeguard all employees and result in safe working conditions and efficient operation will be observed.
- Safety is an operating function and all levels of line management will have a primary responsibility for implementation of safety programmes.
- No jobs will be considered rightly done unless the employee follows the safety precautions and rules to protect himself and his fellow employees.
- All the relevant statutory safety regulations and laws will be complied with in letter and spirit.
- The company will always be on lookout for latest technologies and processes in the interest of safety and environmental quality and will consider them for incorporation.

Exhibit 4.4 Safety Policy of Coal India Fertilizers (RCF)

- Statutory rules and regulations will be continuously implemented so that superior standard of safety is achieved.
- Operations and systems will be planned and designed so that mining hazards are eliminated or materially reduced.
- Working conditions will be made safe and secure so that workers' life is made safe and danger free.
- Material and finance resources needed for smooth and efficient execution of the safety plans will be provided.
- Safety personnel who are otherwise skilled in their jobs are to be deployed wholly for accident prevention work.
- Appropriate forums and workshops with employees' representatives will be organised so that worker's commitment in safety management is ensured.
- Annual long term safety plan is to be prepared unitwise giving effect to by requisite resources; implementation of such a plan is to be monitored closely.
- Internal safety organisation is to be created and made to work so that the safety policy and plans of the company is carried out and observed by safety personnel at the mine, area/group and the corporate level.
- Executives of different ranks and grades are to be integrated so that there is practising involvement in safety towards accident prevention

and achievement of ZAP (Zero Accident Potential) in the work culture of the industry.

- Continuously impart educational training programmes and also organise retraining activities so that proper emphasis may be laid in the operational management and work culture on safety.
- A continuous effort is to be made towards attitudinal change of work persons right from workers upto the highest level of the organisation so that safety is accepted as an inevitable ingredient of work system in the organisation.

Exhibit 4.5 Safety, Health and Environment Policy of Galaxy

Safety

- Safety and health of people is of primary importance.
- The company shall continually strive for eliminating incidents/ accidents at workplace.
- The company shall build and embed safety awareness, consciousness, and attitude in all its personnel.
- All occupational injuries and illnesses can be preventive.
- Safety results from attitude and commitment of people.

Environment Protection

- The company shall ensure safe, healthy and eco-friendly environment at its workplace.
- The company shall continually work towards identification and reduction of risk and prevention of pollution at its plants and surroundings.
- The company shall comply with all the statutory and regulatory requirements relevant to health, safety, and environment.
- The company's safety culture is sum of beliefs, attitudes and practices being followed in various activities of its business.

Safety Culture and Climate

The concept of safety culture and climate is central to contemporary thinking on health and safety management. Accidents such as that at Chernobyl (1988) have been attributed in part, to the safety culture of the organisation. The terms safety culture and safety climate have become almost interchangeable in the literature. Much of the research into the assessment and quantification of culture and climate for safety has centred on the use of attitude surveys. Employee attitudes are one of the most important indices of safety culture and climate. Also attitude towards safety are one of the basic components of a safety culture and climate.

There are four components of a good or positive safety culture:

Competence: recruitment, training and advisory support.

Control: allocating responsibilities and seeking commitment.

Cooperation: between individuals and groups.

Communication: verbal, written and visible.

Safety culture is a sum of beliefs, attitudes, and practices being followed in various activities of business. The principle of any positive safety culture management ensures that the businesses are making sustainable profits.

The creation of a safety system requires that all employees must have a commitment to safety and ownership of the safety system that supports a safe place of work. This commitment requires a cultural change of all employees and this change can be initiated and developed through employee involvement in "work teams".

Every company has a "culture", simply described as "the way we do things round here." As a part of the company culture, safety should start "at the top" from the senior management and pervade the whole organisation. Safety is unfortunately, usually seen as a cost item and a necessary evil demanded by law. But as a part of the company culture, it can actually contribute to profit, just like quality. An unsafe plant or an inferior product, though seemingly cheap, can prove far too expensive in the long run for everyone. Safety manuals are no substitute for a safety culture or a safety-minded workforce. It is people who cause accidents and it is only people who can prevent them. In the absence of a "safety culture," no amount of sophisticated gadgetry, fool proof safety devices and alarms will ensure a safe plant operation.

Safety Standards

One of the significant benefits of having a written safety programme is that it acts as a vehicle for establishing the safety roles and standards for management, supervisors and employees.

Setting and communicating standards is not an easy task. For this purpose, management has to develop specific tools to train employees, monitor activity and document compliance. All these must be done without over-burdening production, maintenance, and quality control efforts.

Communicating safety standards starts day one with the new employees by having them participate in safety orientation class and continue with periodic specific training that applies to their jobs and assigned tasks. Other tools for communicating standards and expectations are company newsletters, posters, immediate on-the-job correction of unsafe behaviours, safety information pamphlets, and informal safety talks by supervisors. Standards that define the periodic evaluation of employee compliance and programme is essential for gathering information on which to base needed changes in standards and procedures. Audits, inspections and task completion documents, such as training forms, should be simple to follow and to complete.

The ILO has adopted more than 40 standards specifically dealing with occupational safety and health, that provide essential tools for governments, employers and workers to establish such practices to ensure maximum safety at work.

Safety Control

Safety control could be synthesized in three phases or stages:

(a) the post control;
(b) the control of the events; and
(c) the pre control.

The post control stage includes all the safety performances and the activities that occur as a result of an accident or injury. Some of these performances are: (a) first aid and medical care that were needed, or that were required, on the part of the injured victim; (b) control of the damages resulting from the accident or injury; (c) emergencies or reactions in the face of a certain situation; and (d) programmes of rehabilitation offered to the accident/injury victim.

The second stage of safety control makes a reference to the control of the events. In this phase, the visible side of safety is observed and all the measures that should be taken are taken so that in the event of the accident, the damages and the occupational injuries are diminished.

The third stage of safety control includes all measures for prevention of accidents and creation of safe work environment.

Safety Promotion

Although safety can to some extent be engineered into equipment and processes, it is still necessary to motivate employees to perform their work safely. Safety promotion is persuasion through motivation. An effective safety programme must be persuasive; it must provide a stimulus to which employees will respond positively. The goal of any such programme is active participation by employees.

Direct safety promotion approach involves intervention into the system, such as:

- Modification of safety training programmes, including retraining.
- Providing short safety talks either at the start of each shift, following lunch, or during coffee breaks.

- Holding safety meetings led by the line manager or the top executive officer in the facility explaining management's concerns over declining safety performance.
- Development of a safety information programme.
- Evaluating safety performance on a regular basis.

Making employees interested in their own safety and well-being is the responsibility of the safety professional who plans the programme and the supervisors who carry it out. The safety professional must realise that a safety promotion programme can succeed only through the combined efforts of management and employees, and must therefore use every available means to maintain their enthusiasm. Interest in safety programme can be created in various ways. In many companies, safety contests have been very effective; usually such contests are held for the purpose of reducing the frequency of accidents. Any good idea about safety promotion can be used as the basis for competition.

Establishing and implementing written safety and health procedures and guidelines should be viewed as an enhancement for any company, not a hindrance. It aims to prevent incidents that cause personal injury, loss of assets or business interruption. It is both a moral obligation and sound business practice. And there is no conflict between operating safely and operating efficiently. Injuries cause both human suffering and financial loss, neither of which is acceptable to a well-managed company. A written process also provides everyone in the organisation with a clear message of what is required to maintain a safe work environment.

It is very important to obtain management's commitment for successful operation of any health and safety procedures. Their commitment provides the motivating force and resources for organising and controlling the activities within an organisation. Without it, the efforts of safety personnel are futile.

Employees at every level of an organisation need to be involved in the safety and health process. Active involvement of employees may be possible by their participation in:

- site safety committees
- incident investigation and inspection
- housekeeping
- recognition committees
- emergency response teams
- safety meetings
- project reviews
- off-the-job safety committees
- development of operating procedures and processes

DUPONT – Safety Pays

DuPont model refers to the approach of the management to health and safety championed by the US multinational chemical company, which is a recognised world leader in all aspects of accident prevention and safety.

At DuPont Corporation, safety experts provide feedback while engineers observe workers and then redesign valves and install key locks to deter accident-causing behaviour. When injuries do happen, the company reports them quickly to workers to provide a sense of urgency, trying to show the behaviour that caused the accident without naming the offender. It also fosters peer pressure to work safely by giving units common goals – the way workers are working together instead of independently. DuPont offers carrots too. Its directors regularly give safety awards, and workers win $15 to $20 prizes if their divisions are accident free for six to nine months. The company's incentive for doing this is not altogether altruistic – it estimates its annual cost savings to be $150 million. According to the president of DuPont safety resources: "When a business culture focuses on safety, it makes a public proclamation of its commitment to caring about the welfare of its people. A workplace noted for safe practices builds both trust and faith across the board."

Safety Committee(s)

A safety committee is most useful mechanism for facilitating the necessary co-operative effort that is essential for success in accident prevention. It should consist of representatives from top management, supervisors and from the workers. It should act as an advisory body and meet regularly.

Safety Committees should draw up terms of reference and a list of agreed objectives, one of which should be the promotion of cooperation between employees and employer in instigating, developing and carrying out measures to ensure the health and safety of all employees. The functions of a safety committee will normally include the following:

1. To consider safety performance.
2. To monitor the effectiveness of the organisation's health and safety policy.
3. To study disease and accident statistics to:
   (a) look for trends,
   (b) highlight unsafe and unhealthy conditions and practices, and
   (c) recommend corrective action.
4. To consider safety inspection and audit reports and recommend action for improvements.
5. To undertake safety audits.

6. To consider reports of factual information provided by enforcing authority inspectors.
7. To consider proposals for future developments put forward by management.
8. To assist in the development of written safe systems of work and work safety rules.
9. To consider revisions of the organisation's health and safety policy
10. To keep a watch on the effectiveness of the safety training programmes.
11. To monitor the adequacy of health and safety communication and publicity in the workplace.

The model rules framed under the Factories (Amendment) Act, 1987 suggest the following functions of the safety committee:

* Assisting and co-operating with the management in achieving the aims and objectives outlined in the "Health and Safety Policy" of the occupier.
* Dealing with all matters concerning health, safety and environment and to arrive at practicable solutions to problems encountered.
* Creating safety awareness amongst all workers.
* Undertaking educational programmes, training, and promotional activities.
* Discussing reports on safety, environmental and occupational health surveys, audits, risk assessment, emergency and disaster management plans and implementation of the recommendations made in the reports.
* Carrying out health and safety surveys and identifying causes of accidents.
* Looking into any complaint made on the likelihood of an imminent danger to the safety and health of the workers and suggesting corrective measures.
* Reviewing the implementation of the recommendations made by it.

Besides the above functions, few more areas can be considered by the safety committees:

- Assisting in enforcement of safety norms by safety department.
- Notification of any exposure that are potentially dangerous.
- Reducing the number of safety related plans without infringing on workers' rights.
- Arrangements for celebrations of safety day in the organisation.
- Need for introducing various types of respiratory and non-respiratory personal protective equipments.

- Modifications and changes to personal protective equipments for elimination of hazards caused by the introduction of new technological process.
- Administration and handling of suggestions and recommendations obtained through the safety suggestions scheme.
- Publicising safety within the organisation members.
- Documentation of health and safety records.
- Establishment of communication channel for improvement of safety and health within the organisation.

It must be remembered that safety committees can only make recommendations on the matters discussed in committee; it is the management's responsibility to take decisions on the implementation of any of the recommendations made. All members of the safety committee should receive training in health and safety and made fully aware of their role.

Safety Specialists

There are a number of specialised safety functionaries known by different designations such as safety officers, safety engineers, safety advisers, safety directors, who are mainly appointed to administer the organisation's safety policies and programmes. Their role and functions are given below:

Safety Officer

Under Section 40-B of the Factories Act, 1948 appointment of Safety Officers in certain factories is a statutory requirement. State governments are empowered to frame Rules prescribing qualifications, experience, etc. for appointment as Safety Officers in factories. In this regard DGFASLI, Government of India, has prepared Model Rules to guide state governments for preparation of rules prescribing qualifications for appointment of safety officers in factory.

The number of safety officers required to be appointed by the occupier of a particular factory is governed by the state government notification issued under Section 40-B(1) of the Factories Act, 1948. Guidelines in this respect are given by DGFASLI in its Model Rules.

The Maharashtra Safety Officiers (Duties, Qualifications and conditions of service) Rules, 1982 laid down the duties of a safety officier, which are as follows.

(i) The duties of a safety officer shall be to advise and assist the factory management in the fulfilment of its obligations, statutory or otherwise, concerning prevention of personal injuries and maintaining a safe working environment. These duties shall include the following namely:

1. To advise the concerned departments in planning and organising measures necessary for effective control of the personal injuries.
2. To advise on safety aspects in all job studies and to carry out detailed job safety studies of selected jobs.
3. To check and evaluate the effectiveness of action taken or proposed to be taken to prevent personal injuries.
4. To advise the purchasing and stores departments in ensuring high quality and availability of personal protective equipments.
5. To advise on matters related to carrying out plant safety inspections.
6. To carry out plant safety inspections in order to observe the physical conditions of work and the work practices and procedures followed by workers and to render advice on measures to be adopted for removing the unsafe physical conditions and preventing unsafe actions by workers.
7. To render advice on matters related to reporting and investigation of industrial accidents and diseases.
8. To investigate selected accidents.
9. To investigate the dangerous occurrences reportable under rule 115 of the Maharashtra Factories Rules, 1963 and the cases of industrial diseases contracted by any of the workers employed in the factory reportable under rule 116 of the said rules.
10. To advise on the maintenance of such records as are necessary relating to accidents, dangerous occurrences and industrial diseases.
11. To promote setting up of safety committees and act as adviser to such committees.
12. To organise in association with the concerned departments, campaigns, competitions, contests and other activities, which will develop and maintain the interest of the employees in establishing and maintaining safe conditions of work and procedures.
13. To design and conduct either independently or in collaboration with the training department, suitable training and educational programmes for the prevention of personal injuries.

(ii) No safety officier shall be required or permitted to do any work which is inconsistent with or detrimental to the performance of duties mentioned in Sub-rule (1).

In fulfilling his role, the safety officer at all times work in close harmony and collaboration with line management executives and with employees and their representatives, with the object of ensuring a safe and healthy workplace in tune with the organisation's health and safety policy.

Safety Director

The director of safety performs a number of significant tasks. They include the following:

1. The formulation and administration of the safety programme.
2. The acquisition of the latest and best hazard control information.
3. The representation of management to the public, employees, insurance companies, and governmental agencies as the company's safety resource.
4. The communication on safety-related issues to managers at all levels.
5. The collection and recording of pertinent data on safety-related operational matters, including work injury causes and statistics.
6. The reporting to top management periodically, on a regular basis, on the organisation's safety effort.
7. The coordination with the organisation's medical department on the safe placement of new or convalescing employees.
8. The inspection of the facilities for compliance with central, state, and local regulations.

Safety Adviser

The functions of the safety advisor is advisory, leaving executive decisions for line managers. The role and function of the safety adviser will normally include:

1. Monitoring the implementation of the organisation's health and safety policy.
2. Advising line management to assist them in meeting some of their health and safety responsibilites.
3. Assisting in the formulation and implementation of safe systems of work.
4. Recommending suitable protective equipment.
5. Checking compliance with all statutory requirements affecting health and safety.
6. Monitoring the necessary safety registers, records and accident books.
7. Promoting health and safety education programmes to develop safety awareness at all levels.
8. Disseminating information on accident prevention techniques.
9. Investigating, reporting and recording injury and damage accidents to: (a) establish the causes, (b) recommend remedial action to prevent a recurrence, (c) monitor performance, and (d) examine trends.
10. Providing meaningful information on accident statistics.
11. Liaising with outside bodies.
12. Keeping abreast of modern techniques and developments in health and safety.

Safety Department

For effective safety and health management, a well organised safety department is very essential. The safety department has to be assigned with the following tasks:

1. To establish the norms and guidelines for the provision of safety of sites, employees, materials, equipment and structures.
2. To prepare checklists, manuals, and other documents for use by the line management in carrying out their functions.
3. To supervise safety at site and within the organisation.
4. To give advice on all safety matters in accordance with the safety policy.
5. To maintain all safety records, prepare reports and monitor the same to all concerned.
6. To conduct safety training in the organisation.
7. To carry out safety audit periodically.
8. To discharge all statutory obligations of the organisation regarding safety, and maintain liaison with the government and other industry associations.
9. To organise competitions, posters, melas and such other activities that promote safety consciousness amongst employees.
10. To carry out plant safety inspections for removing the unsafe physical conditions and preventing unsafe actions by employees.

There are three distinct methods of operating a safety department in a plant.

1. The department may carry out staff functions in which members are required to develop policies and procedures, make periodic inspections, and make recommendations. The line supervisor is totally responsible for safe running of operational activities.
2. The safety department may combine both staff and line functions.
3. The department may hold all safety meetings, investigate all accidents, issue all work permits, and perform any other function considered by any one as a "safety job".

The safety person, whether called safety engineer, safety specialist, or safety director, and so on is merely management's representative. The chief operating executive is responsible for the safety conduct of the organisation. The safety specialist or the department only develops the information needed as a staff member or adviser. The safety position and function varies from one organisation to another. Although the actual responsibility for safety is line management's, and while supervisors are the immediate administrators of safety in an organisation; many industrial concerns have established departments whose members work exclusively to advance safety.

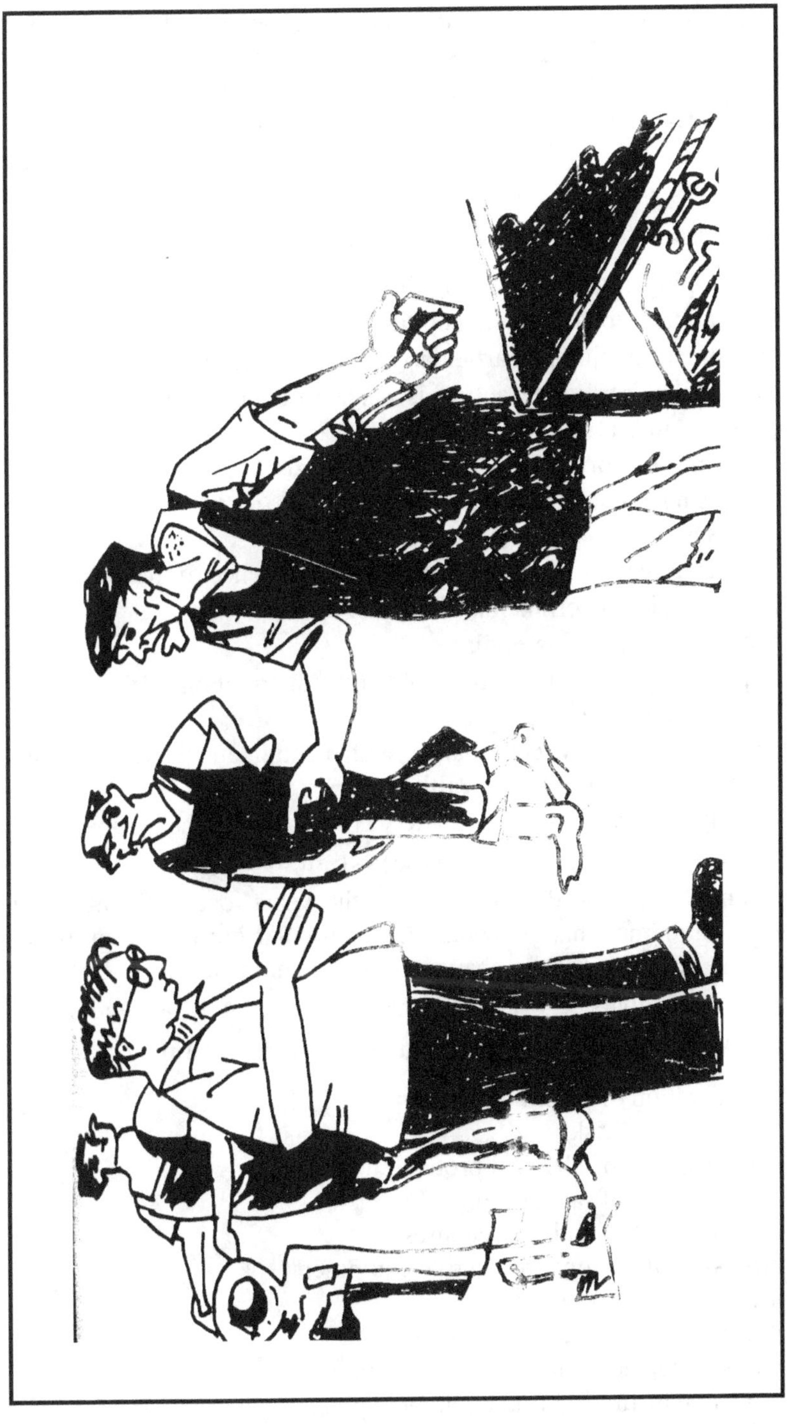

Fig. 4.1 Sound Health Leads to Good Work.

The size of the safety department depends on the size of the company and the nature of its activities. The safety department is more often located within the industrial relations or HR division or department. The safety specialist reports more frequently to the HR manager. The safety department has only advisory power, including that of making recommendations.

To summarise, safety activities usually include:

1. Developing and administering the company safety programme.
2. Inspection to locate unsafe conditions or unsafe practices.
3. Investigating injuries, particularly the more serious ones.
4. Seeing that corrective action is taken to avoid recurrences, or, if possible, to forestall the first one.
5. Maintaining work injury and illness records.
6. Analysing records for clues to prevention of future injuries and illnesses.
7. Preparing reports on the current safety experience of the company.
8. Liaising with governmental and other agencies.
9. Publicising safety materials.
10. Supervising the procurement and distribution of personal protective equipment.
11. Ensuring company compliance with central and state rules and regulations on safety and health.
12. Acting as secretary of major safety committees of the company.
13. Checking on or aiding in safety aspects of training.

Many companies have developed, over the years, occupational health and safety programmes that concentrate in particular on: (a) the problems peculiar to the industry; (b) the changes in the manufacturing process; and (c) the nature of the material being processed.

The safety programmes that normally deal with matters relating to:

- Identification of potential hazards
- Assessment of risk
- Establishment of physical protection
- Development of safe systems of work
- Development of audit procedures
- Provision of personal protective equipment
- Investigation procedures
- Standards for process modification
- Fire fighting activities
- Provision of fire fighting equipment
- Training of employees

Measuring Safety

A measure of performance at the upper and middle management level should meet the following criteria:

1. It should be flexible enough to allow for individual managerial styles and strategies.
2. It should be capable of giving swift and constant feedback.
3. It should be capable of being used in judging promotability.
4. It should be able to get the attention of those in middle and upper management.
5. It should be able to measure the presence as well as the absence of safety.
6. It should be sensitive to change.
7. It should be understandable to both the middle manager and those in top management.
8. It should allow for creativeness.
9. It should be both performance and result-oriented.
10. It should be meaningful.

Good safety performance is always the result of well planned and coordinated efforts on the part of enterprise. The essentials of top rate safety performance in any given establishment may be summarized as follows:

1. There must be forceful and continuous leadership.
2. Plant and equipment must be made safe.
3. Supervision must be competent and intensely safety minded.
4. Full employee co-operation in accident prevention must be secured and maintained.

Measurement of safety performance is necessary in managing safety. It is required primarily to compare and evaluate the results achieved and to plan for the future. Measuring safety, like any other measurement, needs a unit or standard. This unit should cover:

* the number of accidents
* their severity
* the number of employees exposed.

Traditionally, the following measures have been in use to evaluate the safety performance:

***Accident rate per thousand employees*:** This gives an idea as to the number of accidents compared to the number of employees exposed. This is computed by the formula:

$$\frac{\text{No. of accidents} \times 1000}{\text{No. of employees}}$$

***Frequency Rate (FR)*:** This again gives an idea about the number of accidents compared to employees strength and is computed by the formula:

$$\frac{\text{No. of accidents} \times 1{,}000{,}000}{\text{Total manhours of exposure}}$$

***Severity Rate (SR)*:** This is a measure of the severity of accidents expressed in terms of time lost by the injured compared to employees strength. This is computed by the formula:

$$\frac{\text{Total number of days lost} \times 1{,}000{,}000}{\text{Total manhours of exposure}}$$

Average Days charged per injury is computed by the formula:

$$\frac{\text{Total days lost}}{\text{Total number of accidents}} = \frac{\text{Frequency rate}}{\text{Severity rate}}$$

***Disabling Injury Index*:** A unit combining both frequency rate and severity rate computed by the formula:

$$\frac{(\text{FR}) \times (\text{SR})}{1{,}000}$$

Safety Supervision

Supervision of safety performance throughout the organisational hierarchy is a paramount requirement. The immediate supervisors of the workers, more than any others, are the key persons in implementing safety. Just as production and quality rely on supervisor's competent scheduling, training, and leadership; so does safety depend on the supervisor's ability. An indifferent supervisor is soon surrounded by indifferent workers and as a consequence, unsafe work practices are used, guards are removed from hazardous points of operation, and the whole work area become an unsafe place. It is necessary, therefore, that supervisor deligently set an example for safety and insist upon full compliance with the operating rules. The role of a supervisor is vital in all accident prevention programmes. The best supervisor will always look into his men, machines, and materials for having qualitative safe production. Supervisors and not the safety engineer are in a better position to do something about accidents. Because of their close association and frequent contacts with the people in their workgroups, first line supervisors have an indispensable part in the plant safety programmes.

Enforcement of safety measures depends to a great extent on the supervisor, for the workers are in his charge and their behaviour will be powerfully influenced by his behaviour. If the supervisor devotes all his time to production and avoids developing safety measures or feels that he has more important things to do than promote safety, then his department or work area, will have a bad accident record. The supervisor should always be alert for any unsafe conditions of tools, equipment, machinery, materials, structures, or other elements of the working environment that may cause or contribute to accidents. Safety, as well as production and quality control, ultimately are the responsibility of first-line supervisor. It is of utmost importance that supervisory personnel must possess proper training to handle their multifarious responsibilities.

It is felt by many that simpler the approach to safety, the better the safety performance. Nevertheless, the responsibility and accountability of a first-line supervisor must be defined fully.

One of the most difficult activities for the supervisor is to develop and maintain good safety attitude in the minds of all employees. When a good safety attitude is maintained, employees will perform their work safely without close supervision.

It is very important that first-line supervisors be responsible for investigating accidents that occur under their supervision. They are the most qualified to investigate accidents because of their constant contact with jobs, working conditions, and workers. They know most of the details of the jobs, procedures, hazards, environmental conditions, any unusual circumstances, and also their employees' job experience and personal characteristics. They are responsible for training new people, checking for unsafe practices, and looking for unsafe conditions. They must remind men about hazards and act to prevent accidents. By investigating accidents, they will become more proficient in accident investigation.

It is more desirable to train supervisors in safety, so that they may completely discharge their safety responsibility. The following are some areas that should be covered in special supervisory safety training courses.

1. Accident causes
2. How to train employees to work safely
3. Accident investigation
4. Job safety analysis
5. How to conduct a safety meeting
6. Facility inspections

All supervisors face the task of having to train employees to work safely. To a few supervisors it is very easy, to most it is very difficult. Some companies go to specialised trainers for all the training regardless of the type. However, it is felt by many experts in the field that a competent supervisor can train his own employees much better than an outside trainer. However, on some specialised safety programmes on equipment it may be desirable to bring in an expert in that field to do the training.

A job of preventing accidents falls on a supervisor not because it has been arbitrarily assigned to him, but because accident prevention depends upon the normal work, which a supervisor does as matter of course. The supervisor has the following fundamental responsibilities:

(a) Establishing work methods for orderly as well as safe production.
(b) Giving job instructions.
(c) Assigning jobs to people.
(d) Supervising people at work.
(e) Maintaining equipment and the workplace.

These usual job responsibilities of the supervisor are the very actvities through which the work of preventing accidents can also be carried out. These will also help to eliminate accidents caused by 'unsafe methods or procedures'. Making sure that safe procedures are established, is a supervisory responsibility.

Giving job instruction, with the proper accent on safety, will help to eliminate one of the most frequent causes of accidents, viz. 'lack of knowledge or skill'. Supervisors should show their subordinates how to do the work safely and ensure that they have the knowledge and skills to do it exactly in that manner, as required.

When a supervisor assigns persons to various jobs, safety as well as good job performance require, that he be sure, the selected person is qualified, and experienced to do the job and thoroughly understands the work method. Even experienced workers may require the direction and guidance. If workers are left on their own, most of the safety rules and procedures are observed more in breach than in compliance. It is a good practice to correct the fault of workers as soon as those are observed.

Maintaining the equipment and the workplace in safe condition is not different from maintaining them in efficient condition. Accidents very often result from tools and equipment in poor condition, from a disorderly workplace, from make shift arrangements, and so on. Good housekeeping is the foundation of safety.

Safety Responsibility

Having laid down adequate safety as one of the requirements, the responsibility for achieving it rests with the organisation's chief executive.

It should be understood that in the designing of processes, machines, jigs, and fixtures, none should be considered complete and ready for use until it is clear that the operation will be safe to the worker directly involved and will not constitute a hazard to other employees. Workplaces and processes must first be engineered for safety. Wherever possible, machines and processes should be so planned, arranged, or guarded as to preclude the possibility of injury. Even so there still will be plenty of need for everything that can be done to motivate, teach, and control employees so that there will not be any unsafe acts. Even if there are no unsafe conditions, there can be enough unsafe acts on the part of employees to cause serious injuries. The point of emphasis here is the need for providing safe work conditions. Even when hazardous conditions have not been causing injuries, they often are slowing down production.

The work of the maintenance department is extremely important in the prevention of injuries. There are two stages or degrees of safety in which a maintenance department may operate. One is of compliance, in which maintenance employees give full cooperation to the company safety department, readily complying with safety requests. The second, and better, is that of active safety-mindedness. In this stage, the maintenance workers have a considerable amount of safety training and are constantly on the lookout for possible hazardous conditions. Another aspect of maintenance concerns mechanical equipment. Proper lubrication, alignment, and adjustment leads not only to a longer life of the machinery and less downtime but also to the reduction of injuries.

General plant housekeeping is an immediate clue as to the safety rating of a concern. Clean floors, aisles clear of debris, fire-fighting equipment, and all other tools and supplies in their designated places are hallmarks of a safe operation.

Safety responsibilities also include product and process research, accounting department, purchasing and HR/industrial relations department. Usually the safety specialist is located in the HR department. It should be noted, that the HR department has responsibility for minimising the prospect of unsafe acts by employees through its attention to employee selection, placement, training, counselling, and so on.

Each worker has a safety responsibility. Of course it is the workers and their families who suffer most directly from work injuries. It must be made clear to all employees by the HR department, supervisors, top management, and the union that all safety regulations and instructions are expected to be followed just as seriously as any other company directives. In other words, the employees must regard safety measures as part of the requirements of their jobs, not simply as suggestions.

Mention has been made that the union should back up management in the enforcement of safety regulations. Management is generally anxious to promote safety, but it is also seriously concerned with the cost side of the picture. This sometimes leads unions to complain that management is not taking adequate steps to cut down injuries.

A safe place to work and safe work procedures are management's responsibility and cannot be shifted. Nevertheless, any union interested in its members' welfare must be concerned with working safely. Responsible union officials are showing more and more awareness of their responsibility for cooperating to aid management in safety work and responsibilities.

The first-line supervisor is the key person in maintaining day-to-day safety requirements in every organisation, particularly with regard to stemming unsafe acts. A safety-minded supervisor can make a pretty good safety record even if there are other constraints. Supervisors must accept responsibility for all safety failures occurring under their jurisdiction.

The supervisor is the one who directs the workers, says who shall do what, when, where, and how, and therefore cannot evade responsibility for the results, whether in terms of output, spoilage, or injuries. This does not mean that the supervisor is to blame for every injury. In fact, the proper approach to safety is not primarily one of fixing blame but rather of diagnosing hazardous conditions or conduct and taking steps to correct them so that no injurious events will occur.

Thus, everyone is responsible for safety, but everyone has different roles to play to monitor and improve the safety environment. Otherwise, safety becomes static; will fade into the background and again become a bothersome burden.

Safety professionals have customers. As a matter of fact, there are three groups of customers to be served for effectively managing safety programme. They are management, supervisors, and employees.

- The management needs cost control and measurable results – they speak the language of accounting and process management.
- The supervisors need control of their work areas, defined responsibilities and management support – they speak the language of production and quality control.
- The employees need fair treatment, tools for the tasks and a team environment – they speak the language of effort and acceptance.

Safety Inspection

Plant safety is a method of discovering potential risks and is one of the oldest accident prevention techniques. Inspection is an important function of any management and its main objective is to find and remove hazards in jobs/operations before an accident occurs.

Workplace inspections are undertaken with the aim of identifying hazards and promoting remedial action. Many different individuals and groups within an organisation are normally involved in a workplace inspection. The safety inspection should check maintenance standards, employee involvement, working practices, fire precautions, use of guards and adherence to safe working procedures. The elements of a safety inspection procedure are:

1. Identification of possible hazards.
2. Assessment of potential losses from these hazards.
3. Selection of control measures designed to eliminate or reduce the hazards.
4. Implementation of the control measures within the organisation.
5. Monitoring the effectiveness of the newly introduced control measures.
6. Reviewing at frequent intervals to ensure overall compliance with required standards.

Plant safety inspection can bring to light very useful information regarding safety and health status which will include not only unsafe conditions and methods of work but also direct and indirect causes leading to them. This is a very useful tool in the hands of the manager and helps him to take corrective action before harm is done.

It also helps the safety management programme in many other ways.

- Regular inspections at the shop floor level cannot escape the notice of the employees. This in effect, is an indirect way of demonstrating the management's interest in the safety and welfare of employees and thus contributes to better shop floor relations.
- Safety inspection brings to light the areas where waste can be decreased, processes can be improved and productivity increased. This results in better management of resources.
- Safety inspections result in contacts with employees which usually result in better understanding and mutual help which leads to all round success of the safety programme.

Safety inspections should be objective, regular, systematic and backed with sincerity of purpose. Effective procedures, lack of objectivity, lack of regularity, neglect and half-heartedness on the part of the inspection team will make a mockery of the safety management programme in the plant and can even negate achievement by other means.

Safety inspection cannot be done effectively by any one individual or agency. The responsibility should be shared by all those concerned within the plant. It is important that all personnel at the various levels should clearly understand their responsibilities and functions so that management objectives in this respect are implemented through their coordinated efforts.

The persons making inspections should be familiar with the equipment, tools, the processes and operations, and should have adequate knowledge of the hazards associated. They should also be familiar with the company rules and policies and also various statutes dealing with safety, health and environment.

Plant Inspection List

R.P. Blake recommends the use of checklists of items while carrying out a plant inspection.

1. Housekeeping
2. Material handling methods
3. Adequacy of aisle space and working space
4. Guarding of transmission machinery
5. Point-of-operation guards
6. Maintenance
7. Handtools
8. Ladders, portable steps, hoses, etc.
9. Hand trucks, power trucks, wheelbarrows, buggies, etc.
10. Floors, platforms, stairs, railings
11. Cranes, hoists, derricks, plant railways
12. Lighting
13. Electrical equipment, particularly extension cords
14. Elevators
15. Eye protection
16. Other personal protective equipment
17. Dusts, fumes, gases, vapours
18. Pressure vessels
19. Any other explosion hazards
20. Other dangerous substances
21. Oiling methods
22. Inspection of chains, cables, slings and other lifting tackle
23. Access to overhead equipment
24. Exits
25. Yards, roofs, and roadways
26. Any other conditions suggested by the accident records.

Source: Blake, R.P., *Industrial Safety*, P. 77

Safety Tours

This is an unscheduled examination of a work area, carried out by a manager, possibly accompanied by health and safety committee members, to ensure that, for instance, standards of housekeeping are at an acceptable level, fire protection measures are being observed and maintained, and the PPE is being used correctly. To be effective, it is essential that deficiencies noted during the tour are remedied immediately.

Safety Rules

Safety rules are codes of conduct designed for the purpose of avoiding injury and property damage. They should be prepared in realistic and easily understood language. Employees cannot be expected to respect and follow illogical, unfair, or unrealistic safety rules. Whenever safety rules are ignored, other plant rules are usually ignored, too.

The following guidelines are to be observed while preparing safety rules:

1. Safety rules must be clear and easily understood.
2. Keep the number of general safety rules to a minimum.
3. Formulate only those rules that are currently required.
4. Stipulate only those rules that can be strictly enforced.

Safety rules should have the following common characteristics:

1. *Purpose* - A clearly defined purpose for the rule establishes credibility for its issuance and can be an important promotional tool for obtaining employee acceptance and voluntary compliance.
2. *Scope* - Who is covered by the rule? What area(s) of the facility will be covered when the rule is in effect?
3. *Requirement(s)* - Each requirement of the rule must be written clearly, and using words and terms that are easily understood. Arbitrary, ambigious, and confusing words must be avoided.
4. *Date(s)* - The date when the rule was issued and the dates of subsequent revisions are important to ensure that the rule is current.
5. *Signature* - The signature of the person authorising the rule adds credibility and strength to the rule for enforcement.

Safety rules must be accessible to all employees, must be kept up-to-date, and must be reviewed periodically to assure that they are accurate and meet current standards and needs. Once the safety rules are developed, they can be organised in the form of safety and health manual.

Many establishments have developed safety rule books containing rules of general applicability. It is quite common practice to issue a copy to each employee. This course is of little practical value unless the rules and the justification for each rule are really understood by the employee. Necessary efforts are to be made in this direction.

Exhibit 4.6 Health and Safety Rules

Ten Health Rules

1. Aim for a healthy, safe and cheerful workplace.
2. Abstain from over-eating and enjoy meals at a leisurely pace.
3. Keep work clothes clean.
4. Take the initiative in undergoing medical examination.
5. Seek assistance in case of ill-health.

6. Relax during holidays and breaks.
7. Perform warm-up exercises before starting work.
8. Ensure the correct use of personal protective equipment.
9. Do not over exert in relation to posture and movements during work.
10. Take care to avoid occupational injuries such as burns, lower back pain and gas poisoning.

Ten Safety Rules

1. Always use proper protective equipments.
2. Inspect machinery and tools before use.
3. Do work properly in accordance with the instructions.
4. Use a safety belt for work in which there is fear of falling.
5. Do not approach prohibited areas or dangerous locations.
6. Inform the authorities immediately wherever and whenever dangerous work/location is discovered.
7. Perform joint tasks properly according to work procedures and signals.
8. Perform work with electrical machinery and equipment, or with dangerous equipment according to work procedures.
9. Operate movable cranes only after confirming the safety of the surrounding area.
10. Endeavour to keep things in order when working and clean up after finishing work.

Safety Education and Training

Safety education for all levels of management and for employees is a vital ingredient for any successful safety programme. Education in this context refers to the development of proper perspective and attitudes toward safety. Training, on the other hand, is more concerned with immediate job knowledge, skills, and work methods. Safety training of employees, as it is often conducted, has apparently been more effective in arousing safety "consciousness" than in teaching safe job skills.

Top and middle management require education in the fundamentals of safety and the need for an effective accident prevention programme. The supervisors must understand their key role in the safety effort, namely, that they are primarily responsible for preventing accidents. They must conduct safety training programmes for their employees who are directly under their supervision. Formal safety training includes both theoretical and practical that is undertaken in the classroom, workshop or training centre, which follows a definite written syllabus. General safety training should be provided for all employees on an ongoing basis and should include induction courses for new starters, ongoing employee safety training undertaken at regular intervals, safety representative training, supervisory training, and senior/middle management training.

Specific safety training needs can include those relating to safe systems of work for particular operations, first aid training, specific items of plant or equipment, the use of protective equipment, fire precautions, and safety inspections.

Employee safety training should begin on the first day at the workplace and should continue periodically for the length of the worker's affiliation with the company. Training is one of the best methods that can be used to influence human behaviour for the purpose of developing sound and safe work habits. A safety training programme is needed for:

(i) newly recruited employees;
(ii) for employees reassigned to other jobs;
(iii) for employees returning to work after long medical leave;
(iv) when new equipment and processes are introduced or installed in the workplace; and
(v) whenever need arises to improve and update safe work practices and procedures.

Safety training is not difficult if it is handled as a part of ordinary production training. Safety training of new employees, reassigned employees, supervisors and managers, skilled and unskilled workers, and even executives can and should be an important element of the total accident prevention or total loss-control programme.

The training should always be carefully planned and organised, irrespective of whether the supervisor does the training or someone else. Throughout the training session and thereafter the supervisor and the trainer must always set a good safety example.

The following subjects are normally included in a safety course for supervisors and managers:

1. Introduction to accident prevention.
2. Nature, causes, and results of accidents.
3. Accident prevention–principles and procedures.
4. Job safety analysis.
5. Job safety observation.
6. Planned safety inspection.
7. Investigation of accidents.
8. Motivating employees to work safely.
9. Handling of safety problems of employees.
10. Environmental hazards and their control.

In safety training the following key points have to be covered:

- Safety is equally as important as production.
- Accidents are caused and can be prevented.
- Each worker shares responsibility for his own safety and that of his fellow employees.

- Safety rules or regulations will be enforced, and violators of safety rules may be subjected to some form of discipline.
- Supervisors are to be consulted if there is any question in the new employees' minds about any part of their jobs.
- Where required, personal protective equipment must be used as a condition of employment.
- In the event of injury, workers are to report for first-aid or medical treatment and immediately notify their supervisors.

After completion of the safety training programme it has to be evaluated in order to ascertain its results.

Exhibit 4.7 Evaluation of Safety Training Programme

- Name of the programme
- Venue: • Dates:

1. Did you have a clear idea as to the aims and objectives of this programme?

To a great extent	To some extent	Not at all

2. Did the programme meet with your requirements and expectations?
3. Was the coverage of the contents adequate?
4. Was the course presentation interesting?
5. Duration of each session

Too short	Too long	Just right

6. Duration of the programme

Too short	Too long	Just right

7. How will you rate the physical arrangements for the programme?

Excellent	Good	Poor

8. Name some of the topics which you like the most

	Yes	No
(a)		
(b)		
(c)		

9. Do you feel that some more topics are to be added to the programme?

	Yes	No
(a)	________________	________________
(b)	________________	________________
(c)	________________	________________

10. Do you think any part of the programme was superflous and could be deleted?

	Yes	No
(a)	________________	________________
(b)	________________	________________
(c)	________________	________________

11. What difficulties do you anticipate in implementing the new ideas which you came across in the programme?

	Yes	No
(a)	________________	________________
(b)	________________	________________
(c)	________________	________________

12. Do you think this programme will be of use to others connected with your work?

	Yes	No
(a)	________________	________________
(b)	________________	________________
(c)	________________	________________

13. Do you require any further help in implementing your ideas acquired through the training programme?

	Yes	No
(a)	________________	________________
(b)	________________	________________
(c)	________________	________________

14. Do you want to include anything else which you think important?

	Yes	No
(a)	________________	________________
(b)	________________	________________
(c)	________________	________________

Innovative Approach to Safety Training*: *Examples

The General Electric Company once carried out an evaluation of the effectiveness of a safety training program. Its purpose was to reduce the number of accidents and to increase the regularity with which all accidents, major and minor, were reported. The training program consisted of the usual presentations, discussions, and movies, which were very dramatic in describing accidents and their implications. Their evaluation indicated that the program did not achieve its goals. Consequently, a new approach to training was adopted which was oriented to the job relationship between the foreman and each worker. Evaluation of this second approach revealed it to be much more effective. As a result, the information gained from evaluating this program proved welcome and useful to everyone involved in it (Kirpatrick, 1976).

At Ford Motor Company safety is a corporate wide priority. It is the responsibility of the employee to perform his or her tasks in a safe manner and of the supervisor to provide the necessary instruction and equipment. At Ford, there is an emphasis on teams. That is how the company trains and rewards employees for safe work practices - at the team level. Every quarter, the company gives award to the team that has demonstrated the greatest improvement in safe work practices or the most innovative approach to getting employees to be safe on the job. At year-end, the company holds a safety conference and rewards some of its top performing teams (unlike rewarding individuals in some companies). Ford managers emphasise continuous improvement in safe workplace practices.

Safety Communication

Safety communication includes stickers and posters, film shows, talks, competitions with rewards, safety weeks, and so on. As the attitude that accidents must be prevented before they occur has gained ground, people have come to realise that education and training are vital in the promotion of safety. Normally, propaganda seeks to persuade, education seeks to provide information; and training seeks to provide skills.

The following advertising media are commonly used:

Posters: There are all sorts of safety posters and each may help to promote safety in a different way. Safety posters should be displayed in places where workers usually spend some of their time when they are not working, such as the factory entrance and locker rooms. Bulletin boards for posters should be agreeable to the eye and properly maintained. Safety posters can only be an accessory to improving safety; they cannot replace good housekeeping, correct planning, good working habits and suitable guards, but they should help to create a greater awareness of safety among workers.

Posters must keep up with the times, and their messages must be in accord with current conditions, current events, and trends of thought. They must draw the attention of those at whom their messages are directed.

Large Plant Bulletin Boards: To furnish a continuing picture of the accident record and to stimulate a healthy rivalry between different plants or departments, many plants set up large boards at the plant gate or other favourable location.

Plant Newspapers: These are a well-established and valuable media for building up safety morale provided they are interesting, and suitably written. It is essential that the safety messages be timely, pertinent to plant conditions, practical, and carefully written.

Messages in Pay Envelopes: Information on off-the-job and home safety can well be given in this way. Also announcement of special bonuses, awards or prizes, plant outings, and so on, are often made in this manner.

Display of Interesting Objects: Great interest can often be aroused by displaying something that has been a contributing factor in causing or preventing an accident. For instance, a tool with a mushroomed head, broken goggles, a battered safety shoe can make excellent displays.

Signs and Slogans: Properly used, these have considerable value both in giving safety information and in promoting safety interest. All safety signs should be carefully located and be visible to any person approaching the position of hazard.

Slogans can be very valuable, but they must express a worthwhile purpose or goal, and they must be timely. Furthermore, the management must itself, show by its actions that it believes in them and is earnestly trying to live up to them. An outstanding example is the much used, often abused, and much discussed "Safety First".

Films and Slides: A poster gives just one impression of a hazard. A film can tell a whole story of an accident, showing the environment, how the dangerous situation arose, how the accident happened, what the consequences were, and how it could have been prevented. Films made specifically for instruction are more valuable than those made for general propaganda; they are particularly useful for explaining new safety devices or new working methods.

It constitutes a set of queries and checklists relevant to a particular activity which are to be filled in and verified by the site engineer and/or engineer in-charge. Based on this the safety department evaluates the performance of the site or factory in terms of its safety standards. This is an effective administrative tool in the hands of safety department for the effective site safety management.

Safety Appraisal and Evaluation

Appraisal

The basic goals of a planned safety appraisal are:

1. To check machines, devices, and tools.
2. To check materials and chemicals used in laboratories and manufacturing processes.
3. To check on the adequacy and proper use of personal protective equipment and safety guards.
4. To check within the workplace such factors as illumination, ventilation, noise levels, radiation, and so on.
5. To check working surfaces, stairs, ramps, ladders, and fire protection equipment.
6. To give critical attention to sanitation, housekeeping, food handling, waste disposal, and material storage.

Well-planned safety appraisal procedures are an effective means of discovering hazardous conditions. The effort made and time taken in a safety appraisal becomes ineffective unless prompt corrective action is taken following the appraisal. In any safety appraisal system, safety analysis will help keep unsafe conditions and practices to a minimum.

In analysing any safety system, three basic components must be considered: 1. the equipment (or machines), 2. the operators and supporting personnel (maintenance technicians, material handlers, inspectors, etc.), and 3. the environment in which both workers and machines are performing their assigned functions. In studying these three basic components several analysis methods are available, including the following:

Gross-hazard analysis: It is performed early at the design stage considering overall system as well as individual components. It is called "gross" because it is the initial safety study.

Fault-tree analysis: A fault-tree is a pictorial representation of an event(s) and factors affecting the outcome of the event. Fault-trees can be quantified and used to assess overall likelihood or system reliability. It traces probable hazard progression. For example, if failure occurs in one component or part of the system, will fire result? Will it cause a failure in some other component?

Fault-tree is based on the concept of initiating and enabling failure events; initiating events that cause unsafe changes in a process; and enabling events that allow these initiating events to become more serious.

Fault-tree analysis is capable of handling most forms of combinational events. It provides a good basis for and is particularly useful where a small number of major outcomes are concerned and is usually the case in hazard analysis.

Some of the standard symbols used in a fault-tree analysis are rectangles for a failed event including the top event at the top of the tree or branch; circles for basic causes; house for high probability normal event; hexagonal for significant consequences; diamond for fault event not developed to its cause.

Fault-trees are used:

- To assess the effectiveness of a protective system.
- To assess the likely frequency of a desired or undesired event.
- To assess the most reliable number of design options.

Fault-trees are clear means of depicting the contributing factors, and can be used for informed decision making and assess the likelihood of end event. Construction of an event-tree in step-by-step manner will help in identifying possible hazardous events which may result from a faulty event.

Failure mode and effect analysis: It involves reviewing systems to discover the mode of failure which may occur and the causes and effects of such failures. Broadly, this is similar to analysis by a combination of fault and event tree.

Event Tree Analysis: Event tree analysis is usually performed in examining the consequences of failure, because these are greatly influenced by factors like operator intervention and weather conditions which can be expressed most readily in event tree form.

Hazop technique: It is a versatile technique of hazard identification and can be effectively used in new plant design and also in assessing the safety of existing plant operations. The Hazop study involves systematically examining any part of the process, auditing existing plants and operating procedures, and developing new processes. The Hazop study aims to stimulate the imagination of a multi-disciplinary team in a systematic way so that they can identify the potential hazards in the plant.

The procedure for a *Hazop* study is: (a) define objectives and scope; (b) select the team; (c) prepare for the study; (d) carry out the examination; (e) follow-up; and (f) record and results.

Hazard and operability study technique was developed by the petrochemical division of Imperial Chemical Industries to improve safety and reliability of a modern chemical processing plant. The British Chemical Industrial Safety Council Publication defines *hazop* as the application of a formal systematic critical examination to the process and engineering intentions of the new facilities to assess the hazard potential of mall operation or mall functioning of individual items of equipment and the consequential effects on the facility as a whole.

The study is to be carried out by a team of experts.

The key feature of *hazop* is to select appropriate parameters which apply to the designed intention. These are general words such as: Flow, Temperature, Pressure and Composition.

The stepwise procedure to conduct *hazop* studies are:

Select a deviation from normal operating condition with the help of guide words. The guide words used are:

1. NON - No flow, reverse flow i.e., no forward flow than there should be.
2. MORE OF - More of flow, temperature, pressure, viscosity etc. i.e., higher flow, higher temperature and whatever than there should be.
3. LESS OF - Lower flow, temperature, pressure, etc. than there should be.
4. MORE than - Impurities present, i.e. ingress of air, water, acids, extraphase present e.g., vapours, solids.
5. PART OF - Change in composition of stream, e.g., ratio of component different from what it should be.
6. OTHER - What else apart from normal operation can happen, e.g., catalyst change, failure of plant service.

Search for the possible causes, their consequences and action required to control the hazard. Record the *hazop* study findings in a format.

Energy-transfer analysis: It determines interchange of energy that occurs during a catastrophic accident or failure. Analysis is based on the various energy inputs to the product or system and how these inputs will react in the event of failure or catastrophic accident.

Catastrophe analysis: It identifies failure modes that would create a catastrophic accident.

Maintenance-hazard analysis: It evaluates performance of the system from a maintenance standpoint. Will it be hazardous to service and maintain? Will maintenance procedures be apt to create new hazards in the system?

Human-error analysis: It defines skills required for operation and maintenance. It considers failure modes initiated by human error and how they would affect the system. The question of whether special training is necessary should be a major consideration in each step.

Transportation-hazard analyis: It determines hazards to shippers, handlers, and bystanders. Also considers what hazards may be created in the system during shipping and handling.

Evaluation

The most commonly used statistics for the evaluation of safety programmes are frequency and severity rates. Both the frequency and the severity rates give a measure of disabling injuries only. Frequency is concerned with the number of such accidents, and severity with the number of days lost. The frequency rate represents the number of disabling accidents per million man-hours worked; while the severity rate depicts the number of days lost per million man-hours worked.

Organisations need to monitor their safety efforts. Just as a firm's accounting records are audited, periodic audits of a firm's safety efforts should also be made. Accident and injury statistics should be compared with the previous accident patterns to determine if any significant changes have occurred. This analysis should be designed to measure progress in safety management. Another part of safety evaluation is updating safety materials and safety training aids. Safety policies and regulations should be reviewed periodically so as to ensure that they comply with both the existing and new standards. Further, a systematic safety programme requires a consistent effort to maintain a safe working environment. All concerned should continually work to ensure progress in developing a safe and healthy environment for all employees.

A self-inventory checklist for safety is as follows:

1. Is there an organisational policy on safety? Is it written?
2. Are all managerial and supervisory staff fully aware of their safety obligations?
3. Are employees being fully involved in safety programmes? Are there safety committees? Training programmes?
4. Are there written guidelines regarding safety? Are they being consistently and uniformly enforced?
5. Who is ultimately accountable for safety? Are departments charged for accidents?
6. Are safety inspection procedures clearly formulated and enforced?
7. Are all employees indoctrinated regarding proper safety procedures?
8. Are there programmes in place to provide on-going motivation to avoid accidents? Has their effectiveness been evaluated?
9. Are complete and accurate record-keeping procedures in order? Have they been internally audited?
10. Are first-aid and medical facilities available to all personnel at all times? Are all employees aware of the procedures involved in obtaining help?

Safety Audit

Perhaps the best definition of a safety audit is 'a critical examination of an industrial operation in its entirety to identify potential hazards and levels of risk'. Its aims are firstly to ensure that the company is complying with the relevant legislation, and secondly to make certain that the methods of risk control are both effective and economically sound.

Planning an audit programme covers such steps as its objectives, scope, frequency, staffing requirements, reporting, follow-up of recommendations, and audit quality assurance.

Audits are generally undertaken by multi-disciplinary teams. Individual members should have a good knowledge of legal requirements, and understanding of reasonable practice in the industry and, above all, the ability to communicate with all levels of personnel within the organisation.

Health and safety audit systems are designed to cover and, where appropriate, review, an organisation's management system for health and safety. Their objectives are:

- to identify current strengths and weaknesses in the management structure, organisation, planning and control with a view to exploiting and extending the strengths and taking remedial action to reduce or eliminate weaknesses;
- to provide a basis against which overall management effectiveness can be assessed at subsequent audits.

A safety audit subjects each area of a company's activity to a systematic critical examination with the object of minimising loss. Each component of the total system is included, for example management policy, attitudes, training, features of the process and of the design, layout and construction of the plant, operating procedures, emergency plans, personal protection standards, accident records, and so on. An audit–as in the field of accountancy–aims to disclose the strengths and weaknesses as and the main areas of vulnerability of risk, and is carried out by appropriately qualified personnel, including safety professionals.

When carrying out a safety audit, look at:

- frequency of accidents
- accident records
- risk analysis
- accountabilities
- safety information and notices
- training
- consultation
- workplace safety and housekeeping
- accident procedures

- fire equipment
- fire procedures
- handling of loads
- office safety
- hazardous substances
- personal protective equipment

A formal report and action plan is subsequently prepared and monitored. By repeating the audits after a period of weeks or months, a measure of the trend of the organisation's safety standards can be obtained with particular problem areas being highlighted. Safety audit is undertaken by the safety department of an organisation at construction sites or factories at various stages and at regular intervals and may or may not be with prior notice. The audit is for the evaluation of the preparedness of the line functionaries and the workmen in terms of adoption and maintenance of safety norms and practices at workplaces and their role in preventing accidents and dangerous occurrences involving men, materials and machinery.

Health Programme

According to World Health Organisation (WHO), "Health is a state of complete physical, mental and social well-being, and not merely an absence of disease or infirmity." This definition goes beyond the mere absence of disease. It envisages three dimensions or components of health - physical, mental and social–all closely related. A fourth dimension may also be added to it, namely, spiritual health.

A complete health service has five broad stages. These are:

(a) health maintenance or health attainment;
(b) increased risk;
(c) early detection;
(d) clinical; and
(e) rehabilitation.

The purpose of occupational health is fulfilled effectively when all the above services are provided jointly and co-operatively by all the members of occupational health team.

Health care programming among employees at the work place should be an integral element of a company's business strategy. The reason is that a healthy worker is a productive worker. Moreover improvements in employee health result in better work attitudes, higher morale and job satisfaction, and reduced absenteeism and turnover. Health promotion focuses on prevention rather than treatment or cure. Hence, the health programme in an organisation is normally planned around improvement and prevention of controllable risk factors such as smoking, obesity, high level of cholesterol, stress, hypertension, and low level of physical fitness which are responsible for many major diseases.

Planning of a health programme requires a systematic step-by-step process. The organisation needs to identify the various in-house and external health promotion activities, and encourage employee participation therein. The plan should also specify the health promotion policy and procedures.

Once planning is complete, the organisation can proceed to the stage of implementation. However, before the programme can be adopted by the employees, it needs to be marketed to them through various means ranging from circulars, letters and posters.

In many countries, particularly in the developing world, occupational health services are in effect the provision of total health care to the working population. Health promotion at the workplace is an integral component of the health care services to workers, i.e., occupational health service. Health means not only an absence of disease but also the presence of optimal, physical, mental and social well-being.

There are seven content areas that should be included in a comprehensive healthcare (wellness programme). They are:

1. Blood pressure education
2. Cholesterol reduction
3. Weight loss and control
4. General nutrition
5. Smoking cessation
6. Physical fitness and exercise
7. Stress management.

The health promotion programme has to be evaluated in order to ascertain its effectiveness.

Health Services

Normally at the company premises physicians were called upon to treat minor ailments not caused at work, such as colds, sore throats, skin disorders, headaches. But now, industrial health programme of many progressive organisations are taking positive steps to maintain good employee health off the job as well as on the job. Some organisations have also established programmes in the field of mental health. A comprehensive organisation health programme will include the following features:

1. A professional staff of physicians and nurses.
2. Adequate facilities for meeting emergency work injuries.
3. Proper first-aid treatment for occupational injuries and diseases.

4. Pre-employment and post-employment medical examinations and periodic health check-up for those exposed to special occupational hazards.
5. Information and educational services for the personal health of employees.
6. Co-operation with public health authorities in regard to mass inoculations and other measures for the prevention of communicable diseases.

Health Education

Most of the organisations are concerned about the health of their employees. This is true even where there are no clinics or other medical facilities on the premises.

There is nothing wrong with displaying approved health posters. These may caution employees about cough and cold, announce diabetes detection week, or stress on good eating and sleeping habits. Material of this kind is readily available. As a part of their health education efforts, companies may initiate cancer detection programmes for both women and men. During the orientation period or at other times, it is not out of the way to talk about personal hygiene, cleanliness, body odour, overweight, care of the heart, teeth, and other personal matters.

A major health and safety issue in the workplace is HIV/AIDS. AIDS - Acquired Immune Deficiency Syndrome - is caused by a virus that destroys the body's natural defenses against disease and infection. People can become infected with AIDS through intimate sexual contact, contaminated needles, and transfusion of infected blood or blood products. People do not get AIDS by sitting next to a person who has a disease, by eating food prepared or served by an infected person, or through any other casual contact.

AIDS has a significant impact on all workers not just infected people. It affects those who fear "catching" the disease and it is the managers who must defuse problems that arise in the workplace as a result of misinformation and misunderstandings about the disease. Therefore, eliminating the fear and stigma is one of the first and most difficult challenges human resources managers have.

A sound AIDS policy addresses the concerns of equal treatment, legal responsibility, education, and confidentiality rights. Educational programmes that address issues such as working with someone who has AIDS, handling death and bereavement in the workplace, and preventing AIDS are also important parts of creating a healthy work environment.

Industrial Medical Department

The development of industrial medicine in the past three decades may be said to have exhibited three successive phases: (a) surgical, (b) medical, and (c) the preventive with emphasis upon medical and engineering control of environmental hazards. The programme of physical examination now followed in different concerns reflects the changing concepts of industrial medicine in general. Pre-placement examination is vital to both industry and to the employee. One large manufacturing concern states its philosophy thus: "this examination is made not for the purpose of rejecting applicants, but for proper placement of those whose defects might be hazardous to themselves or those around them." There is no uniformity of practice in the organisation and management of medical departments in industry. The character of a given industry, its size, and the conception of management as to what constitutes appropriate medical service to the workers are the factors which determine the nature and the scope of a medical department. There is imperative need of selecting a suitable medical, nursing, engineering, and clerical personnel. They must be adequately trained for the specific duties to be performed and also adaptable to the peculiar needs of the particular industry they serve. There should be a careful disposition of physical equipments and carefully developed system of records.

First Aid

First aid is a post accident strategy vital to prevent loss of life and future deterioration in health following accident injury or sudden illness. First aid is defined as the skilled application of accepted principles of treatment on the occurrence of an accident or in the case of sudden illness, using facilities and material available at the time. It is given:

(a) to sustain life;
(b) to prevent deterioration in an existing condition; and
(c) to promote recovery.

The most important areas of first aid equipment are:

(a) restoration of breathing (resuscitation);
(b) control of bleeding; and prevention of collapse.

Doctor in Industry

It is the aim of the doctor in industry to ensure that employees are healthy and can, therefore, maintain a state of well-being to the benefit of themselves as well as to the industry employing them. The medical duties of a doctor consists of:

1. Impartial advice to individuals, managers and trade unions on all aspects of occupational health within the industry.
2. Advice to those committees within the organisation which are responsible for developing and implementing company's health and welfare policy.

3. Organisation of occupational health services under the care of the doctor.
4. Examination and clinical advice to individuals with problems about their health, particularly those relating to their work.
5. Responsibility for the medical aspects of resettlement and rehabilitation within the industry.
6. Responsibility for providing such primary medical care and/or first-aid facilities in the industry as are considered necessary.
7. Responsibility for health education of all employees with respect to prevention, fitness and hygiene.
8. Assisting with the training of nurses, safety officers and first-aid personnel.
9. Arranging for co-operation with medical services outside the industry.

Occupational Nurse

The functions of a nurse include:

1. helping people in their home and with family problems;
2. providing an opportunity to the workers to discuss their problems;
3. encouraging workers to handle their own problems and situations by offering different alternatives;
4. assisting the physician in pre-placement work and in the periodical medical and health examination;
5. providing therapeutic care for the worker and carry on significant positive health programme.
6. establishing meaningful interpersonal relationships between herself and the worker in order to work out common solutions.
7. organising occupational health and safety education by utilising both individual contacts and group sessions.

To complement nursing profession, it is worthwhile to quote Florence Nightingale (1895). "Nursing is not only a service to the sick; it is a service also to the well. We have to teach people how to live." The nurse's position as a member of the occupational health team is recognised and valued because she believes more in co-operation rather than competition.

Occupational health nursing is the application of nursing practice at the place of employment. The nurse practices her profession not in a hospital or at the bedside but at the workplace. Her nursingcare consists of curative, rehabilitative, and health maintenance and promotional activities.

Mental Health in Industry

In 1950, a WHO expert committee on mental health reviewed the various definitions of mental health and observed, "Mental health, as the committee understands it, is influenced by both biological and social factors. It is not a static condition but subject to variations and fluctuations of degree."

Mental health is not mere absence of mental illness. A mental healthy person has three main characteristics:

(a) The mentally healthy person feels comfortable about himself, i.e., feels reasonably secure and adequate. He neither underestimates not overestimates his own ability. He accepts his shortcomings. He has self-respect.

(b) The mentally healthy person feels right towards others. This means that he is interested in others and loves them.

(c) The mentally healthy person is able to meet the demands of life. He is able to think for himself and take his own decisions.

Mental illness like physical illness is due to multiple causes. Among the known reasons are the organic conditions, heredity, and social and environmental factors. The different preventive measures of mental ill health can be classified as primary, secondary and tertiary.

Mental health services are those services provided to individuals who are mentally ill. Mental health services in a community are concerned not only with early diagnosis and treatment, but also with the preservation and promotion of good mental health and prevention of mental illness. The mental health service comprises of: (a) early diagnosis and treatment; (b) rehabilitation; (c) group and individual psychotherapy; (d) mental health education; (e) use of modern psychoactive drugs; and (f) after care services.

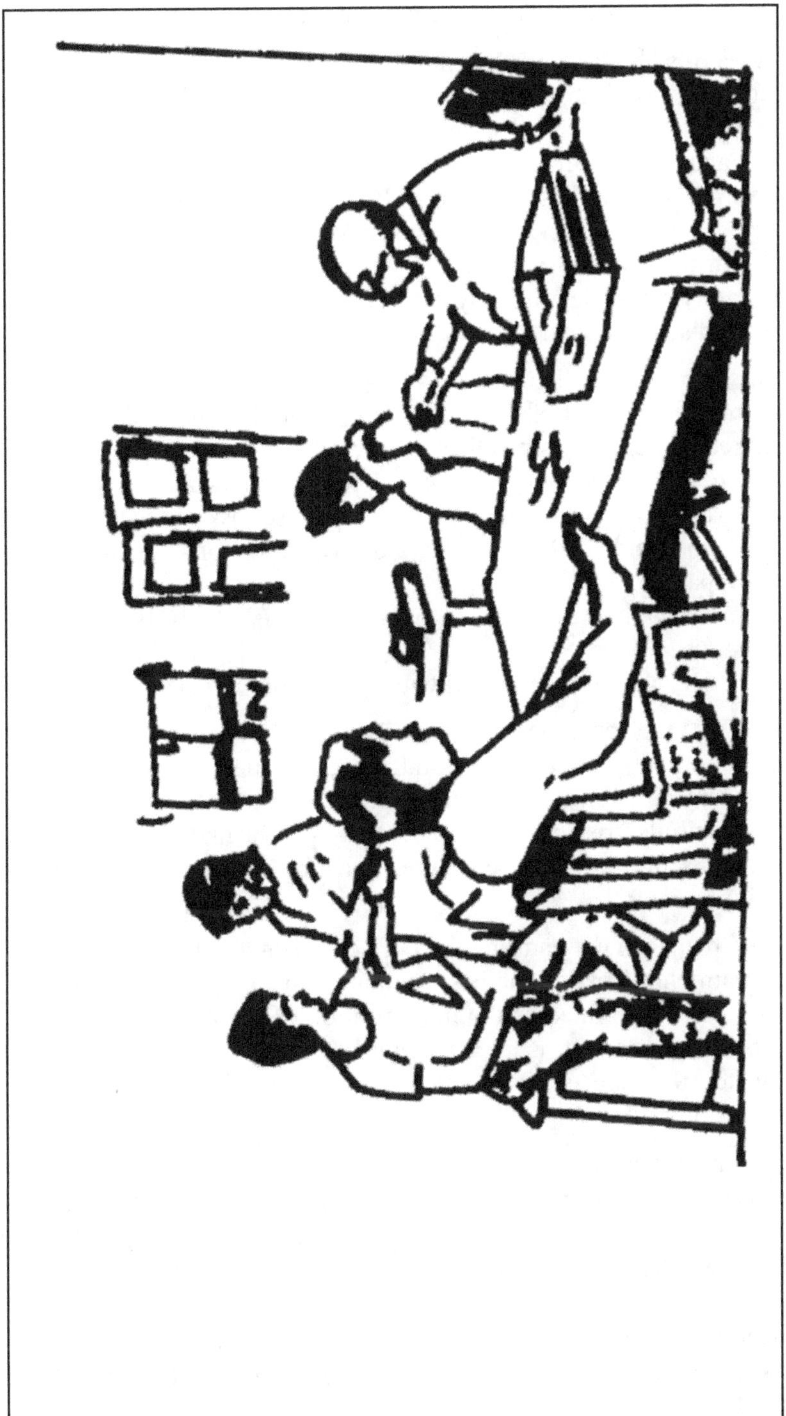

Fig. 4.2 Placement Medical Examination.

Risk Management and Loss Control

The idea of safety as a profit centre is comparatively new. Until a few years ago, indeed, the opposite view was held among most firms - namely, that a high investment in accident prevention was something which only the largest and most profitable enterprises could afford. Such a view is infact still common among small firms. It is now becoming more and more widely accepted that in order to sustain a smooth flow of production and cut down the number of accidents which are costly in every sense of the term, safety can be legitimately regarded as a profit centre, hence subject to business disciplines and techniques like any other activity.

The overall discipline, or control, is the technique now known widely as 'risk management'. Risk management involves four main steps, all of which require action by the client. These are:

1. The identification of all risks attached to the business.
2. Examination of the factors likely to cause loss and extent of the loss if an accident does occur, plus evaluation of the likely financial impact.
3. Investigation of ways in which the risks can be eliminated or reduced.
4. A decision on how to deal with the remaining risks.

Risk is the potential harm which a hazard can cause. It is the combination of the probability and the consequences. An unlikely event with severe consequences (for example, a plane crashing or the building) carries a similar risk to an event which is very likely to occur but with trivial consequences (for example, pricking thumb on a staple while handling papers).

Risk management can thus be defined, in the words of K.J. Duffy, as, "the treatment of risk to the best economic advantage of the organisation". It may also be defined as the minimisation of the adverse effects of risks within a business. It is a management system designed to reduce or eliminate all aspects of accidental loss that lead to a wastage of organisation assets.

Risk management involves the skills and knowledge of many disciplines such as civil and mechanical engineering, industrial hygiene, environmental monitoring, computer fraud, as well as experts in accounting and insurance. It involves the identification, evaluation, and control of the pure risks within a business. Risk analysis and evaluation may call for a blend of managerial experience and actuarial skill to estimate maximum loss from a specific event. Risk reduction or elimination must clearly play a major role in any risk management strategy.

Fundamentals of Loss Control:

1. An unsafe act, an unsafe condition, and an accident are all symptoms of something wrong in the management system.
2. One can predict that certain sets of circumstances will produce severe injuries. These circumstances can be identified and controlled.
3. Safety should be managed like any other company function. Management should direct the safety efforts by setting goals and by planning, organising, and controlling to achieve them.
4. The key to effective live safety performance is management procedures that are accountable.
5. The function of safety is to locate and define the operational errors that result in accidents. This function can be carried out in three ways:

(a) by asking why accidents happen - searching for their root causes;
(b) by asking whether certain known effective controls are being utilized; and
(c) by asking to what extent it helps in improving safety performance.

The activity which has come to be known as 'loss control' is essentially a facet of risk management. Loss prevention may be defined as the application of engineering techniques in order to reduce the occurrence of accidents that result in personal injury; damage to property, product, equipment and buildings and those accidents that have no end result - the near miss accident. It is linked with risk reduction or elimination and concerned with the activities inside the business. It is a practical measure aimed at reducing or down grading incidents and unplanned events in the field of accident prevention, fire protection and plant security. Loss control extends to practically every facet of the business, including injury prevention, damage control, fire protection and industrial security, occupational health and hygiene, pollution control, product liability, and business interruption.

As soon as possible after doing an assessment, the following information must be recorded:

- the date of the assessment;
- the product name or other identification of the hazardous substance;
- the degree of risk to be assessed;
- the control measures for use of the substance;
- the type of monitoring and/or health surveillance, if any, and the intervals at which it must be carried out.

Record Keeping

One of the most important parts of an accident programme is record keeping. Accident records are one of the primary means any company has for measuring the effectiveness of its accident measuring programme. There are essentially two different kinds of accident records - (i) injury records, and (ii) accident investigation records. All companies should keep both types of statistical information. Injury records are kept as a measurement of safety performance. By comparing current injury records with past injury records, a company can see if it has made any progress with its safety programmes. This type of record can also be used to determine problem areas or trends, and it will pinpoint areas that warrant immediate or long-range attention.

Accident investigation records are also a tool for measurement. This form of record should serve two purposes. First, to determine the surface causes and the underlying causes of an accident; and second, to evaluate the supervisor's ability to prevent accidents. Accident records may be used to

1. Create interest in safety among supervisors.
2. Determine the principal accident causes.
3. Point out to supervisors and safety committees the most frequent unsafe practices and unsafe conditions.
4. Judge the effectiveness of a safety programme.

All injuries, however minor, must be recorded giving the following information:

(a) Full name and address of the injured person.
(b) Date and time of accident.
(c) Place where the accident happened.
(d) Cause and nature of injury.
(e) Name, address and occupation of the persons giving the notice, other than the injured person.

Research and Surveys

Research and Surveys in the area of industrial health and safety is of utmost importance to health and safety professionals. Safety issues include safety standards of machinery and equipment, cost benefit analysis of safety measures, and industrial accidents and prevention measures.

Research and survey areas normally include :

- The interactions among worker, machine, and environment.
- Understanding engineering principles of processes and systems.
- The management processes of planning, organising, and controlling.
- Role of learning, motivation, and morale in safety performance.
- Quantification of accidents, injuries, and loss control.

- Safety evaluation and safety audits .
- Fire and electrical safety.
- Industrial ventilation and temperature.
- Industrial hygiene and toxicology.

TQM in Safety, Health and Environment

Total quality management (TQM) is to improve the quality of work of all the people, at all the levels, and in all the functional areas of an organisation. Quality management tools with its focus on customer's expectations, prevention of problems, building commitment to quality in the workforce and open decision making, and continuous improvement can be applied to the practices of safety and health.

Safety cannot be controlled and checked at one point only. It needs to be "built into" and not "inspected into" the industrial cycle like planning, designing, purchasing, manufacturing, supervision, shop operation, inspection and testing, and marketing and service to customers. The need of the hour is to develop a safety philosophy in the organisation and systematic approach to total safety management system.

Safety has commonality of cause and control measures with quality. In attending to problems of quality and safety, the concepts of TQM need to be applied. TQM is not a narrow concept; it encompasses the quality of the environment as a whole. When safety is mentioned, it is not restricted to the safety of the personnel working in a manufacturing industry, but also the safety of the general public as the process and operations inside a factory may spill over the perimeter of the factory. Therefore, the term "safety" is to be understood in its entirety.

TQM demands integration of the requirements of safety, health and environment at the appropriate stages in the closed loop of the quality system. Achieving total quality management is not a limited concept, but demands an overview approach by all concerned at all stages of quality systems cycle. For achieving total quality in safety, health and environment, interface of TQM and involvement of all concerned is essential.

The emphasis should be an overview approach to the safety, health, environment and quality of the product by insisting on the quality of the raw material and by determining the reaction parameters. A safe and eco-friendly product with no health hazard and at the same time conforming to international quality standards cannot be manufactured without having an integrated total quality approach to safety, health and environment at the place of manufacture.

Safety has to be integrated into all aspects of the business as part of the total quality approach. Safety standards and safety objectives are an essential part of ensuring business efficiency in the same way as other technical standards and business objectives.

Take a moment to answer the following key questions:

1. Do you know how well you perform in health and safety?
2. How do you know if you are meeting your own standards for health and safety?
3. How do you know that you are complying with the health and safety laws that affect your business?
4. How great are your losses, whether these are measured in terms of lost working time, sick leave, compensation payments or damage to plant, vehicles or equipment?
5. Do you have accurate records of injuries, ill-health, and accidental loss?

CASE 1

A newly recruited employee was told to clean and arrange the work area of an engineering shop. He set about the task enthusiastically, putting various pieces of surplus metal and scrap on to a four-wheeled trolley. He then pushed the trolley towards the scrap area at a fast walking pace. Unfortunately, on turning a corner the trolley bumped into another employee who was crossing its path. This employee was injured, not only by the impact of the trolley, but also by a sharp piece of metal that slid off the trolley and entered his eye.

Questions

1. Who was to blame for the accident?
2. How should the accident have been prevented?

CASE 2

In Raigarh Petrochemicals, employees are closely involved in every part of the safety process. For instance, they serve on safety committees, develop daily and monthly safety meetings, and conduct job safety analysis.

At the company, safety begins with the company's top management team. Senior representatives serve on the company's health and safety committees. Among other things, this committee meets once a month to

→

review accident reports, establish health and safety goals, review safety statistics, and endorse and sponsor safety programmes. At the departmental level, employee representatives serve on health and safety committees where they similarly monitor overall safety performance.

The company's "safety first" culture is cultivated almost from the day the new employee arrives at work. For example, new entrants are encouraged to participate in the safety process during their orientation. Also they attend a one-day orientation where company officials explain and emphasise the importance of the company's health, safety, and environmental policies and programmes. Employees also participate in monthly departmental training sessions to discuss shop-level safety issues and to discuss safety suggestions. The firm's employees also work with their departmental committees to conduct monthly safety audits, to review and document departmental job safety, and to submit safety suggestions. Employees are actually required to report every incident and near miss of a safety nature. The information is used to analyse accident trends and to enable management and departmental employees to develop safety recommendations.

Question

1. Suggest ways and means of developing safety culture in an organisation based on the above case.

SUMMARY

- Accidents incur not only medical costs and indemnities, but also loss of productive work and damage to property.
- The four basic steps to conventional safety programming are case analysis, communication, inspection, and supervisor's safety training.
- The key elements of safety planning are (a) setting objectives, (b) identifying hazards, (c) assessing risks, (d) implementing standards and targets, and (e) developing a safety culture.
- Good planning is as essential in safety as it is in production.
- It is essential to lay down organisational safety policies and programmes.
- Interest in safety can be created in many ways. for example, formation of safety committees, appointment of safety officers with specific responsibilities, safety inspection, and so on.
- Another aspect is risk management which is designed to reduce or eliminate all forms of accidental loss.

→

- Risk management involves the skills and knowledge of many disciplines.
- Attitudes that cause accidents and injury are (a) overconfidence, (b) laziness, (c) stubbornness, (d) impatience, (e) ignorance, (f) showing off, and (g) forgetfulness.
- Formation of good safety attitudes consist of (a) being informed of safety procedures, (b) being aware of safety procedures, (c) co-operating with safety representatives, and (d) being alert.
- A goal-driven safety programme must be (a) written, (b) measurable, (c) well-defined, (d) achievable, (e) challenging, and (f) realistic.
- The overall responsibility for maintaining and promoting organisational health and safety primarily rests on chief executive.
- In brief, the elements of safety organisation are safety policy, safe work practices, safety training, safety meeting, safety investigation and analysis, safety rules and regulations, safety promotion, safety inspection, safety records, and emergency preparedness.
- Safety standards and safety objectives are an essential part of ensuring business efficiency in the same way as other technical standards and business objectives.
- A strong safety culture begins with visible management's commitment and highly engaged and empowered employees at all levels.

DISCUSSION QUESTIONS

1. Why have the roles of safety personnel changed over the years?
2. What are the functions of a safety officer/department in an organisation?
3. What are the general functions of a plant safety committee?
4. What type of safety training be given to supervisors?
5. What is the role of a supervisor in accident control and prevention.
6. What is risk management? Risk management involves the skills and knowledge of many disciplines. Enumerate.
7. How does safety inspection help in detecting potential causes of injures?
8. How the occupational safety and health policies, and their objectives are aligned with strategic policies and overall direction of the business including integration with the business system?

5

Organisational Role on Safety and Health

Organisational arrangements to reflect the culture of a company are vital if safety performance is to be improved. These will include:

Role of Management

The management collectively and individually has the primary responsibility for providing an effective management system that will ensure a safe and healthy workplace.

A number of negative and positive motivating factors or approaches can be identified in determining the overall approach. All may have a part to play and in practice, a combination of these factors is normally present rather than one factor in isolation.

Employees should not suffer ill health or meet with accidents because of their work. There is no doubt that it can be a complete motivating factor for certain individual managers, but for an organisation as a whole other factors have an important influence on the amount of attention given to safety.

Essentially, the legal approach has its focus on what the law requires, and nothing beyond it. Managers who subscribe to this philosophy will fundamentally do only what is necessary for legal compliance. While compliance of the law is proper, and mostly a necessary prerequisite for a safe working environment, much more has to be done to improve the safety performance of an organisation.

Organisations that subscribe to financial or cost-benefit approaches subscribe to and try to find a balance between, on the one hand, the value of the resources necessary to obtain and maintain high standards of safety and on the other hand, the value of the benefits derived therefrom. While this may sound eminently reasonable and economically sound, the problem with this approach is that it is ultimately confronted with the vexed question of "what is the price of a human life?"

The maintenance of high safety standards requires adequate financial resources, but accidents and damage to equipment also cost money. It is easy to calculate the costs of training or machine guarding, but it is not so easy to calculate the costs of an accident.

Organisations that subscribe to the human factors approach focus on the various factors in the workplace that influence compliance with safety procedures and practices. They take account of human resource considerations and regard people as an important asset to the business. The problem with this approach is that it does not go far enough in actually addressing the issue of what motivates people, what effect this has on their attitudes, and ultimately how this influences their behaviour. However, companies making real progress in this field have turned their attention to and concentrated increasingly on behavioural or cultural aspects.

Whatever approach or combination of approaches is used, sound safety and health management is the foundation upon which the company's approach to managing the business should be based. It requires knowledgeable directors, managers, engineers and employees operating within a management system that not only strives to eliminate accidents and control the risks, but also is responsive to the needs of a business. If we are to make fundamental and sustainable improvement in our business, we must consider health and safety as a core activity and one that is visibly integrated into everything we do from a management and leadership perspective.

Managers are required to carry out the following:

- Establish health and safety as a priority and core value of the company. Senior management must be active in keeping up to this priority and value.
- Establish a world class health and safety programme. The programme must have a strong and clear internal accountability system, in which every employee understands his/her responsibilities.
- Set health and safety workplace standards.
- Visibly demonstrate leadership in the implementation of management's responsibilities.
- Integrate employee well being into all activities.

- Ensure that competent people are identified to perform the required work.
- Provide the necessary resources to support the development and implementation of health and safety programmes as a part of the business improvement plan.
- Reinforce that the proper management of health and safety contributes to improved costs, and enhances its image leading to improved performance of the company, particularly in reliability, quality, and team development.
- Develop a management system that uses both negative and positive reinforcement. Negative reinforcement is the more traditional method employed in obtaining compliance. Positive reinforcement applies the principle of determining what you want people to do and reinforcing them for doing that in a way that is meaningful to the employee.
- Set improvement targets and integrate health and safety into every business decision.
- Communicate health and safety issues as an integral part of the business.
- Foster participation and involvement of all employees in improving health and safety performance.
- Include health and safety performance as a key business indicator.
- Ensure that the right people are in the right jobs.

Role of Employees

In improving safety performance, the importance of the role of all employees need not be emphasised. This is particularly true as companies increasingly recognise the importance of the behavioural aspects of safety. A good employee is a safe employee. All employees should:

- Work safely and follow rules, regulations and procedures.
- Adopt a proactive rather than a reactive role to safety issues.
- Accept responsibilities for safety - their own and their colleagues.
- Report all unsafe situations.
- Participate fully in the development of safe working practices.
- Undertake the necessary training to ensure their competence.
- Apply beneficial learning fully in all activities.
- Take part in the investigation of accidents, incidents and near misses.

- Set Certain examples, in particular, for the young and for new employees.
- Support management in all their efforts to improve safety and health performance.
- Contribute to the work of safety committees wherever necessary.

Role of Trade Unions and Employee Representatives

Where unions are present in the workforce, it is important to ensure that they are given a positive role in the safety programme. In general, the primary responsibility for safety rests with the management of the company. However, in most situations, it is the development of a direct relationship and a shared involvement of the company management and its employees in safety responsibilities. Where they are active, unions can play a significant role in maintaining and improving safety standards.

Unions can provide positive reinforcement of company safety programmes. They can provide effective training programmes, have on-the-floor practitioners as safety representatives, participate in audits, accident investigations and risk analysis, and participate as members of safety committees.

On a broader level, unions can be effective lobbyists with governments where there are companies who are not living up to legislated health and safety obligations or where such roles and regulations are not effective in the workplace. In addition, unions may assist to keep health and safety issues at the forefront of management attention both on an ongoing basis and periodically in the process of contract negotiation.

As with all effective working relationships, there needs to be a high degree of trust between the union, employees and management, in the involvement of safety issues. This requires that the interest of management and the union be aligned and that the common objectives are agreed to. Unions, wherever present, can provide a positive impact on the development and implementation of effective safety programmes.

Role of Safety Organisation

In the professional safety management, safety organisation has to play a vital and integral role in the company's overall approach to safety. To amplify this:

- A safety organisation is necessary for every company/plant.
- Every employee should be aware of that organisation and have access to it, if necessary.

- The safety organisation should include experts from different disciplines within its ranks or have access to them.
- The methods by which the safety organisation operates and its relationship to the rest of the business operations should be clear.

Safety is no different from any other activity carried out by the company. As safety enters almost every facet of company activity, the case exists for a separate organisational structure. Whilst its major impact is traditionally felt to be during the production process, its needs also have to be taken into account in most other activities, e.g., planning for new plants, maintenance of existing plants, research into new materials and processes, purchasing of new materials, sale of finished goods, disposal of waste material, in addition to contact with external bodies and regulatory authorities. Indeed, no part of the business activity is outside its scope.

There is a need for safety experts with varying specialisations that includes engineering, occupational hygiene, medicine, and ergonomics. They should also know in detail the legal requirements, which will vary from country to country. Also, it is recognised that in particular circumstances it may be necessary to use an external consultant to address specific problems.

The safety organisation method of working is primarily through three rules: (i) advising both the management and to all the employees, if necessary; (ii) developing of policies, systems and procedures (both general and specific); and (iii) tracking of performance - specifically by organised auditing, and also by more general discussion and observation.

The safety organisation in determining its method of auditing also needs to take into account a number of other issues, namely the legal requirements, which apply within the company concerned and the way in which the company is operated generally.

A formal structure should be drawn up relating to the organisational arrangements by which the company's safety policy is to be implemented. This should establish clearly defined responsibilities between all employees.

Behavioural Model for Safety

The issue of a behavioural model for safety is not different or distinct from behaviour in general. Behaviour is a function of activators and consequences. An activator is something that precedes the behaviour and influences the behaviour beforehand. A consequence is something that follows the behaviour and discourages or reinforces repetition of the behaviour.

By designing and controlling effective workplace activators and consequences, we can increase key safety behaviour and reduce unsafe behaviour. Effectively managing behaviour by a combination of well planned activators and consequences will result in fewer accidents and injuries. Positive, immediate and certain consequences influence behaviour significantly more than negative, future, uncertain consequences. Safety programmes based on negative consequences rely on enforcement. Rules and policies are established and disciplinary procedures, including, termination of employment are the negative consequences. Negative consequences build compliance whereas positive consequences foster achievement. People work better to achieve success rather than to avoid failure. In fact, using positive consequences is the only way to get continuous improvement. The most convenient, inexpensive, readily available and often the most powerful positive consequence is positive reinforcement or recognition. Positive reinforcement of safe behaviours works best if it is a part of the normal work environment and is part of the values and principles of a company. Praising people for safe behaviour is probably the best thing that can be done on a daily basis to prevent injuries. Positive reinforcement helps to build self-esteem.

Behavioural research has revealed strategies to maximise the impact of recognition. Guidelines for quality recognition are listed below:

- Make it timely - as soon as the behaviour or condition is observed
- Give it on an individual basis
- Be sincere, genuine
- Describe the behaviour or condition that was observed and why it should be recognised and supported
- Explain the importance of the behaviour
- Give personal praise and thanks
- Pay attention to the person's comments, suggestions or reactions
- Reaffirm approval and offer encouragement, acknowledge recognition of a job well done and reinforcement to keep up the good work
- Use material rewards for symbolic value only

Strategic Issues

In the realisation that safety is good business, the management of safety and health should be a strategic issue in every member organisation. The strategy should spell out the beliefs of the organisation, stating the values and principles to which it subscribes.

A set of principles and values is as follows:

- All injuries and damage can be prevented.
- Preventing injuries and damage make good business sense.
- Everybody occupying a managerial or supervisory position is personally responsible for the safety of those working under him or her.
- Everybody in the organisation carries personal responsibility for his or her own safety and also his or her fellow-workers' safety.
- Training is an essential element in ensuring the safety of employees.
- Safety audits are to be carried out regularly and meaningfully.
- Every shortcoming in the field of safety and health is rectified immediately.
- All injuries and damages are investigated thoroughly.
- People are the most important elements in the safety and health programme.

The values and principles mentioned above should be translated into a vision which in turn will lead to a mission statement with specific short and long-term objectives.

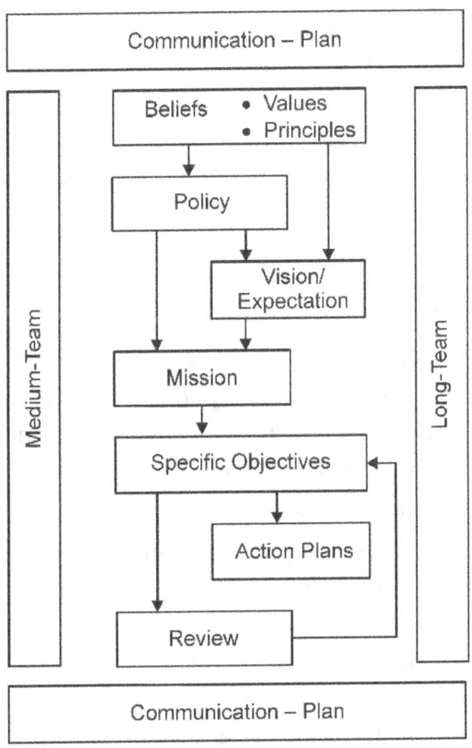

Fig.5.1 A Communication plan.

A formal communication plan is essential, that addresses every issue of the strategy. The strategy should be well designed, properly thought out, creative and innovative. The strategy needs to be driven by a comprehensive communication plan that focuses on the medium to longer term. This plan must clearly identify primary, secondary, and tertiary audiences and must include all stakeholders such as employees, trade unions, all levels of management, shareholders, communities, and families of employees.

Accident/Injury Statistics

Statistics are used by the industry to measure almost all aspects of performance - whether it be 'output', 'delivery', 'plant reliability', 'sales', and so on. It is not surprising therefore, that companies should seek to measure safety performance in a similar way. Measuring safety performance is, however, far from simple, particularly comparisons are attempted. In all companies, individual accidents contribute to the statistics if the incident happens at the company's premises. However, in some companies/countries incidents during home to office travel and vice versa are also included. Similarly, in some companies/countries incidents during business travel are also included.

In the situation where the safety performance of an organisation is poor, measuring safety is a rather straightforward exercise by counting injuries. Once an organisation has made considerable progress in improving safety, counting injuries will not provide enough information to satisfy the demand of the organisation.

Safety Training and Competence

Generally, training has always been an important activity for the industry. If the behavioural and motivational issues are to be followed up and implemented, they will require levels of competence amongst employees to be further enhanced. Training will be a key issue in one of the ways in which competence and personal motivation can be taken forward.

Training gives employees the knowledge of how to work in a safe and healthy way. To keep up standards, training programmes have to be checked and reviewed periodically.

The responsibility of determining and reviewing the training should be in the hands of line management, in cooperation with the safety experts. The most common period for a review is yearly. Managers and supervisors should be responsible in ensuring that necessary training is given.

Engineers and managers should receive training that allows them to fulfil their responsibility for safety. Apart from special technical training, they should attend courses such as risk analysis, health and safety audit, safety management and safety legislation. External experts are also utilised from time to time depending on the required competence, legislative requirement, size of training requirement and availability of training resources.

For different occupational groups, different kinds of training are required. The basic principle for all companies should be that every target group has to undergo training on general health and safety issues. For new employees, a specific training programme should be organised at the workplace by the department concerned.

The focus of training for engineers and managers is more on strategic and managerial aspects. The issues that are more technical like specific courses for first aid, or fire and rescue are also elements of the training programme for engineers and managers.

Wherever contractors are involved, health and safety training should also be organised for them and for their employees.

Safety Audit

Safety audit is a structured evaluation of the application and efficiency of the company's safety policy, safety programmes, and safety systems. It also involves compliance with the national laws and regulations.

There are different kinds of audits such as audits of management systems, audits of safe working procedures and employee behaviour, and audits of contractor activity. There are also audits focusing on specific topics such as:

- Fire protection, means of escape
- Hazardous substances, handling, protection, waste
- Electrical installations
- Temperature conditions
- Noise, lighting
- Working at heights
- Housekeeping

Action plans with a time scale should be agreed after each audit to ensure that the lessons from the audit are learned and actions will be implemented. Follow-up should normally be done by the line management. In regular safety meetings, the outcome should be discussed and corrective actions are assessed.

Safe Working Procedures

For new or existing plants, processes, operations or tasks, safe working procedures should be established by management with assistance from employees and experts in safety, ergonomics, and hygiene. Employee involvement is crucial, as they know the key risk elements in their activity. They are also suitable persons to test and ensure that procedures are clear.

Safe working procedures are communicated in a written form, and should be accessible at shop-floor level during training and meetings. Competence assessment can be done by periodic testing through simple written work. Review of safe working procedures should be done periodically. Modification of equipment, tools or tasks, and incidents or injuries should always trigger a review of procedures.

In routine operations involving contractors, company procedures should apply and be properly communicated to the contractors. In case the contractors have their own procedures they need approval of the company.

The major goal of an accident and incident investigation should be the prevention of a reoccurrence of the same or a similar event and not to identify someone to blame. It can also be used to motivate those participating in the investigation, including shop-floor employees.

An accident rarely results from one single cause, but is the result of the coincidence of several causes or factors. Accident investigations try to reveal and identify these factors. The investigation process involves two phases. The first phase consists of gathering information on the site of the accident, and establishing circumstances and the course of events; and interviewing the victim(s) and eyewitnesses. In the second phase, the accident factors are studied in a broader perspective to reveal so-called 'root factors'. These 'root factors' give clues to why 'things go wrong' and are therefore of fundamental importance. There is a real need to distribute accident investigation reports widely not only to those directly involved, but also to other parts of the organisation which may face similar situations.

Contract System

Contract system is used by some companies predominantly on construction-type projects and on work that requires specific expertise. In other organisations, they normally work on a daily basis alongside the parent company's own employees. Whatever the local conditions, the approach to the management of safety with contractors should be consistent. Companies need to develop a clear policy that establishes their approach to managing the health and safety of contractors. Normally, this is best set out formally in a document so that the company and contractor employees understand what is expected of them. Contractor selection is a vital area in achieving the best results. Only those contractors who can demonstrate good performance and organisation in health and safety should be considered for selection. Auditing of contractors is a tool that can be utilised in different factors such as the safety policy, safety meeting arrangements, and safety training. The training programmes required by contractors will vary depending on contractor activity. Core contractors who are regularly on site should be imparted training on the general hazards that occur site-wide and on specific safety issues relating to the place of work.

The key principles for management of the contract employees health and safety are:

1. Set objectives (targets) for improvement
2. Understand the relationship between company and contractor
3. Establish ownership
4. Check competence
5. Provide resources
6. Communicate
7. Equal weighting
8. Equal health and safety benefits on par with company employees
9. Same treatment

Health-Related Issues

Many companies have defined a policy for the control of substances hazardous to health, which is often based on legislation. Its aim is to minimise a risk to health due to hazardous substances exposure in workplaces. Key elements are identification of the hazard and evaluation of the risks, elimination, substitution and control of the substance.

The health and safety departments should approve products before purchasing. If risk assessment indicates an exposure to a hazardous substance or upon request by management, the occupational health department may carry out special health surveillance. The medical check-up is related to the degree of exposure and its nature. Typical examinations are blood sampling, x-rays, physical examination, urine analysis and biological monitoring questionnaires. The outcome of these surveillance is used for medical treatment and to recommend preventive measures in workplaces, including the use of personal protective equipment.

Some other health related issues are the following:

- The need for risk assessment.
- The health effects of noise.

- Manual handling.
- Repetitive work and vibrations.
- Influence of psychological and social factors.
- Economic problems, if any.

Health promotion as an objective is taken into account in some companies, with concern for smoking, alcohol or ischemic heart disease. Elsewhere, no medical examination is, or no longer is performed. In many cases, the effectiveness of these health promotion activities seem to remain unknown or difficult to measure. Most of the companies have rehabilitation programmes.

Drug and Alcohol Abuse: Many companies feel the need to have a clear policy document which outlines the approach to drug and alcohol abuse on the following lines:

- To ensure a safe working environment for all employees, visitors and contractors.
- To train and inform all managers, superiors, safety representatives, and safety experts.
- To notice and handle abuse problems.

The policy intent of a company is to ensure that whilst at work, all employees, visitors and contractors are provided with a safe working environment free from risk of injury due to the consumption, possession or use of alcohol and/or drugs in the workplace. The aim of the policy is to enable drug and alcohol issues to be treated in a proactive manner, in an effort to eliminate injury or illness related to drugs and alcohol.

Where it is identified that an employee has attended the workplace in a condition that renders him unfit for his normal duties due to the consumption of alcohol and/or other drugs, he will be treated in accordance with laid down procedures.

An organisations drug and alcohol policy will be to:

- Reduce alcohol and drug problems in the workplace through good management of personnel, employment practices, arrangement of work and consultation.
- Prohibit the availability of alcohol or illegal drugs, and control the use of legal drugs in the workplace.
- Reduce alcohol and drug problems in the workplace through information, education and training programmes, relating both to safety concerns and health promotion.

- Identify and assess workers who have alcohol or drug problems.
- Provide confidential counselling treatment and rehabilitation for individuals with alcohol or drug problems.
- Take appropriate disciplinary action where rules related to alcohol and other drugs are breached.

Exhibit 5.1 Behavioural Aspects of Safety Culture

1. Unsafe ways of working will lead to disaster.
2. Safety culture is a business imperative.
3. Reasons that companies do not spend on safety are ignorance, cost, and misuse of regulations.
4. Increase in participation at the bottom line can be linked to a good safety culture.
5. Organisations that implement behavioural safety approach, can ensure that no one is injured or falls sick through their work.
6. Behavioural safety culture is a tool which can be successfully used to mitigate risks.
7. Extend behavioural safety culture programmes to employees families and other key stakeholders.
8. Annual behavioural safety culture reviews are a must, as culture is the key driver.
9. Continually spread the message of behavioural safety culture to keep the industry moving.
10. Supportive safety culture is an essential pillar for building a progressive and sustainable organisation.
11. Business leaders must support safety now more than ever.
12. Measure behavioural safety culture system efficiency as a core value of corporates.
13. Total time spent on safety by each member in an organisation by way of daily observations and spot correction of at risk behaviours is the best leading indicator.
14. People need to know about hazards and their consequences.
15. Spending time and money on safety and health is an investment and not an expenditure.

SUMMARY

- Research indicates that more than 85 percent of injuries at work involve some unsafe behaviour of people.
- The organisations making real progress in safety and health performance have paid their attention to and concentrated increasingly on behavioural aspects.
- Anybody occupying a managerial or supervisory position is personally responsible for the safety of those working under him or her.
- Training on safety and health will be a key issue and one of the ways in which competence and personal motivation can be taken forward.
- The collection of a company's statistics on a traditional basis should not be disturbed.
- Suppliers of materials should provide and keep upgraded material safety data sheets, describing hazardous aspects of materials and remedial actions when problems occur.
- The companies that carry out risk assessment for hazardous substances in general should include the same in the audit process.
- An audit is an useful instrument that can be used for identifying and improving safety and health issues in an organisation.
- Though leadership and management are different, they are co-existing inter-related activities for initiating and sustaining improvement in health and safety.

DISCUSSION QUESTIONS

1. Why should a manager care for providing an effective management system that will ensure a safe and healthy workplace?
2. What is the role of management, employees, and trade unions in improving and sustaining safety performance in an organisation?
3. In the realisation that safety/health is good business, the management of safety should be a strategic issue in every organisation. Elucidate.
4. How do managers and employees perceive the effectiveness and integrity of current safety communication culture?
5. Write your plan of action the way you would like to transform your HSE department into a profit centre or strategic business unit.

6

Fire Prevention and Control

The first major hazards in process plant is fire. It is regarded as having a disaster potential less than explosion or toxic release. There are three conditions essential for a fire: (a) fuel; (b) oxygen; and (c) heat. These three conditions are often represented as fire triangle. Fire normally grows and spreads by direct burning.

Active fire protection measures are provided in the basic plant design; but are effective only when activated in response to fire elements, such as fire warning systems, fire detection systems, fire-fighting agents, fire water supply system, fixed fire-fighting system, and mobile fire-fighting system.

In case of fire, should the occupants just walk out of the area and wait for the firemen to come? The time taken by the fire-fighting team to reach the site may prove too costly. A small fire that could have been tamed with portable fire- fighting equipment, if allowed to grow, may turn the industry/official building into a virtual furnace. Therefore, it becomes essential that apart from installing fire detection systems, the system should also be equipped to fight fire and prevent excessive damage till the fire-fighting team finally arrive.

Chemistry of Fire

The potential of fire or explosion is always present wherever there is a build- up of any flammable vapours or gases. Areas such as sewers where the gases evolve naturally from raw sewage i.e. methane and hydrogen sulphide are confined and cannot disperse easily to open air, so there is a hazard of a fire or explosion.

In other processes, within confined spaces or pits and enclosed chambers, gases may build up over a long period of time and may not be detected until a gas detection test is carried out or an ignition source is applied. To ignite and explode these flammable mixtures, they need to mix with sufficient air and be exposed to an ignition source. This must not be permitted to happen. Safe working practices, permit to work procedures, and exclusion of any ignition source ensure that fire and explosions do not occur.

Classification of Fire

Not all fires can be dealt with by the same method. Pouring water on some fires can be explosive, or it can also spread the fires. Types of fires are grouped to provide different methods of extinguishment and are as follows:

Classification	Material Involved	Extinguishing Agents
A	organic solids, e.g. pulp, paper, wood, wool, etc.	water, dry powder
B	liquids or liquefiable solids, e.g., petrol, oils, waxes, hydraulic fluids, etc.	foam / dry powder, carbon dioxide
C	gases or liquefiable gases, e.g. propane, butane, methane, etc.	foam / dry powder
D	metals, e.g. magnesium	chloride powders

Methods of Fire Extinguishment

Cooling – Removing the heat and bringing the material below its ignition temperature.

Starvation – Remove the fuel from the fire zone or physically removing materials to stop further spread.

Smothering – Remove the air by covering the area with an air tight blanket.

Chemical Reaction Interruption – This method extinguishes fire by stopping the chemical reaction of the flame itself. Prompt use of the correct fire extinguishers has checked many incipient fires. It is desirable, therefore, to have a well-trained and organised firefighting brigade as the first line of defence. Immediately after a fire is detected, all fire doors leading to the area should be closed as well as all windows, blowers, ventilators, and conveyors.

The public fire department should be notified immediately unless it seems obvious that first-aid fire extinguishing equipment can control the fire. For this reason, it is necessary that all first-aid fire equipment, particularly fire pails should be located so that exit facilities becomes easily accessible and a quick retreat becomes easier.

Dust fires require a special technique for extinguishing fire. It is necessary to attack the fire in such a manner that deposits of dust are not disturbed, causing a dangerous cloud of fire particles.

In every instance of a fire, combustion materials should be promptly moved away from the fire and wetted down. After the fire has been extinguished, all debris should be wetted down thoroughly to make certain that the fire will not occur.

When a fire breaks out in adjacent building, the other buildings should be protected by (a) closing every window facing the burning building; (b) stationing people with fire extinguishers at each window nearest to the fire; and (c) stationing employees on the roof of the adjacent buildings with host lines and with appropriate extinguishers to keep the roof wetted down.

When putting out fires near electrical equipment, it is necessary to follow specific precautions. The fire extinguishing compound should be of a type that will not conduct electricity and endanger the operation of the extinguisher.

Fire checklist

- Know your fire extinguishers and their locations
- Know the locations of fire exits and fire doors
- Never wedge open fire doors
- Fit and check smoke detectors
- Be careful while shifting radiators and other equipment
- Be especially careful with cigarettes
- Store flammable liquids and gas containers correctly
- Take fire drills seriously
- Know where the fire alarms are and how to use them
- In the event of an alarm or fire breaking out, leave the area quickly and calmly
- Empty bins and dispose off rubbish promptly

Fire Hazards

Fire at work is probably the single most serious widespread hazard. No one really expects a fire but everyone must know what to do in the event of one. The first priority is to prevent fire and an understanding of the causes of fire at work is essential for taking necessary precautions in future.

Common causes of fire include:

- careless handling or storage of inflammable materials
- misuse or abuse of electrical equipment
- allowing inflammable litter to accumulate
- covering or blocking air vents in equipment

- 'hot work' not properly contained e.g. welding
- smoking, particularly the careless disposal of cigarette butts in prohibited areas.

Management must pay sufficient attention to check on potential fire dangers, appoint suitably trained persons to check on potential fire dangers and also check the premises at the end of the working day.

The risk of fire cannot be entirely eliminated, but much can be done to reduce the risk by education and by good housekeeping. Education is necessary to combat carelessness, ignorance and irresponsibility. Good housekeeping will help in reducing the spread of fire.

***Fire Risks*:** These can be classified roughly under three headings:

(a) Freely burning materials
(b) Highly inflammables
(c) Electrical

Freely burning materials constitute the most general fire risk. This group includes carbonaceous materials - wood, paper, furnishings, and all those materials which when consumed by fire leave an ash deposit. The method of extinction is "cooling down".

Highly inflammable risks include oil, petrol, paint, varnish, solvents, fats, waxes and similar products. All are either normally liquid or like fats, waxes and grease, become liquid on heating.

For electrical risks, consideration has to be given to the use of a non-conducting medium, to safeguard the operator from shock and also avoid damage to the equipment or wiring.

There are three phases of fire prevention:

1. Stopping of fire before it starts – which means no loss.
2. Stopping of fire at the start – which means possibly slight loss.
3. Stopping the fire in midway – which means partial loss.

If we reject all three, the result must be to allow the fire to burn itself out – complete loss.

The primary cause of fire is the human element, such as:

- Carelessness - a lighted match or cigarette butt thrown down.
- Forgetfullness - a naked flame left burning unattended.
- Ignorance - an example of which might be the storage of two reactive substances.
- Miscellaneous - defects in machinery or plant, defective electrical wiring, and so on.

It is not possible to eliminate the risk of fire altogether in hazardous units, and hence plans must be formulated for action in the event of fire. The organisations' fire instructions must be drawn up and displayed throughout the building. In the event of a fire, the first priority is to summon the fire brigade and to warn other occupants of the building of the danger. This is best achieved by means of smoke and heat detectors and break glass fire alarm call points. On hearing the bells, all occupants of the building should close their doors and windows, disconnect all non-essential electrical appliances, and leave the building by the shortest possible route. The only exception to this rule is a hospital, where those staff on duty having responsibility for patients should close the doors and windows, disconnect non essential electrical appliances and then stand by to await further instructions. Patients should not be evacuated from the building unless they are in imminent danger from smoke or fire. To enable people to leave the building, it is essential that all fire exit routes and emergency exit doors are kept open. During the evacuation of the building there should be no running or pushing on staircases. People should leave the building as quietly as possible in order to hear clearly any instructions that might be issued.

The sound of the fire alarm bells or sirens must be known to the occupants of the buildings through regular fire drills. The people must be in a position to distinguish the sound of the alarm bells from that of the other alarms. In the event of the building being evacuated, it is essential to ensure that no one is left behind, especially visitors who may not be familiar with the sound of the alarm system or what is expected of them.

All employers must ensure that their staff receive training in fire precautions and in the use of fire extinguishers. The risk of fire at work can be reduced by adopting a number of simple precautions:

1. Never overload electrical circuits.
2. Ensure all appliances are fitted with the correctly rated fuses and are properly earthed.
3. Electrical leads should never pass through doors or windows or under rugs, and they should be secured as far as possible.
4. All the leads should be examined periodically and those that are worn out replaced.
5. Insulating tape joints and multiple adapters should not be used.
6. Appliances should be switched off at the mains and unplugged after use and not operated in close proximity to ordinary curtains.
7. Avoid radiant electric heaters - converter heaters should be used as far as possible.

Fire Explosion

Fires occurring during working hours constitute a much greater danger to workers. Fires and explosions can also affect the dwelling surrounding the factory in both the short and long term. Sufficient care should be taken at the time of building the factories. The factory workers too have a very high responsibility for ensuring the effectiveness of fire-prevention measures.

For a fire to start, three elements must be present: oxygen, combustible material, and heat. If any one of these components is removed, the fire will not start, or, if the fire is already in progress, it will go out. The methods of fire prevention basically involve reducing or eliminating one of these components.

One of the most common precautions against fire is the "No Smoking" rule. Other precautions are:

(i) fire resistant construction of buildings;
(ii) careful and safe storage and handling of inflammable material;
(iii) provision of adequate fire extinguishing equipment;
(iv) a fire alarm system; and
(v) alert, efficient and well-trained fire fighting squads promptly available at all times.

The first line of defence against fire is in the construction of the building itself. Industrial buildings should be sufficiently fire-resistant. Fire resistant construction should ensure that structural parts do not readily burn and that fires cannot spread, either horizontally or vertically, through walls, floors, doors, elevator shafts, and so on. Exits are more important and must conform to certain general rules.

Fire-extinguishing equipment may range from buckets of water or sand to complete sprinkler systems. The type and amount of equipment needed will depend on the size and construction of the building to be protected and the processes carried on in it. Sometimes, it is sufficient to have portable fire extinguishers, or even a supply of dry sand or some barrels filled with water. Factories where the risk of fire is great, such as refineries, should normally be equipped with sprinkler systems.

Every workplace should have an alarm system to warn people if a fire starts. The alarm system can be automatic, or alarm bells, whistles or sirens can be installed in different parts of the factory. Alarm must be audible everywhere in the factory, including workrooms, storehouses, gangways, locker rooms, lavatories and wash rooms. The following measures have to be undertaken in any fire-prevention programme. Firstly, every undertaking should have trained fire-fighters on each shift; secondly, it is essential that undertakings be inspected at regular intervals for fire risks and to ensure that all fire fighting equipment is in good condition; thirdly, it is desirable that regular fire-drills be organised to make sure that all workers know how to use the fire-extinguishing equipment, and where the nearest exit is; and lastly it is necessary to ensure close co-operation with the local fire brigade.

In some factories, precautions are necessary not only against fire risks but also against the risk of explosions, which are usually very violent and destructive. Explosions may be caused by commercial explosives or by concentrations of certain vapours, gases or dusts in the air. Hence precautions are necessary in the manufacture, handling, storage and use of commercial explosives.

In every place where people work there should be provided means of raising the alarm should the fire break out, means of fighting the fire, and means of escape.

When the fire alarm sounds:

1. Close the windows, switch off electrical equipment and leave the room, close the door behind you.
2. Walk quickly along the escape route to the open air.
3. Report to the fire warden at your assembly point.
4. Do not attempt to re-enter the building.

When you find a fire:

1. Raise the alarm by hitting the fire button and/or ring.
2. Leave the room, closing the door behind you.
3. Leave the building by the fire escape route.
4. Report to the fire-warden at your assembly point.
5. Do not attempt to re-enter the building.

In case of a fire, every employee must know:

1. The nearest fire alarm
2. The nearest fire extinguisher
3. The nearest means of escape
4. How to operate fire alarm
5. How to operate extinguishers
6. How to use means of exit.

Fire Inspection

Due to the extreme consequences that may result from a fire, it is always prudent to perform fire inspections on a regular basis. When an inspection is made, proper attention should be paid to fire fighting equipment and to other fire hazards.

Fire fighting equipment generally falls into five definite types; the hand portable extinguishers, the wheel extinguishers, automatic or manual sprinklers, fixed chemical systems, and hose lines; and in many cases, a sixth type such as a company owned fire truck. Each type of equipment takes a specific type and frequency of inspection.

After the fire fighting equipment check is completed the next obvious part of the facility inspection is to check for fire hazards. Particular attention should be paid to improper storage of flammable liquid. It is not uncommon to find flammable liquids in hazardous areas because proper storage is not available. Precautions must be taken as regards to piling of combustible materials such as paper, packing materials, and scrap lumber. It is wise to never have more than a two-day supply of raw material and a one-day supply of finished products in the manufacturing or process area.

Smoking is a continuous problem in most facilities, and should only be allowed in designated areas. Specific note should be made of smoking in unauthorised areas.

Fire Safety

Building regulations form an integral component in the design framework for fire safety and protection in structures. It is generally accepted that adherence to building regulation can not only control potential fire hazards, but also prevents them to a great extent. But is it just a case of ensuring that these regulations are adhered to at the construction stage? Efficient management and maintenance of building facilities tend to play an equally important role in ensuring that the safety standards and norms are met.

The National Building Code of India (2005) is considered to be the 'mantra' to any development project laying down the base guideline for administrative regulation. Part IV of the code relates specifically to fire protection.

In India, however, the enforcement of stricter norms for structural safety, environmental factors and disaster management still pose a challenge despite having numerous standards for firefighting and firefighting equipment in place. Indeed, the country continues to grapple with cases of gross negligence and violation in the implementation and practice of such standards.

A responsible person must access the fire risks on their premises and take steps to eliminate, reduce or otherwise manage them and implement and maintain a fire management plan covering:

- a means of detection and giving warning
- a means of escape
- a means of fighting fire
- a means of training of occupants in fire safety

This means installing and maintaining appropriate fire safety systems and equipment. Although not an expert on fire safety, the responsible person will be legally accountable for any failure or inadequacy in these systems.

A rigorous and systematic approach to the assessment of explicit levels of fire safety and protection thus requires comprehensive risk analysis to be undertaken for building fire safety and protection. Fundamentally, explicit consideration must be paid to the multiple fire scenarios. The fire risk assessment could be carried out by the building owner, occupier or controller.

Fire safety audits are also found to be effective tools for assessing fire safety standards. It helps identify the areas for improvement and evolve an action plan. Strict adherence and implementation of fire safety standards is just the stepping stone to the creation of a safe built habitat. Periodic fire safety audits can not only prevent losses but also reduce the cost of insurance.

The damage caused by a fire can be catastrophic, if fire safety precautions are not properly observed. Electrical short circuit shares a major part of building fires. Nowadays, various electrical components are available that help us to mitigate the risk of fire.

The major components that are used for fire-fighting system are fire extinguishers, dry and wet risers, yard hydrants, automatic sprinkler systems, manual/automatic fire detection and alarm systems, underground and overhead tanks and fire pumps. Usage of these components, however, depends on type of building occupancy.

All said and done, the most important point in the end is the coordinated operation of all these services. A fire matrix is established wherein the operation sequence of all components is configured so that at the time of fire they operate in a logical sequence.

Fire Prevention – Key Points

- Know what to do and say when calling the fire services in the event of fire.
- Know who your fire wardens are.
- Know your nearest fire exits or escape routes.
- Know the sound of your fire alarm.
- Keep corridors clear.
- Do not prop fire doors open (unless they operate automatically).
- Take fire drills seriously.
- Know where your nearest fire extinguishers are located.
- Know how to use a fire extinguisher (of the sort that you have in your workplace).
- Check when your fire extinguishers and safety equipment were last serviced.
- Make sure that fire extinguishers are accessible.
- Know what to do if you were to discover a fire.
- Know what causes fire.
- Don't empty ashtrays into waste paper bins.

- Keep electric wires away from combustible material.
- Ensure that electrical installation equipment and connections are regularly checked.

First Aid and Emergency Procedures

It is important to have written procedures in connection with fire, first aid and other emergencies. In all work premises, employees should be instructed and trained so that they understand the fire precautions, instructions and drills and what they should do in the event of fire. The instructions, which should be included in written procedures should ensure that the following points are covered:

1. The action to be taken upon discovering a fire.
2. The action to be taken upon hearing the fire alarm.
3. Raising the alarm, including the location of alarm call points, internal fire alarm telephones and alarm indicator panels, if used.
4. The correct method of calling the fire brigade.
5. The location and use of fire-fighting equipment.
6. Knowledge of the means of escape - i.e. escape routes and the need to keep them clear at all times. Under no circumstances should lifts be used.
7. Appreciation of the importance of fire-doors and of the need to close all doors at the time of a fire and on hearing the fire alarm.
8. Stopping machines and processes and isolating power supplies where appropriate.
9. Evacuation of the building.

First aid boxes (which should be easily accessible and properly maintained) should be placed in the charge of a trained first aider who should be readily available during working hours. The contents are prescribed with quantities varying according to the number of employees served by each box. Separate first aid kits should be provided for those who have to work away from their base establishment unless the facilities of other employers can be made available to them. The names of the trained first aiders or appointed persons should be displayed in every workplace. In addition to information in connection with first aid treatment, all employees must be made aware of the procedure to be followed if they sustain an injury.

Every organisation or company should ensure that effective plans are drawn up for all foreseeable emergencies and contingencies likely to arise at each factory or office location. The emergencies could include: fire, bomb threat, chemical spillage, or escape of toxic gas, and in most cases require evacuation of part or all of the premises. Other emergencies that need to be considered include natural disasters such as storms, flood, subsidence; loss of power or lighting; civil disturbances and demonstrations; plane crash; explosions; and out of control production processes.

Essentially the contingency plan should incorporate procedures designed to (a) deal with the cause of the emergency; (b) evacuate all non-essential personnel away from the danger area; (c) include arrangements for the co-ordination of emergency services - for example police, army and bomb disposal; (d) include public relations information service; and (e) restore production or operations as quickly as possible.

Exhibit 6.1 Fire Safety in Industry and Workplace

- Fire safety in industry is a crucial aspect that should not be taken lightly.
- Fire at the industry or workplace can cause harm to human life; damage to property and the environment; and hence it is essential to have fire safety in the workplace.
- Fire safety measures in industry include preventing ignition of an uncontrolled fire; limiting the development and effects of a fire; and establishing a fire prevention plan and emergency procedures.
- Fire safety also involves conducting a hazard analysis; training employees in fire safety; practising good housekeeping and having an evacuation plan.
- Fire safety is everyone's responsibility; hence employees and management in general, departments of safety, production, maintenance, and other concerned departments have to work cooperatively to maintain a safe working environment.

Fire safety becomes everyone's job at a worksite; employers should train workers about fire hazards in the workplace.

SUMMARY

- When a fire occurs, it is necessary to take prompt, definite and correct steps.
- Organising for fire protection necessitates (a) a plan of the grounds and buildings; (b) a location plan of all main control valves; (c) a plan of each available water supply source; (d) a knowledge of first-aid fire extinguishing equipment, alarm systems, and automatic sprinkler, foam, water spray, and so on.
- In order to reduce the fire hazard, there are two considerations to follow: (a) fire safety and prevention, and (b) fire detection and control.
- Fire safety and prevention requires proper installation of equipment and operations and must be designed to ensure that they are as free as possible from causes of fire.
- Fire protection necessitates the development and the use of design and methods for detecting and controlling fires to limit the probable damage from fire, if one does start.
- A basic and extremely important rule in fire prevention is good housekeeping, if fires are to be avoided.
- Waste or rubbish should not be allowed to accumulate.
- Factors like poor housekeeping and unsafe storage of materials have also been responsible for fire hazards.
- Pertinent building codes and all fire regulations must be observed.
- Immediately after a fire is detected, all fire doors and windows leading to the area should be closed and blowers, ventilators, and conveyors should be shut off.
- Proper planning and scheduling of fire preventive measures, both technical and administrative, minimise fire outbreaks and contains the damage.

DISCUSSION QUESTIONS

1. What are the methods of fire extinguishment?
2. What are the precautions and emergency measures to be taken for preventing and controlling fire?

7

Occupational Health and Safety

Occupational health and safety is a diverse subject embracing many disciplines such as engineering, law, occupational psychology, construction, physics, and chemistry. It encompasses the social, mental and physical well-being of workers, that is the "whole person". Successful occupational health and safety practice requires the collaboration and participation of both employers and workers in health and safety programmes, and involves the consideration of issues relating to occupational medicine, industrial hygiene, toxicology, education, engineering safety, ergonomics, psychology, and so on.

The discipline of occupational health has been defined by the American Industrial Hygiene Association as follows:

"That science and art devoted to the recognition, evaluation and control of those environmental factors and stresses, arising in or from the workplace, which may cause sickness, impaired health and well-being, or significant discomfort and inefficiency among workers or among the citizens of the community."

Occupational health is concerned with prevention of accidents and ill-health at work, the treatment in certain cases of injuries and illnesses sustained at work and the promotion, maintenance, and restoration of health.

Occupational health comprises the identification, evaluation, and control of processes and substances that may harm people, cause their discomfort or damage their environment.

While in many parts of the world occupational hazards have not yet been adequately dealt with, new ones are constantly being introduced everywhere.

The practice of occupational health requires a multi-disciplinary approach, both occupational medicine and occupational health being fundamental components. While some hazards are easy to recognise, others are not so evident. Therefore, it is important that preliminary surveys be conducted by experienced occupational hygienists.

Occupational health issues are often given less attention than occupational safety issues because the former are generally more difficult to confront. However, when health is addressed, so is safety, because a healthy workplace is by definition also a safe workplace. The converse, though, may not be true – a so-called safe workplace is not necessarily also a healthy workplace. The important point is that issues of both health and safety must be addressed in every workplace.

Occupational Health

Occupational Health is concerned with health in its relation to work and the working environment. According to the Joint ILO/WHO Committee on Occupational Health, "Occupational health should aim at the promotion and maintenance of the highest degree of physical, mental and social well-being of workers in all occupations: the prevention among workers of departures from health caused by their working conditions; the protection of workers in their employment from risks resulting from factors adverse to health; the placing and maintenance of the worker in an occupational environment adapted to his physiological and psychological ability and, to summarise the adaptation of work to man and of each man to his job."

It is clear from this definition that occupational health covers a wide field. Indeed, it calls for specialised knowledge from many disciplines like medicine, engineering, chemistry, toxicology, psychology, physiology, and statistics.

Occupational health includes studies on all factors relating to work, working methods, conditions of work and the working environment that may cause diseases, injuries or deviation from health, including maladjustment to work. However, occupational health implies not only health protection but also health promotion, a concept which includes everything that can promote the health and working capacity of the worker, such as preventive measures against communicable diseases, improvement of nutrition and general mental health.

The principles of occupational health services were laid down in the ILO's Occupational Health Services Recommendation, 1959 (No. 112), with the following aims:

(a) protecting the workers against any health hazards which may arise out of their work or the conditions in which it is carried on;
(b) contributing towards the workers' physical and mental adjustment, in particular by the adaptation of the work to the workers and their assignment to jobs for which they are suited; and

(c) contributing to the establishment and maintenance of the highest possible degree of physical and mental well-being of the workers.

Legislation is also used in many countries to improve occupational health at the national level. Workers' protection legislation lays down minimum standards of health and safety at work and is usually enforced by the factory inspection service. At the national level, occupational health and safety is currently implemented in three main ways: (a) legislation in establishing compulsory minimum standards of health and safety at work; (b) occupational health services at places of employment or other work places; and (c) research and field surveys in occupational health and safety, as well as education and training by national or regional occupational health institutes.

Healthcare Services

Healthcare service broadly covers all personal and community health services, including medical care and related education and research directed towards the protection and promotion of the health of the community. Healthcare is not synonymous with medical care. Medical care is a subset of a healthcare services system. The term "medical care" refers chiefly to those personal services that are provided directly by physicians or rendered as a result of physician's instructions.

Since health is now viewed as an integral part of socio-economic development, "an acceptable level of health for all", as defined by WHO, cannot be achieved by the health sector alone. It requires an *intersectoral approach*, i.e., the coordinated efforts of the health sector and relevant activities of other social and economic development sectors.

The current social policy throughout the world is to reorient and restructure the healthcare services towards the policy objective of "health for all" by the year 2000, with primary healthcare, as its central function and main focus.

Health services, by their very nature, are complex. The complexities include: (a) the various agencies through which services are provided; (b) the different schemes for financing such services; (c) the broad spectrum of illness in the community; (d) the standard of living and socio-economic status of the community; and (e) the multiplicity of governmental and voluntary agencies actively engaged in planning the organisation and delivery of health services.

In evaluating the quality of care, four dimensions are generally taken into consideration:

(a) Appropriateness
(b) Availability
(c) Accessibility
(d) Acceptability

Health services of a worker include:

- workplace evaluation from an expert's point of view;
- examination of health at the time of assignment, taking into account the potential hazards;
- checking health of a person periodically by a health service team that knows the workplace conditions;
- checking of specific health risks of a worker periodically for possible health changes;
- communication of health check-up results to the persons concerned;
- due respect and protection of the privacy of health data;
- timely access to medical care as and when necessary; and
- recommendation of health services team to the management, safety committee, and worker representatives for follow-up and corrective action.

Occupational health services in places of employment are defined in the ILO's Occupational Health Services Recommendation, 1959 (No. 112). The scope and functions of occupational health services in individual undertakings depend to a great extent on:

(a) the legal provisions concerning their establishment and the degree to which the undertaking is owned or controlled by the government authorities;
(b) the overall personnel policy of the employer and the extent to which these services are an object of agreement between employees and employer;
(c) the extent to which medical services are provided through other agencies, and the availability of different types of health services in the neighbourhood of the undertaking;
(d) the type of production and the main occupational hazards; and
(e) the general health status of the population and the pattern of health organisation of the country.

The occupational health services normally include:

(a) medical examinations
(b) supervision of the working environment – industrial hygiene, occupational safety
(c) advice to management and workers' representatives on working environment and ergonomics
(d) health education and training – education in health and hygiene, training in first-aid
(e) compilation and periodic review of statistics concerning health conditions, maintenance of records, preparation of reports
(f) medical treatment – first-aid and emergency treatment, ambulatory treatment
(g) health counselling-individual

(h) nutrition
(i) family planning-education counselling
(j) research in occupational health
(k) co-operation with other services in the undertaking
(l) collaboration with external services.

The major factors governing the size and type of occupational health services and the health personnel required are:

(a) the extent, scope and functions of services offered;
(b) the number of employees covered, their sex, age and general health status;
(c) the nature and conditions of the work, special occupational hazards, and the number of workers exposed; and
(d) the degree to which the employees understand and accept the occupational health programme.

Occupational Physician

The functions of the occupational physician have been defined comprehensively in the ILO Occupational Health Services Recommendation, 1959 (No. 112), as follows:

(a) Surveillance within the undertaking of all factors which may affect the health of the workers and advice in this respect to management and to workers or their representatives in the undertaking;
(b) job analysis or participation therein in the light of hygienic, physiological and psychological considerations and advice to management and workers on the best possible adaptation of the job to the worker having regard to these considerations.
(c) participation, with the other appropriate departments and bodies in the undertaking, in the prevention of accidents and occupational diseases and in the supervision of personnel protective equipment and of its use, and advice to management and workers in this respect.
(d) surveillance of the hygiene of sanitary installations and all other facilities for the welfare of the workers of the undertaking, such as kitchens, canteens, day nurseries, and rest homes and, as necessary, survelliance of any dietetic arrangements made for the workers;
(e) pre-employment, periodic and special medical examinations – including where necessary, biological and radiological examinations – prescribed by national laws or regulations, or by agreements between the parties or organisations concerned, or considered advisable for preventive purposes by the industrial physician. Such examinations should ensure particular surveillance over certain classes of workers, such as women, young persons, workers exposed to special risks and handicapped persons:

(f) surveillance of the adaptation of jobs to workers, in particular handicapped workers, in accordance with their physical abilities, participation in the rehabilitation and retraining of such workers and advice in this respect.
(g) advice to management and workers on the occasion of the placing or reassignment of workers;
(h) advice to individual workers at their request regarding any disorders that may occur or be aggravated in the course of work;
(i) emergency treatment in case of accident or indisposition, and also, in certain circumstances and in agreement with those concerned (including the worker's owned physician), ambulatory treatment of workers who have not been absent from work or who have returned after absence;
(j) initial and regular subsequent training of first-aid personnel, and supervision and maintenance of first-aid equipment in cooperation, where appropriate, with other departments and bodies concerned;
(k) education of the personnel of the undertaking in health and hygiene;
(l) compensation and periodic review of statistics concerning health conditions in the undertaking.
(m) research in occupational health or participation in such research in association with specialised services or institutions.

The wide-ranging functions of the occupational physician can, infact, be grouped into three categories: industrial hygiene, health supervision, and first-aid. His functions imply supervision of both the worker and his work. He should acquaint and concern himself with the environmental factors. In the course of initial and periodic medical examinations, he should acquaint himself with the worker's general state of health, aptitudes, degree of adaptation and, of course, watch out for any signs of incipient occupational disease. In many cases, observation and examination will need to be backed up by instrument and laboratory testing and the industrial physician may require to call upon the assistance of specialists such as radiologists, oculists, ear nose and throat specialists, dermatologists, gynaecologists. He may also need the assistance of industrial hygienists, safety engineers, chemists, industrial psychologists, social welfare officers, and so on. The occupational physician should establish effective liaison with other services or activities in the undertaking involved in promoting the safety, health and welfare of the worker. The occupational physician should be a member of the joint safety committee and should present at every meeting and give an account of all injuries dealt with by the first-aid room. Liaison with the welfare officer and industrial psychologists will also prove of value.

The occupational physician should perform his duties competently, conscientiously, and should tolerate no interference in his work. To carry out his work effectively, he requires the active collaboration of both the employer and the workers.

The occupational physician is concerned primarily with man and the influence of work on his health; on the other hand the occupational hygienist is concerned primarily with man's environment. Both, in conjunction with other members of the team, have responsibilities for recognising, measuring, and controlling health hazards in the psychological environment.

There is great shortage of occupational hygienists everywhere, particularly in developing countries. This situation jeopardises the success of many occupational health programmes. Various requirements for the practice of occupational hygiene are: recognition of hazards, evaluation of hazards, control of hazards. Understanding of the basic principles of occupational health is also very important for both workers and managers, as well as other professionals such as production engineers, administrators and industrial planners.

Occupational Health in Developing Countries

The state of a man's health has a great influence on his work performance. Nowadays, therefore, an occupational health programme may be considered not only as aiming at "the promotion and maintenance of the highest degree of physical, mental and social well-being of workers in all occupations, but also as an important means of achieving higher productivity" (Joint ILO/WHO Committee on Occupational Health).

The various functions of the occupational health service and its minimum requirements if it is to function efficiently within an undertaking, are outlined in the ILO's Occupational Health Services Recommendations (No. 112). Certain problems, however, may be encountered when the establishment of occupational health services in developing countries is being considered. For instance, some countries may have difficulty in providing the basic necessities of life – food, clothing, shelter – for their population, and the prevention of malnutrition and endemic diseases in such a situation might not be looked upon as one of the top most priorities. The National Commission on Labour (1969) in its report pointed out that the loss of life in accidents and agonising process of an occupational disease may not stir the community as much as it would in countries with chronic labour shortages, although to the affected ones it is a tragic occurrence. Relief gets organised soon after the event, but prevention gets side-tracked.

In most industrialised countries, the health needs of the gainfully employed are met mainly by the national authorities or by organisations not connected with the undertaking. This considerably reduces the load on the service to be provided at the place of employment itself. In developing countries, however, well organised public health and medical care services may be conspicuous by their absences, and almost all the health problems of the workers may have to be dealt with through the place of employment. The magnitude of the problem is, therefore, much greater than in the industrialised countries. It becomes greater still when the high prevalence of endemic diseases and the poor sanitation and overcrowding of slum areas are taken into account. Thus, occupational health service have to tackle a wide range of infectious diseases, e.g., tuberculosis, leprosy, dysentry, diseases of parasites, e.g., malaria, filariasis, and epidemic diseases e.g., smallpox, cholera.

Most developing countries are situated in the tropics and the hot, dry and humid climate create additional physiological hazards. Heat stress caused by exposure to high temperatures, or to high humidity rates will often create additional health problems that are not easy to solve. The wearing of personal protective appliances and clothing and of safety appliances may cause additional problems in view of these high temperatures.

When a country embarks on a program of industrialisation, the newly industrialised areas often call for far more workers than can be provided by local manpower. Thus, workers flock to the industrial zones from far and wide, and the abrupt social and cultural changes which these migrant workers are obliged to undergo and a result may give rise to considerable stresses and new health problems. These changes are particularly marked when underdeveloped countries undertake rapid industrialisation programmes as large groups of people move from agrictulture to industry and from rural to urban areas. These cause basic health problems like occupational disease and injury, fatigue, unsatisfactory man/machine relationships, and undue physiological stresses. Workers find themselves in an atmosphere totally different from that which they have been used to in the past - living in slums. Since they cannot easily severe their roots in the villages, they try to keep contacts with their place of origin. The result is disruption between their new life in the city and their old life at home. Such social disorganisation can have far-reaching effects on community health and lead to behavioural disorders such as delinquency, crime, prostitution and alcoholism and contribute to an increase in the incidents of mental illness.

Migrant workers who have neither vocational training nor previous industrial experience are predisposed to high risk of accidents or occupational diseases. Illiteracy and ignorance make it difficult for such workers to understand or appreciate written or oral instructions. A lack of sufficient skill and knowledge for the job is the most common cause of accidents. Under such circumstances the apathy of these migrant workers towards safety equipment can be readily understood. To them, accidents are a matter of fate or an inevitable part of the game.

The ILO's recommendation No. 112 and the 6th Report of the Joint ILO/WHO Committee on Occupational Health state clearly that occupational health services should be the responsibility of the undertakings themselves or should be attached to an outside body. Since such services will ultimately reduce the load on the general medical and public health service, the government of a country may step in to provide the necessary support. In many countries, the government has set up social security organisations or health insurance schemes. The funds are contributed by employers, employees and government in varying proportions.

A better method may be to encourage the employers to provide occupational health services at the place of employment under an agreement with the workers or with their unions and attend them to have a say in their running. A limited scheme along these lines is in operation in the Scandinavian countries, and the idea is also being put into practice in a slightly modified form in some Eastern European countries.

The existing organisational patterns often show a wide disparity from country to country. Some large organisations do employ a full time doctor whereas medium and small-sized firms engage them on a part-time basis. The position regarding safety and accident prevention in medium and small scale industries cannot be considered satisfactory.

One of the most outstanding features of the last few years has been the massive adoption of legislation imposing the obligation to establish occupational health services. Legislation on prevention and periodic medical examinations for workers exposed to special hazards was adopted by many developing countries as well. Obviously organising occupational health service is a tremendous task. Although legislation requiring the employers to provide for the treatment of injuries and for dealing with medical emergencies at the workplace exists in almost all countries, the implementation of statutory measures varies widely. Instead of introducing new legislation, governments should concentrate on plugging loopholes in existing legislation.

Developing countries might profitably copy the industrialised countries by setting up occupational health institutes or special university and hospital departments run on commercial lines. These can render effective service both to workers and to industry as a whole. There is also a need for a programme of education for the workers to inform them of the benefits that accrue from the supervision of their health. The participation of the workers themselves in such a scheme is obviously essential, and the results may well be a reduction in the misuse of social security benefits and in absenteeism due to sickness and injury. In this way, the occupational health service may come nearer to achieving its object of "total health".

Occupational Safety

Safety is an organised and conscience effort to prevent accidents, minimise risks. There is a general tendency to forget or overlook the aspect of safety. This may be due to the fact that accidents do not happen everyday. As a matter of fact when safety measures work, accidents do not occur; and in absence thereof, accidents take place. So as a common human tendency, one quite often neglects 'safety'. One looks at it only when an accident takes place.

No undertaking is an entity on its own. It is unrealistic to imagine that safety problems can be resolved exclusively and solely within an individual undertaking. However, one can say that everything begins and ends with the undertaking. Everything begins with the undertaking because an occupational safety and health policy that is not based on the day-to-day realities of life in industry may be academically satisfying but cannot be successful in practice. And everything comes back to the undertaking because an in-plant occupational safety and health policy would remain a non-starter if it were not practically applied and constantly adjusted to each individual plant or undertaking.

The basis of all safety organisation is awareness of accident risks and hazards both quantitative (i.e. their frequency and severity rates) and qualitative (i.e. according to the nature of the causes of actual or potential accidents). Plant managements should have the means to access their own degree of risk, enabling them to make a comparison with other firms. Qualitative analysis is the most important branch, because it examines accident causes directly. Plant managements should not be content merely to investigate and analyse their own accidents. They should draw on the experience of trade organisations, employers' and worker's associations, specialised institutions and other firms.

In general, the greater the employee participation in the safety programme, the more effective it will be. Such activities include:

1. Safety campaign and contest
2. Safety meetings and safety stunts
3. First-aid training
4. Plant fire brigades
5. Plant inspection
6. Accident investigation
7. Job safety analysis
8. Safety suggestion systems
9. Safety inventories
10. Safety committees

Worker participation in plant safety organisation can be planned in many ways, depending on national law and practice. In order to ensure their effective participation, the workers and their organisations should be made aware (a) that safety regulations are enacted to protect workers' life and limb and prevent health impairment; (b) they are closer to the production process and have to spend many hours daily in the working environment; and (c) in the last analysis, it is for their safety that safer design of materials and equipment and works safety rules are developed. It is obvious, therefore, that workers and their organisations should be associated with in-plant occupational safety and health organisation, and that they should be given every facility for this cooperation. It is desirable that they or their representatives should participate in all systematic investigations of accident casuality, inquiries following accidents, and policy making and implementation of decisions concerning measures to reduce or eliminate hazards.

Among the many ways and means of providing for worker and trade union participation in plant safety organisation, the most frequently adopted are safety and health committees. Details concerning the membership, powers and duties of these committees are usually laid down in laws and regulutions varying from country to country. Another form of participation undertaken to achieve enhanced safety and all round improvement of working conditions is the establishment of research teams.

A minimum plant safety and health programme should be based on the following ten principles:

1. Safety is not a separate subject. It is just another aspect of industrial progress, and should be closely and fully integrated in all production questions and decision making.
2. Promotion and organisation of occupational safety and improvement of working conditions should be part and parcel of all top-level management policy, as a matter of course. Top management should communicate its policies and decisions in this field preferably in writing, to all plant personnel at every level, through normal channels of communication.
3. Every representative of top and middle management, engineers and supervisory staff, down to shop floor level should be responsible for plant safety, i.e. at all production levels. It should receive equal importance with other professional skills and performance, and taken into account in weighing up promotion possibilities.
4. The safety department should be considered a kingpin service, reporting directly to the highest management echelons. Its role consists essentially in safety promotion, documentation, information and general training, advisory tasks, coordination and inspection. The department should participate fully in all in-plant occupational

safety and health activities, but in no case takeover any of the basic responsibilities of management and supervisors in this field.

5. The occupational health service should devote a considarable part of its time to workplace inspections, with special attention to erogonomic and "health and safety engineering" aspects.
6. The workers and their accredited representatives should be associated in occupational safety and health policy making, planning and action. This cooperation is certainly one of the best ways of ensuring optimum employee response and collaboration in implementing safety programmes.
7. All action taken in the field of occupational safety and health should be coordinated as part of an over-all rationalised programme embracing the following stages: preparation, application, checks and controls regarding implementation, assessment of the results, administration, enforcement and application of the results.
8. A safety programme of this kind should be based not only on the analysis of past accidents, but also on indepth studies – carried out before accidents occur – of the risks and hazards in production departments. Total loss control methods (employment injuries, material loss or damage, financial aspects) are highly recommended.
9. The improvement of workplace and working conditions, which includes occupational safety and health aspects, can be achieved most effectively by workplace studies, leading to improvement or better design or machinery installations and production processes; drafting of work routes, and so on.
10. Safety training at all levels should be an integral part of vocational training.

Occupational Safety in Developing Countries

A study of safety and health conducted by the Trade Union Congress in a developing country in 121 establishments showed that hazardous conditions existed in 102 (84%) of those surveyed, serious accidents or fatalities had occurred in 81 (67%) in the previous year or so. Developing countries present a panorama of vast contrasts as far as industries are considered, viz.

- chemical, explosives, and refineries with a high standard or standard of safety and accident prevention activities;
- organised mechanical and other industries with some in the above category, but in most of which hardly any reward or at most lip service is paid to safety;
- a large number of small or unorganised factories with practically no knowledge or activity in the field of safety.

Some of the reasons for the present state of affairs are summarised below:

Workers: The majority of workers in developing countries are illiterate, come from a rural or agricultural background and are neither trained to handle machines nor appraised of the hazards or risks involved. Facilities for the training of workers, particularly in accident prevention, are inadequate.

Management: Managers are rarely trained or even aware of their role in accident prevention. With the exception of a few enlightened employers, the attitude for provision of safe working is at most to comply with the letter of the law rather than its spirit. There is hardly any knowledge of accident costs, as insurance schemes scatter for compensation as well as medical care. Penalties for violation of labour laws are usually very light and hence do not act as a deterrant.

Plant and Equipment: One estimate mentions that nearly 70% of factories in developing countries have obsolete machinery, with inadequate guards. Appropriate guards cannot be procured either because no one manufactures them or for lack of funds. Machinery and plant has to be imported from developed countries who design them for their own workers and modes of operation. Personal protective equipment is rarely provided and wherever it is, it is often unsuitable for the prevailing conditions of heat and high humidity. Maintenance of machinery and guards is neglected or of a poor standard.

Labour Legislation: Responsibility for safety is a statutory responsibility of employers. Safety rules and regulations are elaborate and mostly copied from the developed countries. But in most cases their execution is most ineffective. Besides the labour laws generally apply only to the bigger undertakings. Non-compliance with safety regulations resulting in serious or fatal accidents at most attracts a fine, which neither deters nor compels the employer to accord proper attention to the safety of his workers. In addition, legal action for violation of safety generally involves protracted proceedings.

The number of factory inspectors employed by the governments to enforce labour standards are few and hence not adequate for the tasks. They are generally lowly paid and hence the eligible or the properly qualified are not attracted to government service. Training facilities for inspectors are limited and this affects their ability to inspect, spot the hazards and render suitable advise for remedial measures. Many factories remain uninspected for years because of the shortage of inspectors. Accurate accident statistics are not available because of non-reporting to the labour inspectorates.

Working Conditions and Environment: It is not uncommon to notice inadequate or ineffective lighting, extremely hot or cold working environments, and inadequate ventilation. High level of noise without provision of use of ear plugs is noticeable in many cases. Housekeeping does not pay proper attention. Many factories, particularly the small factories, do not have any first aid or immediate medical care facilities.

Trade Unions: Trade Unions are active in agitating for higher wages, social security benefits, and so on. But they have hardly directed their attention to or clamoured for improved working conditions and safety requirements. No effort is made to train new and other workers in the field of safety.

Compensation Schemes: Most developing countries provide insurance cover for medical treatment and payment during absence from work for the injured worker and / or his family in the organised sector of industry. A large number of workers in the small or unorganised sector are not covered and hence are entirely at the mercy of the employers even for medical care and treatment.

Occupational Safety and Health in Small Undertakings

There is no universally accepted definition of a small undertaking and, for a variety of reasons, different countries have adopted different definitions.

The question of definition is an important one, since in some countries, occupational safety, health and social security legislation does not cover small undertakings. They are, consequently, not subject to inspection by labour, factory or industrial health inspectors and also not required to report industrial accidents or occupational diseases to the competent government authority.

One reason for exempting these small undertakings from any legislative requirements is the shortage of trained occupational safety and health inspectors. In some of the more advanced industrialised countries, all industrial undertakings irrespective of the employees employed are subject to the same labour laws but, in practice, the occupational safety and health inspectorate concentrates more on the large and medium-sized factories.

In certain countries, the number of persons employed is a deciding factor in order to be a small undertaking. Of course, this figure varies from country to country. However, the number of workers is not necessarily a true indication of the size or importance of the undertaking for the purpose of occupational accident and disease prevention. A very highly mechanised or automated factory such as a flour mill or an electric power station may employ only a small number of workers but it may have a large capital investment and its occupational safety, health and welfare problems may be more important than a larger factory employing several hundred workers.

In some countries, a small undertaking is defined as one employing less than ten workers and not equipped with power machinery; such undertakings are not subject to occupational safety and health legislation and are not visited by factory inspectors. However, the introduction of a single power-driven machine brings the factory under the labour laws and it becomes subject to inspection.

The classification of a firm as a small undertaking on the basis of the number of persons employed is quite arbitrary as far as safety and health is concerned. An appreciable number of small factories have very low standards of occupational safety and health, especially in developing countries and countries where no labour legislation is applicable to them. In such undertakings, the hours of work are usually very long, and young children are often employed in the most unsatisfactory conditions. The factories are in many cases housed in old slum buildings or ramshackle structures with rough earthen floors. The standards of safety, hygiene, fire precautions, lighting, first-aid, sanitary facilities, and protective clothing are low.

Safety studies have highlighted that small firms (those with less than 50 employees) have poorer accident records than large organisations. This is made even worse by the large number of accidents that go unreported by small companies. Ignorance about health and safety measures and contempt for basic safety standards are the characteristic features of small firms.

The reasons for the low standards of health and safety in small undertakings are: (a) the absence of labour legislation with a system of enforcement by an inspectorate; (b) the lack of technical knowledge by the owners; (c) obsolete machinery and plant; and (d) unsanitary or overcrowded factory buildings which are sometimes a fire-trap. Some of these small enterprises are short of capital and cannot afford to rent better factory accommodation, buy new machinery, or install new safety devices.

In a number of countries governments are making efforts to deal with the problem of health and safety in small undertakings. However, it is not possible for a small undertaking to provide suitable health and safety services, but a group of such undertakings can do so, or alternatively a government department or a public authority can establish a centre providing these services for an industrial area containing small undertakings. It is not practicable for all developing countries to immediately introduce legislation to enforce the same standards of labour law in the small undertakings as in larger ones. There is an inadequate inspection staff to enforce such legislation and even if it were enforced, many of the small undertakings would be pushed out of business and the workers thrown out of employment. One useful step would be for the governments in such countries to provide low-rent factory accommodation in new buildings replacing some of the older ones and extend the benefits of workmen's compensation and medical services to workers in small undertakings irrespective of the number of persons employed.

Promoting Occupational Health

Some suggestions for promoting occupational health and safety include:

1. Make the work interesting by using job enrichment and by curbing boredom, fatigue and stress.
2. Ensure that all workers are provided with a safe and healthy work process.
3. Establish bipartite safety committees and ensure their effective functioning in collaboration with all concerned.
4. Encourage safety contests and awards amongst employees who submit good accident-prevention ideas.
5. Publicise accident statistics.
6. Identify the causes of accidents and the conditions under which they are most likely to occur.
7. Carry out regular inspections and audits in order to eliminate risks.
8. Organise continuous education and training programmes on accident prevention and safety.
9. Maintain records listing all dangerous occurrences, their causes and the remedial actions taken.
10. Enforce vigorously safety legislation mechanism to curb work hazards and the penalties on convictions must be sufficiently severe to convince most employers that it is more sensible to implement safety precautions than risk the consequences of failure to do so.
11. Provide for rehabilitation of physically-disabled in any legislation dealing with health and safety at workplace.
12. Assign occupational health and safety a greater priority at the national level and a concerted government, employer, and union action is needed to improve the quality of the working environment.

ILO Strategy

ILO has published its "Global Strategy on Organisational Safety and Health" in 2004. The following are some of its important findings and recommendations:

- Significant efforts have been made at all levels to come to terms with the problem of occupational accidents and diseases, as well as major industrial disasters. Nevertheless, over two million workers die each year from work related accidents and diseases and globally this figure is on the increase.
- In addition to established measures to prevent and control hazards and risks, new strategies and solutions need to be developed and applied

both for well-known hazards and risks as well as for emerging issues, such as biological hazards, psycho-social hazards and musculo-skeletal disorders. Furthermore, as organisational safety and health (OSH) is an intrinsic part of social relations it is affected by the same forces of change that prevail in national and global socio-economic contexts.

- Although effective legal and technical tools, methodologies and measures to prevent occupational accidents and diseases exist, there is a need for an increased general awareness of the importance of OSH as well as a high level of political commitment for effective implementation of national OSH systems.
- The fundamental pillars of global OSH strategy include the building and maintenance of the national preventive safety and health culture and the introduction of a systems approach to OSH management. A national preventative safety and health culture is one in which the right to a safe and healthy working environment is respected at all levels and where the principle of prevention is accorded the highest priority.
- The launch of a global knowledge and awareness campaign focused on promoting the concept of "sound management of safety and health at work" is the most effective means for achieving strong and sustained preventative safety and health culture at both the national and enterprise levels.
- It is important to provide technical advisory and financial support to developing countries and also countries in transition for the timely strengthening of their national OSH capacities and programmes.
- In the field of OSH, adequate capacities to develop, process and disseminate knowledge that meets the needs of governments, employees and workers are a prerequisite for identifying key priorities, developing coherent and relevant strategies, and implementing national OSH programmes.

Let us ask a few questions to ourselves:

- Are we aware of the occupational health risks to which our workforce is exposed to?
- Do we have a system of periodically assessing the hazardous factors prevailing in the work environment?
- Have we evolved a system of proper pre-placement and periodic medical examination of our employees?
- Do our factory medical personnel undergo sufficient training and expertise towards occupational health?
- Have we made sufficient efforts to reduce the hazards as such?

- Have we complied with the statutory provisions related to health and safety prescribed under the Factories Act, 1948?

If the answers to all these questions are positive, then we have given an honest beginning to the solution of certain problems associated with safety and health.

CASE

Health and safety in a department store

A year ago, you were appointed by the store in-charge as the competent person responsible for assisting the health and safety programme in the department store. The store has a turnover of 600 crores per annum and a staffing establishment of around 300. This case relates to the general services, warehousing and maintenance (GSWM) department. There are 40 staff in the GSWM department who undertake work associated with cleaning, general maintenance and repairs, transport and storage of goods and store security. The GSWM working environment contains many hazards and is characterised by significant risks to health and safety.

The GSWM department has traditionally had a poor health and safety record. The demands of supporting customer-focused sales departments have resulted in health and safety concerns being subordinated to operational demands. There is a carefree and fatalistic attitude towards health and safety among the staff, which is evident from high accident, injury and absence rates. Other problems include:

- unsafe manual handling
- irregular use of personal protective equipment
- untidy working areas
- poor storage of goods.

In a cooperative and supportive way, you pointed out these health and safety problems to the GSWM department manager some months ago. The manager perceived it something of an encroachment on his responsibilities. However, he reluctantly agreed to do something about your concerns. He stated that his primary focus was on operational efficiency in a climate of tight resources and incessant demands from sales departments where the 'customer was king'. The GSWM manager prided himself on his direct management style and his ability to train and educate his staff and he had no time for presentations from occupational health professionals, video tapes or poster initiatives. He decided on a strategy of dealing immediately and severely with any incidents of unsafe practice, as he believed in the 'red hot stove'

→

approach punishment of undesirable behaviour. He briefed his supervisors accordingly and they began a campaign of warning and punishment in relation to unsafe practices, injuries and accidents. The punishments and sanctions included giving out unpleasant jobs, restricting access to desirable overtime, and causing embarassment to health and safety transgressors by different means.

A monitoring of accident and injury statistics revealed a downward trend in the first few months. This had to be offset by some deterioration in the relationship between the GSWM and the sales departments. Sales department managers reported that GSWM staff were more concerned with health and comfort than they were with keeping the business competitive. It also became apparent that the decrease in accidents and injuries was illusory, as informal discussion with GSWM staff revealed that accidents and injuries were being under-reported because of a fear of being punished. It was also clear that the staff and the supervisors were uncomfortable about the adversarial relationship that was being created by the GSWM manager's policy on health and safety.

You were left with little alternative but to discuss this unsatisfactory picture with the GSWM manager. He was distressed by his apparent failure to resolve the problem and admitted that he appeared to have made things worse. You agreed that you would work together to put things right.

Questions

1. Comment on the style and approach of the GSWM manager in attempting to resolve the health and safety problems in his department. Why did his approach fail?
2. What strategy would you adopt to resolve the problems?
3. How can an active and positive health and safety culture be developed, and what organisational processes or mechanisms will be required?

SUMMARY

- Occupational health and safety is a discipline with a broad scope involving many specialised fields.
- The occupational health plays a key role of prevention, treatment of health problems and adaptation of work to the worker.
- Occupational health and safety is concerned with prevention of accidents and ill health at work; the treatment in certain cases of injuries and illnesses sustained at work; and the promotion, maintenance and restoration of health.
- Hygiene is concerned with the science of preservation of health whilst at work.
- More and more developing countries are giving increased attention to occupational health.
- The occupational health service varies according to the size of the work group, the hazards involved, the location of the plant, and many other factors.
- Occupational safety is a community problem, because the whole community suffers from the cost of occupational accidents and diseases, both in terms of compensation costs, pain and suffering, and loss of productivity.
- Occupational health and safety must strive for a co-operative approach and in pursuing such an approach, there is need for improving training, workplace safety standards and work processes.
- The object of occupational health service should be to set-up a preventive occupational service tailored to the needs of the workplace and its particular hazards.
- Occupational health and safety (OSH) management can and should be viewed in monetary terms as part of a business system.
- Developing a good safety culture in the organisation should be on top of the agenda for all forward looking organisations to ensure long term business sustainability.
- Many companies developed occupational health and safety programmes over the years that focused mainly on: (a) the problems peculiar to the industry; (b) the processes in manufacturing; and (c) the nature of the materials being processed.

DISCUSSION QUESTIONS

1. How do you classify the occupational health hazards?
2. List some of the common types of mechanical and chemical hazards?
3. What are some of the well-known occupational diseases?
4. Select an important work area, activity or project. Make a list of hazards likely to be encountered in it. Who could be affected by them and how?
5. What is the role of occupational health service in an industry?
6. What are the various ancillary health services which an organisation can provide to its employees? Give examples.
7. How does a supportive safety culture act as one of the essential pillars for building a progressive and sustainable organisation?

8

Industrial Hygiene

The word hygiene is derived from *hygeia*, the goddess of health in Greek mythology. She is represented as a beautiful woman holding in her hand a bowl from which a serpent is drinking. In Greek mythology, the serpent testifies the art of healing, a symbol which is retained even today. Hygiene is defined as the "science of health and embraces all factors which contribute to healthful living."

Hygiene is a close relative of epidemiology. It aims not only at preserving health, but also at improving it. The purpose of hygiene is to allow man to live in healthy relationship with his environment. Air, weather, soil, waste, bodily cleanliness, disinfection, and nutrition are the widely differing concerns of the hygienist.

Hygienic working environment plays an important part in the maintenance of good health among the workers. The term "hygiene" includes not only the material environment but also personal hygiene. The importance of personal hygiene should continually be brought home to the industrial workers. Occasional health campaigns may emphasise the point that health is a concern of every single individual, and not merely a job which can be left to the doctor. All the employees in a factory should become health-minded.

Industrial hygiene has been defined as "that science and art devoted to the anticipation, recognition, evaluation, and control of those environmental factors or stresses arising in or from the workplace, which may cause sickness, impaired health and well-being, or significant discomfort among the workers or among the citizens of the community." Industrial hygiene is an applied science and a profession like other applied sciences and professions, such as medicine and engineering. It is founded upon basic sciences such as biology, chemistry, mathematics, and physics. Drawn from these is a foundation of theory that supports the practice of industrial hygiene. The rationale for industrial hygiene practice is based on the profession's purposes, that is, the recognition, evaluation, and control of work-related hazards. Industrial hygienists use environmental monitoring and analytical methods to detect the extent of worker exposure and employ engineering, work practice controls, and other methods to control potential health hazards.

Genesis

There has been an awareness of industrial hygiene since antiquity. As far back as the period of Babylonian civilization, occupational diseases were known, for tablets found refer to lead poisoning and pulmonary diseases of the workers. The environment and its relation to worker health was recognised as early as the fourth century BC when Hippocrates noted lead toxicity in the mining industry. In the first century AD, Pliny the Elder, a Roman scholar, perceived health risks to those working with zinc and sulphur. He devised a face mask made from an animal bladder to protect workers from exposure to dust and lead fumes. In the second century AD, the hazardous exposures of copper miners to acid mists has been recorded.

In the middle ages, guilds worked at assisting sick workers and their families. In 1556, the German scholar, Agricola, advanced the science of industrial hygiene even further when, in his book *De Re Metallica*, he described the diseases of miners and prescribed preventive measures. The book included suggestions for mine ventilation and worker protection, discussed mining accidents, and described diseases associated with mining occupations such as silicosis.

Industrial hygiene gained further respectability in 1700 when Bernardo Ramazzini, known as the "father of industrial medicine", published in Italy the first comprehensive book on industrial medicine, *De Morbis Artificum Diatriba (The Diseases of Workmen).* The book contained accurate descriptions of the occupational diseases of most of the workers of his time. Ramazzini greatly affected the future of industrial hygiene because he asserted that occupational diseases should be studied in the work environment rather than in hospital wards.

Industrial hygiene received another major boost in 1743 when Ulrich Ellenborg published a pamphlet on occupational diseases and injuries among gold miners. Ellenborg also wrote about the toxicity of carbon monoxide, mercury, lead, and nitric acid.

The U.S. Congress has passed three landmark pieces of legislation related to safeguarding workers' health: (1) the Metal and Nonmetallic Mines Safety Act of 1966, (2) the Federal Coal Mine Safety and Health Act of 1969, and (3) the Occupational Safety and Health Act of 1970 (OSHA Act). Today, nearly every employer is required to implement the elements of an industrial hygiene and safety, occupational health, or hazard communication programme and to be responsive to the Occupational Safety and Health Administration (OSHA), and its regulations.

Industrial Hygienist

Industrial hygienist can help us to:

- Identify, evaluate and bring under control all the chemical, physical, mechanical, biological and psychological agents that are known to be hazardous.
- Ensure the physical and mental demands of the above agents at work and the jobs performed by the workers are matched with their individual anatomical, physiological and psychological capabilities, needs and limitations.
- Provide effective measures to protect those who are vulnerable to hazards.
- Discover and improve work situations that may contribute to the overall ill-health of workers.
- Educate managements and workers to fulfil their responsibilities relevant to health protection, promotion, and stress factors.
- Elevate morale of workmen and management by defining the actual work-place environment and removing unhygienic factors.

The primary responsibility of the industrial hygienist is as follows:

- To protect the health of employees.
- To examine the work environment.
- To maintain an objective attitude towards the recognition, evaluation, and control of industrial health hazards.
- To counsel the employees regarding the industrial health hazards and necessary precautions to be taken to avoid its adverse health effects.
- To conduct programmes for the education/training of the workers and the public regarding prevention of occupational diseases.
- To direct the industrial hygiene programmes.
- To make specific decisions as to the need for effectiveness of control measures to protect the health of the employees.
- To prepare rules, regulations, standards, and procedures for the proper conduct of work and prevention of nuisance in community.
- To act responsibly in the application of industrial hygiene principles.
- To report findings and recommendations accurately and honestly.
- To conduct research for enhancing the knowledge concerning the effects of occupation upon health and means of preventing occupational health hazards and related problems.

To be effective in recognising and evaluating on-the-job hazards and recommending controls, industrial hygienists must be familiar with the hazards' characteristics. Potential hazards can include air contaminants, and chemical, biological, physical, and ergonomic hazards.

Occupational Health Hazards

A hazard is a source of potential harm. The consequence of a hazard can be harmful to people, to the environment, or to the plant and assets. Hazards can be physical (from things) or behavioural (from people).

Some of the occupational health hazards are:

***Air Contaminants*:** These are commonly classified as either particulate or gas and vapour contaminants. The most common particulate contaminants include dusts, fumes, mists, aerosols, and fibres. Dusts are solid particles generated by handling, crushing, grinding, colliding, exploding, and heating organic or inorganic materials such as rock, ore, metal, coal, wood, and grain. Any process that produces dust and remain in the air long enough to be inhaled or ingested should be regarded as hazardous until proven otherwise.

***Chemical Hazards*:** There is hardly any industry which does not make use of chemicals. The chemical hazards are on the increase with the introduction of newer and complex chemicals. Chemical agents act in three ways: local action, inhalation, and ingestion. The ill-effects produced depend upon the duration of exposure, the quantum of exposure, and individual susceptibility.

Some chemicals cause dermatitis, eczema, ulcers and even cancer by primary irritant action; some cause dermatitis by an allergic action. Harmful chemical compounds in the form of solids, liquids, gases, mists, dusts, fumes, and vapours exert toxic effects by inhalation (breathing), absorption (through direct contact with the skin), or ingestion (eating or drinking). Airborne chemical hazards exist as concentrations of mists, vapours, gases, fumes, or solids. Some are toxic through inhalation and some of them irritate the skin on contact; some can be toxic by absorption through the skin or through ingestion, and some are corrosive to living tissue.

***Mechanical Hazards*:** Mechanical hazards are often obvious to the trained safety professional who will readily spot the unguarded sections of factory machinery. Ergonomic problems may be anticipated by observation of the way in which posture and repeated movements of limbs and joints occur in the course of work. Injuries caused by mechanical hazards have a high potential severity since often they are apt to result in permanent partial disability.

***Electrical Hazards*:** Electricity like fire is a useful servant when it is under control. But, it may also create a hazard or damage when a person becomes a part of an electrical circuit which may result in electric shock. Also various other hazards are associated with the use of electricity. A person may fall after losing his balance if subjected to a violent jerk after an electrical shock and it may be fatal. Sometimes, while working in confined space in the presence of flammable vapour, a person may be exposed to fires and explosions. A deafening noise produced by the blowing of a high-tension fuse or an explosion may rupture the eardrum. Some of the insulating oils used in electrical equipment are also found to have carcinogenic effects. Certain fumes and vapours from volatile liquids used in the electrical equipment may have adverse effects on the health. A person can receive an electric shock if he interposes any part of the body into an electric circuit.

***Biological Hazards*:** These include bacteria, viruses, fungi, and other living organisms that can cause acute and chronic infections by entering the body either directly or through breaks in the skin. Occupations that deal with plants or animals or their products or with food and food processing may expose workers to biological hazards. Laboratory and medical personnel also can be exposed to biological hazards. Any occupation that result in contact with bodily fluids pose a risk to workers from biological hazards.

In occupations where animals are involved, biological hazards are dealt with by preventing and controlling diseases in the animal population as well as properly caring for and handling infected animals. Also, effective personal hygiene, particularly proper attention to minor cuts and scratches especially on the hands and forearms, helps keep worker risks to a minimum.

In occupations where there is potential exposure to biological hazards, workers should practice proper personal hygiene, particularly hand washing. Hospitals should provide proper ventilation, proper personal protective equipment such as gloves and respirators, adequate infectious waste disposal systems, and appropriate controls including quarantine in instances of particularly contagious diseases such as tuberculosis.

***Physical Hazards*:** These include excessive levels of ionising and non-ionising electromagnetic radiation, noise, vibration, illumination, and temperature.

In occupations where there is exposure to ionising radiation, time, distance, and shielding are important tools in ensuring worker safety. Danger from radiation increases with the amount of time one is exposed to it; hence, the shorter the time of exposure the smaller the radiation danger.

Distance also is a valuable tool in controlling exposure to both ionising and non-ionising radiation. Radiation levels from some sources can be estimated by comparing the squares of the distances between the worker and the source. For example, at a reference point of 10 feet from a source, the radiation is 1/100 of the intensity at 1 foot from the source.

Shielding also is a way to protect against radiation. The greater the protective mass between a radioactive source and the worker, the lower the radiation exposure.

Noise, another significant physical hazard, can be controlled by various measures. Noise can be reduced by installing equipment and systems that have been engineered, designed, and built to operate quietly; by enclosing or shielding noisy equipment; by making certain that equipment is in good condition and properly maintained with all worn or unbalanced parts replaced; by mounting noisy equipment on special mounts to reduce vibration; and by installing silencers, mufflers, or baffles.

Substituting quiet work methods for noisy ones is another significant way to reduce noise - for example, welding parts rather than riveting them. Also, treating floors, ceilings, and walls with acoustical material can reduce reflected or reverberant noise. In addition, erecting sound barriers at adjacent work stations around noisy operations will reduce worker exposure to noise generated at adjacent work stations. It is also possible to reduce noise exposure by increasing the distance between the source and the receiver, by isolating workers in acoustical booths, limiting workers' exposure time to noise, and by providing hearing protection.

***Psychosocial Hazards*:** The psychosocial hazards arise from the workers' failure to adapt to an alien psychosocial environment. Frustration, lack of job satisfaction, insecurity, poor human relationships, emotional tension are some of the psychosocial factors which may undermine both physical and mental health of the workers. The capacity to adapt to different working environments is influenced by many factors such as education, cultural background, family life, social habits, and what the worker expects from employment.

The health effects can be classified as: (a) psychological and behavioural changes including hostility, aggressiveness, anxiety, depression, tardiness, alcoholism, drug abuse, sickness, absenteeism; and (b) psychosomatic illhealth including fatigue, headache, pain in the shoulders, neck and back, propensity to peptic ulcer, hypertension, heart disease and rapid ageing.

***Ergonomic Hazards*:** The science of ergonomics studies evaluates a full range of tasks including, but not limited to, lifting, holding, pushing, walking, and reaching. Many ergonomic problems result from technological changes such as increased assembly line speeds, adding specialised tasks, and increased repetition; some problems arise from poorly designed job tasks. Any of those conditions can cause ergonomic hazards such as excessive vibration and noise, eye strain, repetitive motion, and heavy lifting problems. Improperly designed tools or work areas also can be ergonomic hazards. Repetitive motions or repeated shocks over prolonged periods of time as in jobs involving sorting, assembling, and data entry can often cause irritation and inflammation of the tendon sheath of the hands and arms, a condition known as carpal tunnel syndrome.

Ergonomic hazards are avoided primarily by the effective design of a job or jobsite and by better designed tools or equipment that meet workers' needs in terms of physical environment and job tasks. Through effective worksite analyses, employers can set up procedures to correct or control ergonomic hazards by using the appropriate engineering controls (e.g., designing or redesigning work stations, lighting, tools, and equipment); teaching correct work practices (e.g., proper lifting methods); employing proper administrative controls (e.g., shifting workers among several different tasks, reducing production demand, and increasing rest breaks); and, if necessary, providing and mandating personal protective equipment. Evaluating working conditions from an ergonomics standpoint involves looking at the total physiological and psychological demands of the job on the worker. Overall, the benefits of a well-designed, ergonomic work environment can include increased efficiency, fewer accidents, lower operating costs, and more effective use of personnel.

***Nuclear Hazards*:** Careful design, effective use of equipment, and other stringent precautions make the nuclear reactors inherently safe; but, however, nuclear power programmes have been beset with problems. These problems reached a climax at Chernobyl in Ukraine in 1986. The accidental explosion at the nuclear power plant was a catastrophic event of the atomic era that posed serious potential global health consequences. It released radioactive material into the atmosphere, causing major concerns and environmental health issues. Most seriously poisoned by radiation were workers at the scene, including those who fought with the fire despite dangerously high radioactive levels. The radioactive cloud that was released in the blast and the ensuing fire was responsible for several deaths in the coming years, mostly in the neighbouring countries.

Recognition of Hazards

There are a number of different methods that can be used to recognise hazards:

1. ***Accident or injury reports***: In terms of recognising hazards that cause traumatic injury, a study of accident or injury reports can be useful. Group statistics from such reports can indicate areas of the plant and processes that are involved in large numbers of accidents or injuries.
2. ***Physical injuries***: Pre-employment examinations along with periodic physical examination can help to identify chronic conditions that may be a result of contact with a hazard in the work environment.
3. ***Employee notifications***: In some cases, the employee will recognise a health or safety hazard before it is recognised by any other plant personnel. Given the right management atmosphere, the employee will bring this problem to the attention of those responsible for rectifying the problem.
4. ***Required inspections***: Certain pieces of equipment are on a schedule of required inspections. These required inspections can indicate problems before they become a hazard to the health or safety of the worker.
5. ***Literature and discussion with other professionals***: It is the responsibility of the professional to keep abreast of changes that are occurring within the occupational safety and health field.
6. ***Sampling and spot inspections***: Spot inspections should be conducted to detect potential hazards, if any.
7. ***Fault-tree analysis***: Using fault-tree analysis, a probabilistic model of the system events is constructed. With the application of fault-tree analysis, it is possible to determine the likelihood that a given event will occur and that a given series of events will cause this event to occur. Though it is mainly used for safety analysis, it can also be used as a tool to recognise potential health hazards.
8. ***Critical incident techniques***: Using this technique, a number of workers from a given location or plant are interviewed to determine unsafe practices or errors that have occurred while they have been on the job.
9. ***Job safety analysis***: Using this technique, each individual job is broken down into tasks that must be performed and elements that are required to perform these tasks. Each task and element is reviewed to detemine if a potential hazard exposure exists. Action is taken to modify either the procedures used, the equipment involved, or the protection afforded to the worker to eliminate the exposure.

Table 8.1 Hazardous Factors and their Effect on Health.

Hazardous Factors	Adverse Health Effect or Other Outcomes
1. Mechanical risk	Occupational accidents and injuries factors
2. Physiological strain and heavy physical work	Musculoskeletal disorders, Strain injuries, Low-back pain
3. Ergonomic factors	Strain injuries, Mental stress, Lowered productivity and quality of work
4. Physical factors e.g., noise and chemical hazards	Noise-induced hearing loss, traumatic vasospastic disease vibration Intoxication, fibrosis, cancers, allergies, nervous system injuries
5. Biological factors	Infections, allergies
6. Psychological strain	Psychic stress, work dissatisfaction, burnout, depression
7. Psychosocial aspects of work	Conflicts, lowered productivity, lowered quality of work, mental stress

Source: Rantanan, J.R., *Role of Occupational Health in Industrialised Countries: Strategies and Principles*", Proceedings of Symposium on Occupational Health - An Essential Component of Social and Healthcare Policy, Estonia, 14-15 January, 1993.

Occupational Diseases

Occupational diseases cover all pathological conditions induced by prolonged work, e.g. by excessive exertion or exposure to harmful factors inherent in materials, equipment or the working environment. There are diseases which practically affect workers only. These diseases can very clearly be related to an occupation or to a determined occupational exposure.

Occupational disease occur as a result of exposure to physical, chemical, biological or psychological factors in the workplace.

It is possible to identify a specific cause for certain diseases, whereas others are due to several harmful factors. Occupational disease, is a potential hazard linked to the exercise of an occupation. The diagnosis of occupational diseases rests on specific knowledge and multiferious information gained from clinical examination of the patient, investigation of the working environment and epidemiological data.

On the other hand, the above medical considerations are different from legal considerations. The legal definition of occupational disease is a rather complex problem. It may be stated that the criteria to be considered in the definition of occupational disease vary according to whether such definition is to be used for statistical purposes or with an eye to preventive measures or for the purposes of compensation. In addition, in the last case the importance of the definition varies according to the social security system adopted in each country. In India occupational diseases are treated on a par with accidents.

The Causes of Occupational Diseases may be classified as under:

Physical Causes

Examples include:

(a) Heat – heat cataract, heat stroke
(b) Lighting – miner's nystagmus
(c) Noise – noise – induced hearing loss
(d) Vibration – vibration – induced white finger
(e) Radiation – radiation sickness
(f) Dust – silicosis, coal worker's pneumoconiosis
(g) Pressure – decompression sickness

Chemical Causes

Examples include:

(a) Acids and alkalis - dermatitis
(b) Metals-lead and mercury poisoning
(c) Non-metals-arsenic and phosphorus poisoning
(d) Gases-carbon monoxide poisoning, arsine poisoning
(e) Organic compounds - occupational cancers, for e.g., bladder cancer
(f) Dusts-mercury poisoning

Biological Causes

Examples include:

(a) Animal-borne – anthrax, brucellosis, glanders fever
(b) Human-borne – viral hepatitis
(c) Vegetable borne – aspergillosis (farmer's lung)

Ergonomic Causes

Examples include:

(a) Job movements-cramp (in relation to handwriting or typewriting)
(b) Friction and pressure-bursitis, cellulitis

Some of the well-known occupational diseases are as follows:

Dermatitis**:** The term 'dermatitis' simply means an inflamation of the skin. When the condition is due to contact with substances at work it is called 'occupational' or 'industrial dermatitis'. Some people are more susceptible to skin damages than others, particularly the young, those with soft, sweaty skin, the fair complexioned and those with poor personal hygiene. Occupational disease can affect any part of the body, but the hands, wrists and forearms are most commonly involved.

Anthrax**:** It is a highly infectious disease of ruminants: goats, cattle, sheep and horses, and is due to a baccillus. Man can be affected by direct contact with the animal, or indirectly by contact with the animal products. At risk particularly are those engaged in tanning, wool sorting, manufacture of brushes, bone-meal, fertiliser and glue. Also at risk are dockers and agricultural workers.

***Pneumoconiose*:** The word pneumoconiose literally means dust retained in the lung, with no implication as to whether disease is or is not present. In more common usage, however, the word has become a general term for any of the dust diseases of the lung and is used here with that meaning.

Though it means 'dust in the lung', it medically refers to the reaction of the lung to the presence of dust. The most important cause include mineral dust, silica and asbestos. Symptoms of cough and breathlessness develop usually after many years of exposure, only in the latter stages of disease.

The dust to which a coal worker may be exposed is complex in nature. The coal miners' pneumoconiose apparently caused by coal dust itself, is an established entity and is pathologically different from silicosis.

***Silicosis*:** Among the occupational diseases, silicosis is the major cause of permanent disability and mortality. It was first reported in India from the Kolar Gold Mines (Mysore) in 1947. Ever since, its occurrence has been uncovered in various other industries, e.g., mining industry (coal, mica, gold, silver, lead, zinc, manganese and other metals), pottery and ceramic industry, sand blasting, metal grinding, building and construction work, rock mining, iron and steel industry and several others.

Silicosis results from inhalation of silica dust i.e. crystallized silican dioxide. Silica is present in many rocks. In addition to coal workers, workers involved in quarrying, boaring, stone cutting, sand blasting, boiler scaling, and certain foundries are exposed to silica dust.

Silicosis may be either of an acute or of a chronic nature. The disease is most often reported in manufacturers and packers of abrasive soap powders, in sand-blasters working in enclosed tanks, and in high-power drillers of tunnel rock. After development of symptoms, the survival time is likely to be very short. Chronic pulmonary silicosis, the type usually encountered in industry, is surfaced only after years of silica dust inhalation. The disease is reported to occur most commonly in the mining industries but is also seen in numerous other industries such as potteries, foundries, stone cutting and finishing, tile and clay producing, and glass manufacturing.

There are no symptoms of the disease in the early stage. Later the initial complaint is of a dry morning cough. Next occurs some breathlessness. As destruction of lung tissue proceeds, breathlessness worsens. In the process of gross lung destruction the blood circulation from the heart to the lung may be affected and result in heart failure.

***Asbestosis*:** It is a reaction of the lung to the presence of asbestos fibres. Among those at risk are persons engaged in milling the ore, the manufacture of asbestos products, asbestos spraying, building demolition. Symptoms develop slowly after a period of exposure which varies from a few to many years. In some cases exposure may have begun so long ago that it cannot be recalled. Breathlessness occurs first and progresses as the lung loses its elasticity. There may be little or no cough and chest pain seldom occurs. The chest should be x-rayed every two years and special lung function tests are helpful. Diagnosis depends on history of exposure, chest x-ray, lung function testing, symptoms and physical signs.

***Byssinosis*:** Byssinosis occurs in individuals who have experienced prolonged exposure to heavy air concentrations of cotton dust. The earliest manifestations of byssinosis may become noticeable after several years of exposure to cotton dust. In the most advanced stages of the disease, cough, chest tightness, and dyspnea may be so severe that the worker is forced to leave the cotton industry.

***Bagassosis*:** Bagasse is the fibrous material remaining after the sugar-containing juice has been extracted from sugarcane. It is used in insulating and acoustic materials as well as in the manufacture of paper, fertiliser, explosives, animal feed, and refractory brick.

Bagassosis is a lung disease produced by the inhalation of dust attendant with the opening of bagasse bales which have been stored for several months or years and have subsequently become very dry. Physical examination may reveal dyspnea, cyanosis, and so on.

***Lead poisoning*:** Lead is a relatively common metal, which has a great variety of uses. Early features of lead poisoning include fatigue, loss of appetite, and metallic taste in the mouth. Constipation is the commonest complaint and is sometimes associated with abdominal pain. In addition, other metals like mercury, chromium, arsenic, manganese, nickel, cadimium are also associated with occupational diseases.

***Occupational Cancer*:** It is a form of delayed toxicity, usually serious in clinical course and outcome, due to exposure to chemical agents in the workplace. Biologically and clinically, cancer due to occupation is at present undistinguishable from cancer due to other causes. Cancer is not a single disease but a family of different diseases each characterised by its own aetiological, morphological, patho-physiological and clinical profile. Therefore, the main diagnostic assessment remains a detailed occupational history of the cancer patient.

Work-Related Diseases

The WHO categorises work–related disease as "multifactorial" in origin. These are diseases in which workplace factors may be associated in their occurrence but need not be a risk factor in each case. Such work-related diseases are:

- hypertension
- ischaemic heart disease
- psychomatic illness
- musculoskeletal disorders
- Chronic non-specific respiratory disease/chronic bronchitis

Difference between Work-related and Occupational Diseases

The main differences between work-related diseases and occupational diseases are shown in the Table 8.2.

Table 8.2 Work-related and occupational diseases

Work-related Diseases	Occupational Diseases
1. Occurs largely in the community	Occurs mainly among working population
2. "Multifactorial" in origin	Cause specific
3. Exposure at workplace may be a factor	Exposure at workplace
4. May be notifiable and compensable	Notifiable and compensable

STUDY ASSIGNMENT

Collect the following information while visiting a chemical factory:

- Details of machinery and equipment
- Process details
- Chemicals used, by-products if any, and final products
- Toxicity of chemicals
- Production details – number of shifts, duration of each shift, and number of workers in each shift
- Jobs carried out by exposed persons, their duration of exposure to chemicals and frequency of operation
- Manner in which chemicals are stored and handled
- Existing control measures
- Personal protective equipment and manner of selection, use and maintenance
- Housekeeping
- Personal hygiene measures in practice
- Frequency and type of medical examination

- Educational programmes
- Workplace compliance

(Findout answers for the following questions):

1. Do you have an active safety and health programme in operation?
2. Whether one person is clearly responsible for the overall activities of the safety and health programme?
3. Do you have a safety committee or group made of management and workers' representatives?
4. Do you have a working procedure for handling in-house employee complaints regarding health and safety?
5. Have you considered incentives for employees or work groups who have excelled in reducing workplace injuries/illnesses?
6. Whether all required training plans are in place?
7. Whether all standard operating procedures are in place?

CASE

Your secretary, who normally has an excellent standard of accuracy and speed of working has recently been making small errors and falling behind her work schedule. You are also aware of her timekeeping that is slipping.

When you talk to her about this, she mentions that during the day she often suffers from headaches and finds it hard to concentrate at the computer. She suggests that it might be something to do with her computer or office environment.

It is obviously important not to dismiss this out of hand, and you need to look more closely at the situation.

Questions

1. Describe in detail how you would go about solving this problem?
2. Would you assess her work situation and office environment? If so, create a checklist that you might use.

SUMMARY

- Hygiene is primarily concerned with the science of preservation of health whilst at work.
- Industrial hygiene is primarily concerned with the appraisal and control of occupational health hazards that arise from various industrial processes and operations.

→

- Industrial hygiene is an applied science encompassing the application of knowledge from a multidisciplinary profession.
- For the effective implementation of industrial hygiene programmes, multidisciplinary approach is required which may involve the areas of industrial medicine, industrial physiology, industrial engineering, chemical toxicology, industrial psychology and ergonomics.
- The combination of several disciplines will play a vital role in the application of industrial hygiene principles.
- The industrial hygienist effectively provides information about the manufacturing operations of the company to its medical department, which provides insight into the health background of the employees.
- The various environmental factors are stressors that can cause sickness, impaired health or significant discomfort to workers and these can be classified as chemical, physical, biological or related to ergonomics.
- An important aspect of industrial hygiene programme is that it can recognise potential health hazards and assures a healthy workplace for employees.
- Industrial hygiene includes all the functions needed in the recognition, evaluation, and control of occupational health hazards associated with production.
- Industrial hygiene surveys cover both the aspects of air and biological monitoring.
- Occupational diseases are those arising out of and in the course of employment, and occur as a result of exposure to physical, chemical, biological and psychological factors at the workplace.
- The common occupational health problems encountered in Indian industry are occupational dermatitis, dust diseases, asbestosis, silicosis, coal workers pneumoconiosis, byssinosis, gas poisoning, noise hazards, heat hazards, and so on.

DISCUSSION QUESTIONS

1. What can a company do to encourage neatness and cleanliness among its employees?
2. How occupational diseases occur as a result of exposure to physical, chemical, biological and psychological factors at the workplace?

9

Safety in Construction

The term "construction" refers to the buildings and civil engineering activities. An ILO publication includes the following types of works in the scope of construction activities. Building and civil engineering may be divided into four main parts: work above ground, work in open excavations, underground work, and under water work.

Construction industry is the core industry providing basic infrastructure necessary for the economic development and well-being of people. Almost 50 per cent in any plan outlay goes towards the construction and building industry. Progress in building industry means progress in all sectors. Construction industry is infallible indicator of a country's economic and social health.

The investments in building and construction industry are mammoth compared to any other sector and the effects of success or failure of a construction project go beyond the owner and contractor. They, more than often, affect the entire society over a long and sustained period of time, and even can result in man-made disasters or accentuate natural calamities.

The construction industry in India has registered an enormous growth in the last four decades. It encompasses a large area of activity of varying complexity such as huge river valley projects, construction of off-shore structures, construction of ordinary buildings, high rise buildings and tall structures.

Notwithstanding the technical advances in civil engineering, construction remains a labour intensive industry with diverse groups of skilled, semi-skilled and unskilled workmen working in harsh and complex environment, their number being more than 6 million.

Construction industry is inherently more hazardous as compared to other industries from the point of view of safety. A construction site has may hazards and many pit-falls.

Construction is a high risk industry for clients, contractors and workers alike. Safety in construction is a very complex phenomenon. The risk of injuries in the construction industry is very high, mainly because of the nature of the work carried out by different agencies employing largely unskilled workers. In India, a little over 16 percent of the population depends on construction industries for its livelihood, while about 18.7 million persons are directly engaged in this sector. It is estimated that one out of every five persons employed in this industry suffers employment injury, as adequate attention is not given to safety and health.

As per the report of a survey conducted by ILO, 165 per thousand workers get injured during construction activities. In addition to that, the workers are exposed to a host of hazardous substances, which have a potential to cause serious health and occupational diseases such as asbestosis, silicosis, and lead poisoning.

Its other observations are:

- Employment relations are contractual, and they exist for the project duration only. That is why permanent or semi-permanent systems or arrangements are not thought of.
- Many small contractors do not want to invest in training the workers employed by them particularly regarding safety measures because of the short-term relationship with them.
- The workers themselves do not ask for any safety measures lest this may affect their employment opportunity.
- There is no institutional framework to impart training at the worker's level, barring a few initiatives taken by Construction Industry Development Council (CIDC), National Safety Council (NSC) and some major construction companies.
- Till today, many construction companies do not have safety departments at their project site.
- There is a shortage of safety professionals for the sector as many of the professionals are getting absorbed in factories and some are getting attracted by good salary and joining firms abroad.

Characteristics

Following characteristics of the construction industry affect in some way or the other the implementation of safety measures:

- Large number of small firms involved.
- A chain of contractors ranging from the principal contractor to sub-contractors and petty contractors.
- Competitive tendering.
- High turnover of contract labour.
- Seasonal and migrant labour.

- Unorganised labour.
- Low level of awareness and poverty resulting in their exploitation.
- Working in the open and all weather conditions.
- Difficulties in the absorption of imported technology.
- Accident frequency and severity rates are well above the averages of all other industries.

The construction industry has to ensure the safety of workers as well as of the public. Therefore, the vital requirement for any successful construction enterprise is designing and constructing safe structures and providing safe work environment to the personnel engaged in construction. The concern for safety starts from the conceptual and design stage and continues through the construction process. The designer, consultant, client and constructor and subsequent maintenance and upkeep agencies share equal responsibility for safety.

Safety has to be a vital function in the management of construction projects. The concept of safety has to be built into the work values and systems. Internationally accepted safety standards are to be applied to create a safe work environment.

The major safety issues that are associated with the construction phase of the project can be summarised as:

(i) Client and Contractor (Construction Team) safety systems
(ii) Client and Contractor (Construction Team) compliance with their agreed safety plan
(iii) Level of Client (or Client's agent) supervision
(iv) Client and Contractor safety management structures
(v) Construction safety audits
(vi) Project execution and safety performance monitoring
(vii) On the job safety training
(viii) Construction safety requirements
(ix) Promotion of safety to employees
(x) Duty of care to the general public
(xi) Sub-Contractor compliance with contractor safety plan

Exhibit 9.1 Ten Principles on Construction Safety

1. Use safety helmets and clothing correctly.
2. Inspect machines and tools prior to their use.
3. Follow all instructions of managers/supervisors and carry out operations with accuracy.
4. Use safety belts where there is any risk of falling.
5. Never enter any areas posted with signs of keepout from dangerous places.

6. Immediately inform your supervisor on locating any unsafe work conditions.
7. In joint operations, follow established safety procedures and signals.
8. Handle electrical machinery and hazardous materials under established operational procedures and with care.
9. Operate vehicles, construction machines, and mobile cranes only after confirming safety in terms of surrounding areas.
10. Ensure that work sites are always well arranged and cleaned after operations

Exhibit 9.2 Safety Measures During Construction

- Provide appropriate protective equipment like safety shoes, safety helmets, safety belts and eye goggles depending on the nature of work being carried out like welding, gas cutting, and working at heights.
- Provide safe means of access to the place of work.
- Secure the place of work using suitable guarding like handrails and midrails.
- Barricade excavated areas.
- Provide steel scaffolding and avoid usage of bamboo/wooden scaffolding.
- Ensure proper storage of hazardous materials.
- Ensure proper storage and handling of gas cylinders.
- Examine the suitability of various construction equipment, tools and tackles from safety point of view, scrutinising fitness and test certificates.
- Examine the credentials of operators of construction equipment prior to engaging them on the job.
- Ensure proper earthing of portable electrical tools and temporary electrical installation for construction power.
- Provide portable fire extinguishers and running waterlines while carrying out hazardous activities like hook-up joints.
- Provide first-aid facilities at site.
- Ensure regular housekeeping to facilitate safe work environment.
- Prevent environment pollution during disposal of construction waste.
- Constitute safety committees involving senior level participation on the part of owner, contractors and employees.
- Conduct periodic safety meetings and recommend preventive measures.
- Create and sustain safety awareness at all levels among construction staff and workmen.

- Report accidents, near misses and recommend preventive measures whenever necessary.
- Ensure prompt medical attention to accident victims.

Construction Hazards

Hazard means any existing or potential condition at the job site, which by itself or by interaction with other factors, can result in an accident.

In order to ensure safe construction and to avoid accidents, sincere efforts must be made to eliminate hazardous conditions and to take adequate precautions at every stage of construction. The essential components of hazard control programme are:

(a) hazard identification.
(b) ranking hazards by risk.
(c) establishing preventive and control measures.
(d) monitoring and
(e) evaluating programme effectiveness and feedback.

Hazard control is not the responsibility of just one individual or one department. Planning, design, purchase, operation and maintenance are the departments associated with the construction of the project. It is the collaborative effort of all these departments. The activity of hazard control is a continuous and on-going one.

Every accident has a cause. If hazards can be detected then accidents can be prevented. There are many potential hazards in the construction industry. Some such hazards are listed below:

1. exposure to asbestos dust in demolition and construction;
2. exposure to quartz dust in state quarries;
3. exposure to harmful ground treatment chemicals, or to potentially harmful chemicals used on construction sites at waste tips;
4. potentially harmful chemicals in land reclaimed for construction (on the site of waste tips);
5. exposure to radiation;
6. use of explosives in excavating, quarrying, and demolition;
7. fire;
8. collapse of uncompleted structures;
9. exposure to excessive noise;
10. air or gas pressure in underwater or underground work;
11. risks to children and others owing to unauthorised access to construction sites or plant yards;
12. risks to the public including the effects of noise, dust, contamination, pollution, collapse of excavations or structures, vehicles, and so on.
13. high loads in transit coming into contact with overhead electric power lines or with bridges;

14. hazards in offshore construction work including ionizing radiations, fire, hypothermia, etc.

There will be a continuing need for the industry to be on its guard against such hazards and to identify new hazards resulting from the introduction of new materials or methods of construction or from changing circumstances.

Safety measures and the hazards in construction activities are very closely related. As a matter of fact, safety measures can be adopted only after studying the hazards involved in the construction activity. Accidents of minor and major nature take place at construction sites quite often causing injuries or loss of human life, damage to property, and delays in project commissioning.

Accidents do not just happen, they are caused. Somebody commits a mistake somewhere, which results in accident and its subsequent consequences. It is any existing or potential condition at the job site, which by itself or by interaction with other factors, can result in an accident.

Causes of Accidents

- Defects in technical planning.
- Fixing unsuitable time-limits so that the workforce may have to work excessive overtime to keep to schedule.
- Assigning work to incompetent contractors.
- Insufficient or defective supervision of the work.
- Lack of cooperation between different trades.
- Construction defects.
- Use of unsuitable materials.
- Defective processing of materials.
- Lack of equipment.
- Unsuitable equipment.
- Defects in equipment.
- Lack of safety devices or measures.
- Inadequate preparation of work.
- Inadequate examination of equipment.
- Inprecise or inadequate instructions from supervisors.
- Unskilled or untrained operators.
- Inadequate supervision.
- Irresponsible acts.
- Unauthorised acts.
- Carelessness.

Accidents may occur in a number of ways. Some examples are:

- through the collapse of walls, parts of buildings (particularly during demolition), stacks of materials or stockpiles of excavated material;
- through the collapse or overturning of ladders, scaffolds, stairs or beams;
- by falling of objects, tools or pieces of work;
- by falling of persons from ladders, stairs, roofs, scaffolds or buildings through hatches and windows or through other openings;
- during loading, unloading, lifting, carrying and transporting loads on or in connection with vehicles of all kinds;
- at power plant and power transmission machinery;
- in the operation of railways;
- on lifting appliances;
- on welding and cutting equipment;
- on compressed-air equipment;
- on combustible, hot or corrosive materials;
- by dangerous gases;
- during blasting with explosives;
- when using handtools;
- by stepping on sharp objects.

The basic causes of accidents for persons in civil engineering works are the following:

(a) Slitting–falling from height.
(b) Struck or trapped by objects in motion.
(c) Stepping on or striking against on moving objects.
(d) Handling objects in such a way so as to cause injury.
(e) Using of unsafe materials, handtools and apppliances.

Equipments causing accidents are:

***Cranes*:** The common accidents occurring out of operation of cranes fall under the following categories:

- Non-availability of load ratings.
- Overloading of the crane.
- Operating the crane at unsafe radius.
- Travelling with heavy loads.
- Failure of lifting ropes.
- Travelling with wrong boom position.

***Tower Cranes*:** The cranes are sensitive to winds and should not be operated if winds exceed 40 Km/hr. Only trained operators should be allowed to operate the tower cranes. In case of rail mounted cranes proper stopper should be provided to prevent accidents from occurring due to movement of the crane while handling loads. While not in operation, it should be left free to align itself in the direction of the wind.

***Lifting Gear*:** The lifting gear used in cranes and other appliances should be in proper condition to avoid accidents. The lifting gear includes wire ropes, chains, fibre ropes, etc. They should be tested once in six months to ascertain their mechanical condition.

***Hoists and Lifts*:** Following provisions should be carried out for the safe operation of hoists and lifts:

- Provision of enclosures filled with gates.
- Marking the safe carrying limit inside the lift.
- Provision of over-running limit switches.

***Pulley Blocks*:** The accidents arising out of pulley block have been attributed to axle metallurgy and smaller diameter. To avoid this, diameter should be greater than 20 times that of the diameter of ropes, and hardly worn out sheaves and blocks should be discarded.

Prevention of Accidents

It is absolutely essential to take all precautionary measures to prevent accidents in construction industry. It is very important that the manager of a project enforces the safety measures in his area and devotes his attention to the promotion of safe working conditions at worksites. Safety must be a basic component of the management philosophy just as profit making because the cost of an accident itself may not only drain away the profits but may affect the very survival of the company.

Preventive measures should start at the planning and designing stage itself. The planners, architects and engineers should keep in mind the safety of the works at all stages of the project and the contractors should be made to familiarise themselves with safety regulations and safe working practices. The machinery and equipment which are to be used on the works should also conform to specific safety requirements. The internal organisation should provide for regular inspections and maintenance of all machinery, tools and equipment being used on the works. The site management should take necessary safety precautions at all stages of work. Some of the preventive measures are:

Mechanical Maintenance

1. All machinery such as cranes, dozers, dumpers, tippers, trucks, winches, material hoists, are to be maintained in absolutely up-to-date condition.

2. Upkeep of regular periodical maintenance of machinery and adequate stock of original spares.
3. Lifting load capacity should be clearly marked on the machine.
4. Operators should not leave the machine in running condition with the load on, in their absence.
5. Operators should be able to see the lifted load or the signal from the signal man very clearly.
6. Nobody should stand under the load being lifted by the crane or the hoist.
7. Passengers and material cages should be provided with proper enclosure.
8. Drivers of passenger and material transport vehicles should undergo periodical alcohol test.

Electrical Installation

1. Switch room should be adequately spacious, easily accessible, and regularly attended by responsible staff.
2. Electrical materials should be of good quality and staff should be properly trained.
3. Proper insulators should be used for supporting the electrical live wires.
4. Overhead lines should be supported with poles of adequate height.
5. Fire extinguisher should be maintained in the control room.
6. Proper illumination should be ensured at the job sites.

Housekeeping and Fire Protection

1. Discipline and housekeeping are the key words for successful safety management.
2. Access roads at project sites should be properly maintained for safe movement of vehicles.
3. Exit facilities must be maintained without any obstruction. Fire protection system should be installed and kept always in working condition.
4. Natural calamities have to be taken care of by proper risk analysis, adequate provisions in design, and providing for disaster management.

Personal Protective Equipment

The working conditions in construction are in most cases such that, despite all preventive measures in project planning and work design, some personal protective equipment, such as a helmet, hearing and eye protection, boots and gloves, is needed to protect workers. However, there are certain disadvantages in using PPE:

- Wearing some forms of PPE may involve discomfort to the user and slow down the work.
- Extra supervision is called for to see that PPE is used.
- PPE costs money.

Whenever possible, it is better to try to eliminate the hazard rather than providing PPE to guard against it. However, some PPE such as safety helmets and footwear should be used on all construction sites. Safety helmets protect the head effectively against head injuries. Foot injuries can be minimised by wearing protective footwear.

Hands are extremely vulnerable to accidental injury, and in construction more injuries are caused to hands and wrists than to any other part of the body. They are largely preventable by better manual handling techniques and equipment, and by wearing suitable protective equipments such as gloves and gauntlets.

In construction industry many eye injuries occur as a result of flying material, dust or radiation, spillage, leakage or splashing or corrosive liquids. Some of these hazards can be removed to a great extent by proper machine guarding, exhaust ventilation, work design, personal eye protection, and so on.

On construction sites there are often tasks where harmful dust, mist or gas may be present in operations such as rock crushing and handling, blasting, dismantling buildings, welding or cutting materials and paint spraying. Whenever there is doubt about the presence of toxic substances in the atmosphere, the respirator must be worn. The correct type of respirator will depend upon the hazard and the work conditions.

The majority of fatal accidents in construction are due to falls from heights. Safety harness may be used normally in maintenance work on steel structures such as bridges and pylons. There are many types of safety belt and safety harness available. A full safety harness should always be used in preference to a safety belt.

Checklist

1. Housekeeping.
2. Access to working at height and depth.
3. Adequacy of working space.
4. Illumination.
5. Material handling equipments, cranes, winches, slings and shackles.
6. Fork lifts and industrial trucks.
7. Hard and portable power tools.

8. Electrical equipments-temporary connections and fuses.
9. Welding equipment.
10. Work permit procedures.
11. Storage of explosives and detonators.
12. Precautions during blasting.
13. Fire extinguishers.
14. Personal protective equipments – safety belts, safety shoes, helmets, goggles.
15. First-aid facilities.

Safety Legislation

There is no safety legislation, *per se*, applicable to the construction industry. Repair of workshops of contractors are those attached to construction sites are covered under the Factories Act, 1948. Quarries operated by a contractor are included under the Mines Act, 1952. Vehicles plying to and fro construction sites are covered by the Motor Vehicles Act, 1951. A contractor's office is subject to the Shops and Commercial Establishments Act. The Contract Labour (Regulation and Abolition) Act, 1971 regulates the employment and working conditions of contractors' labour. The migrant labour is covered under the Inter-State Migrant Labour Act, 1973. The Employees' Compensation Act, 1923 covers all construction sites throughout the country. The Employees' State Insurance Scheme applies to the workers employed for the work of construction of building for the expansion of factory. Child labour and maternity are dealt with under separate statutes.

The Building and Other Construction Workers' (Regulation of Employment and Conditions of Service) Act, 1996 regulates the employment and conditions of service of building and other construction workers and provides for their safety, health and welfare. Under Section 38(2) of the Act, appointment of safety officers in construction establishments is a statutory requirement. Qualification and experience of safety officers are prescribed by the Central Government in Schedule VIII under Rule 209(1) of the Building and Other Construction Workers' (Regulation of Employment and Conditions of Service) Rules, 1998.

Every establishment employing more than five hundred building workers shall appoint safety officers, as laid down in the scale given below:

1. Up to 1,000 building workers – one safety officer.
2. Up to 2,000 building workers – two safety officers.
3. Up to 5,000 building workers – three safety officers.
4. Up to 10,000 building workers – four safety officers.

For every additional 5,000 building workers or part thereof – one safety officer.

Accident Reports

Reports of accidents might be obtained through the section supervisor or engineer. These reports would generally identify the department in which the injured person worked, the extent of the injury, the machine (if any), that was involved, and the cause. The report will be of value to the safety department as a source of hazard data and the basis for suitable remedial action.

Table 9.1 Accident Report (To be filled even in case of no injury).

Name of the job site / job no.		:
Exact location of the accident		:
Brief description of the accident		:
Whether resulted in damage of property	Yes/no	:
If yes, briefly state extent of the damage		:
Whether resulted in injury to personnel	Yes/No	:
If yes, names of the injured		:
Age, sex and employee number		:
Name of the subcontractor		:
Date and time of accident		:
Witness–name, employee number		:
Briefly state the unsafe act/condition that caused the accident		:
Safety appliances used by the injured at the time of accident		:
Remedial measures taken		:
	Section Supervisor/Engineer	

Safety Unit

A separate safety unit headed by a safety engineer should be set-up for monitoring all safety activities in a construction site. The functions of safety unit includes:

1. To look into all procedures and practices and examine all civil engineering works.
2. To inspect the works regularly and advise the contractors, departmental officers, and the staff concerning the measures to be taken to ensure safety of the works and persons employed therein.
3. To organise and implement training programmes in the application and observence of safety norms and practices.

4. To collect accident data, analyse, and submit reports of all accidents and initiate corrective measures as may be warranted by the situation.
5. To initiate a suggestion box scheme and encourage suggestions from all concerned.
6. To conduct safety education and create awareness through safety posters, cartoons, leaflets, bulletins, and through various other activities.
7. To carry on safety surveys within and outside the organisation to keep abreast of up-to-date safety practices.

The organisation of safety on the construction site will be determined by the size of the work site, the system of employment and the way in which the project is being organised. Safety and health records should be kept to facilitate the identification and resolution of safety and health problems on the site. In construction projects where subcontractors are used, the contract should set out the responsibilities, duties and safety measures that are expected of the subcontractor's workforce.

Training should be conducted at all levels including managers, supervisors and workers. Subcontractors and their workers may also need to be trained in site safety procedures.

Role of Parties

The role of various parties in safety are:

***Designer*:** The designer while designing should give due consideration to the safety of the workers who will be subsequently employed on the erection of such structures. He should exercise proper care not to include any thing in the design which would require the use of dangerous structural procedures and undue hazards. These could properly be avoided by proper modifications in the design.

***Employer*:** The employer should provide and maintain buildings and work passages and should organise work in such a manner that workers are protected against risk of accidents. While buying machinery he should ensure that they confirm to the various safety regulations. The employer should ensure that proper supervision is provided to workers to perform their work. Proper instructions should be given to all employees about the safety requirements. Buildings and equipments in which dangerous defects have been detected should not be used till they are rectified.

***Contractor*:** Safety is the responsibility of every contractor and his employees.

The contractor should at all costs avoid the following:

1. Use of poor quality of steel and cement.

2. Use of wrong mix of concrete.
3. Use of wrong water/cement ratio.
4. Any change in the design.
5. Any change in the erection procedure.

The contractor should provide the following basic gear to his construction labour and ensure their proper use.

1. Helmets
2. Boots
3. Goggles
4. Gloves
5. Proper lighting
6. First-aid kit
7. Safety harness and
8. Safe temporary structures.

The safety equipments are to be provided to workers, either by the principle employer or contractor, as the case may be. It is also important to ensure that workers make use relevant safety equipments such as rubber boots, hard toe safety boots, hard hats, safety belts, goggles, aprons, respirator shields, gloves, flash-lights, battery lamps, magazine shoes, safety nets, helmets, and so on.

The contractor should ensure adequate number of fire extinguishers at the worksite to meet the risk of fire hazards.

The contractor, unless otherwise contracted upon, shall abide by the provisions of Factories Act, State Factories Rules, Employees' Compensation Act, Employees' State Insurance Act, and Contract Labour (Regulation and Abolition) Act.

***Workers*:** The following are the obligations of the workers

1. Workers should not be bare footed at a construction site.
2. A helmet must be worn at all times during construction.
3. Whenever required the workers should wear safety belts and they must be inspected carefully and periodically for their soundness.
4. Safety goggles, safety spectacles and face shields and masks must be used wherever required.
5. Gumboots, rubber hand gloves should be worn where there is risk of injury.

Trade unions: Trade unions exist to represent, protect, advise, educate their members, and serve their interests. One of the major responsibilities of the trade unions is to look after the safety, health and welfare of their members at the workplace. They have to play the following roles in an organisation regarding occupational safety, health and environment protection:

- Trade unions can co-operate with government agencies and with employers in the investigation of causes of accidents and diseases.
- Once the cooperation of the unions has been sought and secured, they can persuade their members to accept changes which are in their interest and educate them to be safety conscious.
- They can take necessary steps to ensure application of legally defined minimum safety standards at the workplace.
- The management can take active help and participation of trade unions in overhauling the procedures, formulating the safety policy, and organising safety training classes.
- Representatives of trade unions should meet the representatives of management regularly and discuss freely every aspect of accident prevention.
- Trade unions can play vital role in smooth running of safety committee(s) and should generate positive will to accept the safety promotion attitude.
- Trade unions must motivate members for proper safety consciousness in the industry and accept safety as a way of life.

Safety Management

Safety cannot be fully ensured in the construction industry by mere legislation and enforcement. This is true not only in India but in other countries as well. There must be a provision for checklist of all safety measures in each and every contract document. Therefore, the modern trend in occupational safety envisages the following:

1. Integrated multi-disciplinary approach of a team consisting of specialists drawn from academics, government, management, and workers.
2. Training and information to all concerned and actions at the national, enterprise and unit levels.
3. Safety management for building and construction project and systematic approach to certain fundamental issues like management policy on safety, objectives of a safety programme, organisation for safety implementation, evaluation of safety systems and procedures, and ensuing safety operations. Achieving safety is not accidental one, it is the result of deliberate well-directed efforts on all matters in relation to its planning systems, training and supervision, and each one of these is to be based on sound concepts.

The operational management of safety in construction industry has to be concentrated on five aspects, namely (a) design safety, (b) construction safety, (c) safety while executing/constructing, (d) fire safety, and (e) safety against natural calamities.

Design safety

(i) Conceptual/architectural planning.
(ii) Correct appreciation and evaluation of forces and loads during execution and erection.
(iii) Following the bye-laws and statutory regulations.
(iv) Correct structural analysis and detailing.

Construction Safety

(i) Following strictly the design analysis outputs of dimensions and other parameters of structure and the services.
(ii) Using right quality materials in right quantity and following the right specifications.
(iii) Selection, use and up-keep of proper construction plant and machinery.
(iv) Trained manpower.
(v) Proper supervision at all stages of construction.
(vi) Construction safety can be ensured by vigorous training and skill upgradation of persons involved in construction.

Safety Execution

Safety in execution is different from construction safety, as much, as this is mainly to take care of safety on temporary facilities provided in executing the job for all works. The safety in execution centre round on house-keeping at work site, neatness and cleanliness, proper scaffolds, personal attire like helmets, gumboots, apparel, and so on.

Fire Safety

The construction industry, like all others, faces risk of fire and explosion which destroy property and endanger lives. Compared with most other industries and businesses, construction seems ill prepared for such risks in its own work, although it may be constructing buildings with high built-in standards of fire safety.

During construction, improper storage/disposal of inflammable waste, inflammable liquids and explosives, loosely hung temporary electrical wiring and fittings, compounded with proximity of these to the source of fire like welding operations could invite fire hazards. Fire fighting precautions and first aid fire fighting measures is needed to be incorporated in finally completed building or any other civil engineering facilities to ensure safety of the occupants and users.

The planning, design and construction of any building should be such as to ensure safety from fire. For multi-storeyed, high rise and special buildings additional fire precautions are to be taken.

Every building meant for human occupancy shall be provided with exits sufficient to permit safe escape of its occupants in case of fire or other emergency. Some buildings are to be equipped with fire alarm system.

The following fire safety arrangements are to be made:

- Fire extinguishers and fire buckets, painted red should be provided at all fire hazardous locations.
- Extinguishers should be inspected, serviced and maintained in accordance with manufacturers' instructions.
- Excavation facilities and fire exits should be provided at all locations featuring fire hazards.
- Sirens or other suitable fire alarm arrangements should be provided at all hazardous locations.
- Employees should be conversant with the use of different types of fire extinguishing apparatus.

Safety aganist natural calamities

Weather is an important factor in construction work. Apart from this, temperature, wind, rain, hurricanes, tornadoes, and earthquakes, are some of the natural calamities in construction industry.

Duties of Safety Engineer

Deathrage suggests the following duties of the safety engineer:

1. Takes charge of, directs and supervises all job-safety measures and their implementation.
2. Instructs supervisors and workmen on safe practices.
3. Secures, distributes and hands out safety posters and materials to all concerned.
4. Holds weekly safety meetings for supervisors and key personnel.
5. Heads safety committee composed of supervisors and key personnel.
6. Creates and maintains first-aid stations and supplies.
7. Provides for medical, ambulance and hospital services.
8. Investigates all accidents and their causes and arranges to rectify them.
9. Checks all equipment for mechanical defects and safe practices.
10. Takes charge of storage of inflammable liquids and blasting supplies.
11. Assists in estimating costs of safety programme for future cost estimating.
12. Maintains all safety records, files, statistical reports and other data as is necessary.

First Aid

First Aid is an immediate and temporary care given to the victim of an accident or sudden illness until the services of a qualified medical doctor can be obtained. It is important for the first aiders to know not only what to do but also what not to do. Improper and careless moving of the victim may increase the severity of the injury/illness and may even cause death. Great haste in giving first aid is usually unnecessary and sometimes harmful. However, in construction work there are two cases where great speed is necessary. They are (a) cases of severe bleeding, and (b) cases where breathing is suspending, requiring artificial respiration.

A fully equipped first aid box should be available as near the site as possible with adequate supply of various types of bandages. An artificial resuscitator will be of great help while dealing with electric shock cases where the victim stops breathing immediately. In these cases, with the use of resuscitator, breathing can be revived in almost all cases. Contents of the first aid kit should be replenished as and when required, as listed out below:

1. Sterilised dressings
2. Sterilised cotton wool
3. Centrimide solution or suitable antiseptic solution
4. Bottle of Mercurochrome solution in water
5. Bottle containing sal-volatile
6. Scissors
7. Adhesive plaster
8. Sterilised eye-pads
9. Aspirin or any other analgesic tablets
10. Polythene wash bottle for washing ice
11. Roller bandages
12. Triangular bandages
13. Tourniquet
14. Splints
15. Safety pins
16. Snake-bite lancet
17. Potassium permanganate crystals
18. Ointment for burns.

Safety Control and Regulations

Safety control and regulations aims at;

1. Elimination of unsafe conditions by means of:
   - (i) Safeguarding all machines, equipment and workspace.
   - (ii) Rectifying or prevention defective conditions.
   - (iii) Suitable and safe design and construction.
   - (iv) Safe processes and methods of work.

(v) Adequate and suitable illumination.
(vi) Adequate and suitable ventilation.
(vii) Safe dress and personal protective equipment.

2. Discovering causes by means of:
   (i) Job safety analysis.
   (ii) Investigation of accidents.
   (iii) Inspection of plant and equipment.
   (iv) Recording and tabulation of data.
   (v) Analysis of data.
3. Eliminating of unsafe actions by means of:
   (i) Discipline.
   (ii) Safety education.
   (iii) Supervision.
   (iv) Personal adjustment.

The following precautions are needed to carry out safe construction activities at the work place.

1. Safety in developing, planning and arrangement of work place.
2. Safety of the plant which includes machinery, equipment, appliances and other technical devices used or positioned in the work place.
3. Planning safety work methods and safety systems.
4. Maintaining and monitoring a safe and healthy working environment by using proper industrial hygiene.
5. Adequate safety precautions and better house-keeping to prevent injuries from falls, slips and other objects.
6. Suitable arrangement for rendering prompt and efficient first-aid to injured persons should be maintained under the guidance of the medical officer in charge of the project.
7. Restricting and minimising the use of substances liable to cause ill health, and ensuring that conditions for maintaining safety and health exists at all times.
8. Adequate precautions against fires, explosions and electric shocks.
9. Providing adequate health, hygiene, welfare and first-aid facilities.
10. Providing and maintaining in proper condition all the personal protective equipment and clothing.
11. Ensuring proper selection, inspection and certification of materials to be used for temporary works.
12. Providing information, education and training to workers during placement or as early as is reasonably practicable to ensure their safety and health.
13. Providing information and protection of workers against toxic or dangerous substances and articles in use, manufacture, handling, storage and transportation.

14. Ensuring that plant and machinery used at work sites are designed, constructed and installed in such a way that they are safe from risks to health when properly used.

Accidents occur mostly due to human error and observing safety is the function of all those who are associated with the project. Safety culture is required to be developed at the job site and the job site manager has to play a very important role in developing safety culture. He should treat productivity and safety as two related parts of high job performance. He should demonstrate by his actions the importance he attaches to safety culture. Further, he should reach and watch every worker's action at the project sites through his supervisory staff, including sub-contractors and convey the message of safety. He should have 'safety eyes' and should be quick to spot safety hazards during the walk around the job.

Most accidents are the results of carelessness, lack of thought, ignorance or failure on the part of the worker to sense danger and hazard. The worker should realise that one careless or reckless act on his part can cause an accident. Hence, safety culture has to be injected in them. Supervisors, foreman and engineers who are in day-to-day contact with the workers are in a best position to do it. It is in fact their major responsibility.

Construction environment is filled with potential hazards, no two objects are alike and each project is constantly changing. All workers with experience should be treated as new workers when employed, and should be given necessary guidance, orientation and advice about possible hazardous conditions.

Safety Programme

The safety programme of a construction and civil engineering organisation is a compact package that comprises a safety policy, safety department to implement the policy, and trained personnel to man it. Workers should be closely involved in the safety programme through tripartite safety committees and other participating measures. Safety record, monitoring and training as well as periodic safety audit should be essential aspects of such a programme. The same are discussed briefly in the following paragraphs.

***Safety Policy*:** A client or a contractor who is serious about establishing high standards of safety at work sites as well as throughout the organisation should have a safety policy that forms an integral part of the overall policy of the organisation. The safety policy comprises of a statement of the organisation's objectives regarding safety of employees, in operations, and at work sites. The safety policy should deal with the subject more positively and state that the top management wants risk-free and zero accident work environment.

Safe and healthy working conditions do not happen by chance. Employees need to have a written safety policy for their enterprise setting out the safety and health standards. The policy should name the senior executive who will be responsible for seeing that the standards are achieved, and who has authority to allocate responsibilities to managers and supervisors at all levels and to see that they are carried out.

The safety policy and programme should deal with the following matters:

- Arrangements for training at all levels.
- Safe methods or systems of work for hazardous operations.
- Duties and responsibilities of supervisors and key workers.
- Arrangements by which information on safety and health is to be made known.
- Arrangements for setting up safety committees.
- The selection and control of sub-contractors.

***Safety Records*:** The statutory safety records required to be maintained at sites are the following:

1. Accident Register.
2. Dangerous Hazards Register.
3. First-Aid Register.
4. Register for the inspection notes by government officers.
5. Copies of all accident reports and hazard reports filed to the prescribed authority. Non-statutory records may be such as prescribed by the head office or the safety department.

Role of Entrepreneur: The entrepreneur can assist in the prevention of accidents and thereby improve overall contract performance through the following methods:

1. ***Effective communication*:** Communicating effectively with the workforce on accident prevention is often the key to a successful approach to safety improvement.
2. ***Record keeping*:** Keeping records of the types of accident that occur most frequently, and why they occur, put you in a better position to prevent them as you know what you are fighting against.
3. ***Motivation of the workforce*:** Besides general steps such as providing information on accidents and their causes and consequences, some special motivation measures can be introduced, such as organising a safety bonus for the employees with the best accident record.
4. ***Use of safety equipment*:** Make sure that safety equipment is available when and where it is needed, insist that it is always used, and take disciplinary action against employees who refuse or frequently forget to use the equipment.

Construction Team

The main features of a construction team are :

(i) A senior construction team member should be assigned overall responsibility for construction safety.

(ii) The role, authority and reporting lines of the safety advisers should be clearly defined, down to all sub-contractors.

(iii) The construction team should install appropriate systems to allow effective safety management.

(iv) These management systems should address:
- Safety training
- Ongoing monitoring and inspection of construction work
- Auditing and periodic inspection
- Construction organisation including lines of responsibility and communication
- Enforcement of safety rules and procedures
- Feedback of safety issues to all employees
- Safety co-ordination between the client, the client's representative, the design team, contractors, sub-contractors and suppliers
- Investigation and reporting of accidents and incidents
- Liaison with regulatory and statutory authorities.

(v) The senior members of the construction team should clearly demonstrate leadership, commitment and support for safety, leading by example.

(vi) The construction team should demonstrate appropriate access to expertise and advice on safety issues; including hazard assessment, legislation and risk prevention.

(vii) The construction team should establish and maintain overall site safety policy including the provision and use of common facilities, plant equipment, emergency procedures and services.

(viii) The construction team should maintain good working relationships with appointed safety representatives and/or safety committees

Exhibit 9.3 Safety, Health and Environment Measures at Afcons Infrastructure Limited

- Afcons is an engineering and construction company based in Mumbai, Maharashtra.
- The company consistently delivers the landmark mega projects in the areas of marine, highways, bridges, rail and metro tunnels, and so on.

- Afcons is a Safety Health and Environment (SHE) conscious and ISO 45001 and ISO 14001 Standards Certified Company.
- The top management has fortified to communicate the steps of effective SHE management systems and strategies to achieve zero incident record of work environment in top-down, bottom-up approach.
- The company constantly seeks new and innovative ways to reduce work-related injuries and occupational health problems, conservation of natural resources, and safety for neighbouring people.
- The company has established Afcons safety council for evaluating overall SHE performance, review of SHE audits, and management review for further improvement and development.
- The project management is committed for excellence in safety, quality and execution as a culture in the organisation.
- The management conceptualised and developed the Total Satisfaction Model to maintain highest safety standards, quality delivery, and performance of integrated quality management systems.
- The top management understands the value of developing a high safety culture and maintaining an incident free work environment and thereby creating a good image in business.
- The motto of the company clearly depicts that OHS is everyone's responsibility for achieving continuous improvement and highest standards

Safety Programme Audit

One of the most important elements of a good safety and health programme in construction is the initial and ongoing programme to identify all potential hazards in the workplace. To ensure that the safety programme itself is carried out as designed; it is necessary to conducts periodic regular audits of the safety activities. This should be done by staff other than those responsible for the implementation of safety programme. It is important to remember that this audit is not a safety inspection, it is an audit of the programme.

(A) Safety for Work at Height

1. Whether ladders are of adequate strength and sound design. Yes No Being done
2. Whether scaffolds are self supported, adequately designed, and provided with hand rails. Yes No Being done
3. Whether elevated floors are clean, provided with hand rails, and cut-outs in floors are guarded/barricaded. Yes No Being done

4. Whether no loose materials are stacked on floor beams, and or flangers of structurals like crane girders, etc. Yes No Being done
5. Whether gas cylinders are maintained vertically and stored separately– full or empty. Yes No Being done
6. Whether there is adequate space for the safe movement of the cranes and trucks, dumpers, etc. Yes No Being done
7. Whether aisles and walkways are clear. Yes No Being done
8. Whether workmen are using protective equipment like safety belts, helmets, safety shoes, etc. Yes No Being done

(B) Safety Audit Queries for Electrical Activities

1. Whether all electrical tools, distribution boards, are earthed. Yes No Being done
2. Whether all electrical tools are connected through earth leakage circuit breakers (ELCB). Yes No Being done
3. Whether earthing test of tools are made at 500 volts. Yes No Being done
4. Whether earthing circuit is tested for continuity. Yes No Being done
5. Whether 42 volts/48 volts supplies are provided for electrical tools through step-down transformer. Yes No Being done
6. Whether earthing point for the plug is correctly inserted to socket. Yes No Being done
7. Whether power transmission cables are taken over minimum height of seven feet. Yes No Being done

(C) Safety Audit Queries for Blasting Operation

1. Whether explosives and detonators are transported by road in separate vehicles. Yes No Being done
2. Whether explosives and detonators are stored separately in different magazines. Yes No Being done
3. Whether drilling holes for the explosives/ detonators have been checked as per sequence of firing. Yes No Being done

4. Whether authorised and licenced foremen are employed in handling explosives/detonators. Yes No Being done
5. Whether blasting operation is being done during lean period like lunch time. Yes No Being done
6. Whether proper siren-signal is used before and after the blast. Yes No Being done
7. Whether unblasted/fired detonators are checked and cleared after the blast. Yes No Being done

(D) Safety Audit Queries for Working Conditions

1. Whether floors, passages and stairs are solid, clean and free from obstacles. Yes No Being done
2. Whether floor openings are covered or guarded. Yes No Being done
3. Whether steps, corners, and obstacles are marked clearly. Yes No Being done
4. Whether fire escapes are unlocked and unobstructed. Yes No Being done
5. Whether dangerous machines and equipments are fenced properly. Yes No Being done
6. Whether protective clothing and equipment are provided for users. Yes No Being done
7. Whether safe work practices are adhered to. Yes No Being done
8. Whether equipment and machinery are checked regularly. Yes No Being done
9. Whether they are repaired or replaced, as required. Yes No Being done
10. Whether a trained first-aider is on the premises. Yes No Being done
11. Whether accident and emergency procedures are tested regularly. Yes No Being done
12. Whether a fully stocked, first-aid box is kept and maintained properly. Yes No Being done

SUMMARY

- Construction sector plays an important role in the country's economic development.
- The construction industry is characterised by the predominance of migratory and unskilled workers.
- The industry has registered enormous growth world-wide during the last few decades.
- Construction is a high accident prone industry mainly involving both buildings and civil work projects, and a high risk for clients, contractors and workers alike.
- Despite rapid and latest technology, construction work is still labour intensive, and many workers are exposed to the physical hazards everyday causing injuries.
- Constructing safe structures and providing safe work environment to the personnel is a vital factor in successful construction business.
- There are many factors which make safety an important function in the management of construction projects.
- The concern for safety starts from the design stage and continues till the facility is delivered to the owner.
- Health and Safety in construction is a very complex phenomenon.
- Technological, organisational and behavioural aspects play a key role in any construction safety and accident prevention programme.
- It is necessary to institute a comprehensive programme for the implementation of health measures, and safety rules and regulations during the construction phase, and also thereafter.
- Safety rules have been framed for construction workers in the USA, Australia, Great Britain, and some other countries.
- The designer, consultant, client and contractor should share equal responsibility for safety and health of workers; a culture to be built into their work values and systems
- The three main hazards faced by construction workers are: (a) physical hazard, (b) chemical hazard, and (c) fire and explosion hazard.

DISCUSSION QUESTIONS

1. What are the construction hazards?
2. How do we prevent accidents in construction industry?

10

Health and Safety at Workplaces

Perhaps the one area that has received the greatest safety attention is the workplace. Work environment and the job is responsible for majority of industrial accidents and injuries. Legal requirements do not in themselves optimise safety, however, at best they create a climate for the study and enhancement of means to attain the desired objective. In other words, merely knowing the law and its performance requirements will not necessarily optimise safety. It is necessary to assure that the spirit and the letter of the law are fulfilled for that to take place.

This chapter covers health and safety at factories, mines, ports and docks, plantations, offices, and buildings.

Factories

Types of factory premises vary enormously. At one end of the scale are giant factories employing huge workforce with stringent safety and health regulations. At the other extreme are small undertakings housed in slum buildings or improvised huts. The contrast is seen most vividly in developing countries.

Special precautions are required in the design of factory buildings in which there is a high fire risk such as in oil refineries. Many countries have most stringent regulations regarding the storage, handling and processing of dangerous substances. The most important safety and health factors are the provision of safe storage of such materials, their distribution to the processes, the absence of open lights, flames or any source of ignition.

The Factories Act, 1948 provides for health and safety measures for the workers employed in factories, and their statutory compliance by the occupier of the factory.

Health

Every factory must provide appropriate health measures in accordance with the provisions of the Act:

- To keep clean and free from effluvia;
- To dispose of wastes and effluents;
- To maintain adequate ventilation and reasonable temperature;
- To prevent inhalation of dust and fumes and their accumulation in any workroom;
- To ensure proper standards of humidification where humidity in the air is artificially increased;
- To avoid overcrowding;
- To provide sufficient and suitable natural or artificial lighting;
- To provide for sufficient supply of wholesome drinking water;
- To provide separately for male and female workers sufficient latrine and urinal facility of prescribed types;
- To provide for a sufficient number of spittoons and maintain them in a clean and hygenic condition.

Safety

Every factory shall, in accordance with the provisions of the Act, take appropriate safety measures:

- To fence dangerous parts of the machines; for example, prime movers, fly wheel, electric generator, rotary converters;
- To prohibit employment of women and children near cotton openers;
- To protect workers from repairing machinery in motion;
- To maintain all chains, hoists and lifts in good mechanical construction, of sound material, and adequate strength, and free from patent defect and certified as to its safe working load;
- To keep all floors, steps, stairs, passages and gangways in good condition;
- To prohibit any person from carrying or moving any load so heavy as to be likely to cause him injury;
- To protect workers from injury to eyes from particles or fragments thrown off in the course of the manufacturing process;
- To protect workers from dangerous fumes, inflammable dust, gas, such other materials; and
- To protect workers from fire and provide for precautionary measures; for example, safe means of escape for all persons in the event of a fire; fire resistant construction of buildings; adequate fire extinguishing equipment; and alert, efficient and well-trained fire fighting squads.

If it appears to the Chief Inspector of Factories that any building or a part of a building or any part of the ways, machinery or plant in a factory is in such a condition or a state of disrepair as to be dangerous to human life or safety or is detrimental to the health and welfare of the workers, he may serve the occupier or manager or on both an order in writing, specifying the measures which, in his opinion, should be taken and requiring the same to be carried out before a specified date (Sections 40 and 40-A).

In every factory wherein 1,000 or more workers are ordinarily employed; or wherein, the opinion of the State Government, any manufacturing process or operation involves risk of bodily injury, poisoning or disease, or any other hazard to the health of the persons employed in the factory, the occupier shall, if so required by the State Government by a notification in the Official Gazette, employ such number of safety officers as may be specified in that notification. The duties, qualifications and conditions of service of safety officers shall be such as may be prescribed by the State Government (Section 40-B).

The Factories (Amendment) Act, 1987 added certain new provisions on health and safety. The amended Act:

- Provides that every occupier shall ensure so far as is practicable, the health and safety at work of all workers while they are at work in the factory.
- Stipulates that every person who designs, manufactures, imports or supplies any article for use in any factory shall ensure, so far as in reasonably practicable, that the article is so designed and constructed as to be safe and without risks to the health of the workers when properly used.
- Contains certain provisions regarding grant of permission of the initial location of a factory involving a hazardous process or for the expansion of any such factory. The State Government may appoint a Site Appraisal Committee for the purpose.
- Lays down that the occupier of every factory involving a hazardous process shall disclose to the chief inspector and also to the local authority in the manner prescribed all information regarding dangers, including health hazards and the measures to overcome such hazards arising from the exposures to or handling of the materials or substances.
- Specifies that every occupier of a factory involving any hazardous process shall (i) maintain accurate and up-to-date health records or, as the case may be, medical records of the workers who are exposed to any chemical, toxic or any other harmful substances.
- Authorises the Central Government to appoint an enquiry committee to inquire into the standards of health and safety observed in the factory.

- Provides that the Central Government may lay down emergency standards of safety in hazardous processes and also measures for their enforcement.
- Lays down that the maximum permissible threshold limits of exposures of chemical and toxic substances in manufacturing processes in any factory shall be of the value indicated in the second schedule.
- Stipulates that the occupier shall in every factory where a hazardous process takes place, or where hazardous substances are used or handled set up a safety committee consisting of equal number of representatives of workers and management to promote co-operation between the workers and the management in maintaining proper health and safety at work.
- Gives right to the workers employed in any hazardous process to warn about imminent danger to their lives or health due to any accident.

The Factories Act basically places complete responsibility for ensuring the safety, health and welfare of all workers on the management of the factory. It is the top management's responsibility to have an adequate and appropriate organisation to implement safety. This responsibility which is very wide ranging is specified in the General Duty Clause, under Section 7 of the Factories Act. Some of the specific duties included are:

- Provisions and maintenance of safe systems of work.
- Safe use, handling, storage and transportation of hazardous chemicals.
- Provisions of information, instruction, training and supervision of safety and health.
- Provision, maintenance and monitoring of safe working environment.

A statement by the top management, reflecting the company's policy regarding safety and health of workers, as well as the organisation and arrangements for its implementation is to be issued and publicised among the workers.

As per the Act, all workers have the right to:

- Obtain information from the management regarding workers' health and safety at work.
- Get trained regarding health and safety at work, either in the factory or at an outside training centre.
- Make a representation to the Factory Inspector, either directly or through his union, in case of any inadequacies in safety and health protection in the factory.

Workers employed in factories where hazardous processes are involved have following additional rights:

- Receiving information regarding the hazards and precautionary and control measures while handling hazardous chemicals.
- Receiving reports of health check-ups. The management has to maintain up-to-date medical reports, and make these available to workers.
- Worker's participation in joint safety committees consisting of equal number of management and worker's representatives, who should be elected by workers.
- Right to warn regarding imminent danger to life and health. In such case, workers can approach the management, either directly, or through the factory inspector, or through the safety committee.

The following are the special provisions to be implemented in all hazardous process factories:

- Special approval by site appraisal committee.
- Compulsory disclosure of information by management regarding hazards of chemicals to workers, local public and factory inspector.
- Preparation of emergency plan.
- Periodic health examination of workers.
- Work environment standards i.e. maximum permissible concentrations of chemicals in the work environment.
- A complete list of the Threshold Limit Values (TLVs) and Short Term Exposure Limit (STEL) for all hazardous chemicals given in Schedule II of the Act.

Abstracts and Notices

Some dangerous occurrences do not result in any worker sustaining a physical injury, usually because no worker happens to be in the vicinity at the time. These occurrences are, however, of a very dangerous nature and may recur in different circumstances, resulting in loss of life or serious injury to workers. In many countries certain dangerous occurrences must, by law, be notified to the competent authority in the same way as occupational accidents. Examples of dangerous occurrences are:

(a) collapse of crane, hoist, or other appliances used for raising or lowering persons or goods;

(b) explosions or serious fire resulting in complete stoppage of ordinary work;

(c) electrical short-circuit or failure of electrical machinery or plant attended by fire or explosion or causing structural damage thereto;

(d) explosion of a pressure vessel used for storage at a pressure greater than atmospheric pressure of any gas or gases (including air) or any solid or liquid resulting from the compression of gas;

(e) structural collapse of a building.

A notice containing abstracts of the Act and of the rules made thereunder, in English and in a language understood by the majority of the workers, shall be displayed in every factory at some conspicuous or convenient place at or near its main entrance. A notice containing the name and address of the inspector and the certifying surgeon shall also be displayed in every factory in the same manner. The manager of a factory is required to send to the appropriate authorities:

(a) A notice of certain accidents resulting in death or bodily injury which prevent the injured person from working for a period of 48 hours or more.

(b) In case of any dangerous occurrence in a factory, whether causing bodily injury or disability or not, the manager of the factory is required to send notice to the authorities concerned.

(c) If a worker in a factory contracts any disease specified in the Third Schedule, the manager is required to send notice to the appropriate authorities.

In the case of an accident, if the injured person dies immediately or subsequently, information of his death whenever known shall be sent by the manager by telephone, special messenger or telegram within 24 hours of the occurrence to: (i) the chief inspector of factories, (ii) the district magistrate, and (iii) the officer-in-charge of the nearest police station.

In case, the disease is covered under the Employees' State Insurance Scheme, information of the death should be sent to: (i) the local ESIC office, (b) the insurance medical officer with whom the deceased was insured.

Obligations of Workers

Various obligations of a worker are:

(i) A worker shall not interfere with or misuse any appliance, convenience or other thing provided for the purpose of securing the health, safety or welfare of the workers.

(ii) A worker shall not wilfully and without reasonable cause do anything likely to endanger himself or others.

(iii) A worker shall not wilfully neglect to make use of any appliance or other thing provided for the purpose of securing the health or safety of the workers.

If any worker violates the above provisions, he shall be punishable with imprisonment upto 3 months, or with fine upto Rs. 100 or with both.

Obligations of Employers

Various obligations of an employer are:

(i) Obtain approval of the government regarding the location, plan and construction of the factory, and also licence and registration certificate for operating the factory.

(ii) Implement all the provisions concerning health, safety, and welfare.

(iii) Send a written notice to the chief inspector at least 15 days before occupying or using any premises as a factory, containing the name and address of the factory and occupier and manager, nature of manufacturing process, number of persons to be employed, and nature and quantity of power to be used.

(iv) Comply with all statutory requirements pertaining to hours of work, leave with wages, weekly holidays, and extra wages for overtime.

(v) Display notices, maintain registers and records, and submit returns as required under the Act.

(vi) Report fatal and other accidents, and occupational disease contracted by any workman, to the government or its specified authority in such form or manner as may be prescribed.

Obligations of Enforcement Authorities

Various responsibilities of an enforcement authority are:

(i) Enter premises at any reasonable time.

(ii) Examine and investigate health and safety measures.

(iii) Inspect relevant documents.

(iv) Take samples, measurements, photographs.

(v) Dismantle or test any dangerous article or substance.

(vi) Take possession of articles or substances.

(vii) Require people to give information and affix signature.

(viii) Issue notice for omissions and commissions.

Mines

Developing civilisation throughout the world, requires the development of more and more advanced production technologies so as to fulfill the expanding needs of mankind. These needs, coupled with the resultant growth and complexity of industrialisation have greatly increased demand for minerals and their products. The search for and the extraction of minerals are conducted not only worldwide but also deeper and deeper below the surface of the earth.

Accident risk in mining industry in comparison with other sectors of the economy has led to work in this industry being considered as one of the most dangerous of occupations. Mining differs from virtually all other industrial, agricultural and service activities in that the environment changes continually as the work progresses. As each tonne of mineral is extracted, different conditions of working surfaces, sides and surroundings are created. When the work is underground, it is performed in an alien environment in which light and ventilation must be provided and in which constant attention must be given to the overhead and surrounding ground. The continually changing environmental conditions introduce new and different work hazards and constitute a safety situation radically different from that found in a factory where the workplace surroundings remain constant for long periods at a time and can be gradually modified and improved using well known principles and practices.

Among the principle health hazards in a mine are those related to the poor quality of the mine air to the presence of dust, poisonous gases or other airborne impurities. One of the most common impurities in mine air is dust. Rock dust is released into the air from operations such as drilling, blasting, loading, haulage, shovelling, and so on. Blasting gives rise to heaps of broken rocks which contain a considerable amount of fine dust. Carbon monoxide is the most common dangerous gas encountered. The various environmental hazards are darkness in underground worksites, heat, dampness, radiation, gases, atmospheric pressure, and the like. In contrast to underground mining there are no special occupational hazards in opencast mines.

The general safety of the workers in a mine depends directly on the mining technology and policies adopted to cope with the geological and physical conditions of the mine workings. However, the individual safety of a mine worker is also dependent to a large extent on the effects of his own work activities and also on the action of his fellow workers, including some working in other parts of the mine. Accident prevention in mining requires an extensive knowledge of safe principles and practices which have been established through experience. This depends not only on sound mining technology to provide safe working conditions, but also upon enlightened experience, thoughtfulness and responsibility on the part of each worker in the mine to work safely.

The prevalence and severity of the hazards encountered in this industry vary with the kind of mineral and the type of mine. The following are some of the common hazards in mines:

1. ***Materials handling***: Material handling accidents are those which occur when the worker is moving, lifting, carrying, loading or storing materials, supplies, ore or waste, and are initiated primarily through the use of unsafe work practices or faulty judgment. Intensified safety training and education of workers in the proper techniques for lifting, loading or carrying are the most effective means of avoiding injury during these operations. Although these injuries are not

generally as severe as those from most other causes, a good many of them do result in permanent disabilities.

2. ***Slips and falls of persons***: Accidents in this category usually originally originate from uneven, cluttered, or slippery working surface, and from poorly guarded or constructed elevated walkways or working surfaces. Safety education in working procedures in high places, together with inspection practices and the safeguarding of elevated workplaces would undoubtedly prevent many of these accidents.
3. ***Machinery***: These accidents occur while operating, moving, traming or working around machinery. This hazard is both prevalent and serious and with increasing mechanisation of the mineral industries and the introduction of new or larger and more powerful machines, it requires constant assessment and reassessment. Safe work procedures around machinery must be maintained, and the workers must be carefully trained in these procedures. Technical inspection and maintenance must assure that any exposed moving parts are completely guarded and that no parts of the machine are defective. Increasing mechanisation undoubtedly has a strong influence on safety conditions and practices.
4. ***Haulage and transport***: There are accidents involving the haulage and transport of ore, waste, materials, supplies, and men in and around the mines and minining property. Haulage accidents occur frequently and may often result in fatal injury or permanent disability. They occur in a wide range of work activities with all types of transport equipment such as belt and chain conveyors, rail equipment, winding equipment, trucks and others. Education of the worker in safe haulage practices and through technical inspection and maintenance of equipment are required for control of this hazard.
5. ***Slipping of ground***: This is the most serious hazard in underground mining and one of the most prevalent; the same hazard is present in opencast mines and quarries although to a smaller degree. In underground workings it causes by far the largest proportion of fatal injuries and the danger increases as the workings desend deeper and deeper below the earth's surface. Safety education, technical inspection, and sound engineering practices of roof support are certain important measures for reducing accidents in this category.
6. ***Gas and dust explosions***: The hazard of inflammable and explosive dusts is ever present in underground coal mining but may occasionally be encountered in the mining of other mineral deposits, principally the sedimentary deposits. Dust in mines is becoming an increasingly difficult problem owing to expanding mechanisation. Control is primarily through use of water sprays, improved ventilation systems, and better clean-up or housekeeping practices.

7. ***Mines fires*:** Fires in underground workings constitute a disaster-type hazard and may be one or two types–those in which the timber supports or combustible mineral itself (e.g. coal) is burning; and those in which mining equipment or material is burning. Owing to the ventilation currents, mine fires are difficult to extinguish, and the generated noxious gases spread throughout the mine workings.
8. ***Inrushes*:** Inrushes of water or of unconsolidated material into mine workings may also produce catastrophic accidents. They may be caused by mining into abandoned water-filled workings or into bodies of surface waters, or be due to surface floods entering the mine workings. Accurate, up-to-date mine mapping and knowledge of surrounding abandoned workings are the primary controls of this hazard.
9. ***Explosives*:** By their very nature explosives are extremely hazardous substances and, when brought into the mine environment of unstable ground, dangerous atmospheres and limited and precarious means of escape, their hazardous nature is intensified. Each mine should consequently draw up a complete schedule of safe working practices for explosives based on national regulations, and covering such matters as the selection of suitable explosives and ignition appliances; the conveyance of explosives to the magazine; magazine location and design, and so on.
10. ***Electricity*:** The use of electricity and electrical equipment in an environment which may be damp or contain an explosive atmosphere greatly enhances the hazards of this form of energy. In some cases it is necessary to design electrical equipment and installations to meet special safety requirements.
11. ***Contaminated atmospheres*:** The quality of the air in mines is dependent on the efficiency of the ventilation employed. The gases that may contaminate the mine atmosphere and cause intoxication or asphyxiation include carbon monoxide, hydrogen sulphide, and so on. Efficient ventilation of these gases must be supplemented by regular determination of the atmosphere and the establishment of emergency procedures for the evacuation of areas invaded by dangerous gas mixtures.
12. ***Miscellaneous hazards*:** The remaining hazards generally do not result in the more serious injuries, but they do cause a relatively sizable number of the less serious injuries. They involve accidents from the improper use of hand tools or the use of defective hand tools by the injured worker or a co-worker, from falling objects or materials other than falls of ground, from stepping or kneeling on sharp or loose objects, from acetylene and electric welding and cutting activities, from burns by acid or caustic materials, from flying particles, and so on.

First-aid and Rescue

Every mine should make adequate contingency provisions for the safe conduct of rescue and repair work in mines following an explosion, fire, fall of ground or other accident; where the atmosphere of the mine is subject to contamination, provision should be made for rescue operations in an atmosphere dangerous to life. Rescue equipment, including respirators should be provided in readily accessible places and kept in a good state of maintenance, and a certain number of mine workers chosen for their coolness, powers of endurance and general suitability for the work should be selected for comprehensive rescue training and for inclusion in a permanent rescue corps.

Each mine should be equipped with adequate first-aid facilities in accordance with the number of persons employed. Where the number of mine workers is sufficiently high or where national legislation so requires, the mine should have a first-aid room with a qualified first-aider.

Since mining has many inherent hazards, safety precautions have been laid down in detail in the Mines Act, 1952 and the rules and regulations framed thereunder to guard against dangers in mines. It is the responsibility of the mine management to comply with all mandatory provisions. The statutory provisions under the Mines Act are as follows:

Health and Safety

- **(i)** ***Drinking Water*:** The owner or agent of a mine shall make effective arrangements to provide and maintain, at suitable points conveniently situated, a sufficient supply of cool and wholesome drinking water for all persons employed therein. In the case of persons employed below ground, any other effective arrangement shall be made for supply of drinking water. All such points shall be legibly marked 'Drinking Water' in a language understood by a majority of all the persons employed in the mine.
- **(ii)** ***Sanitation*:** In every mine, a sufficient number of latrines and urinals of prescribed types, separately for males and females, shall be so situated as to be convenient and accessible to persons employed in the mine at all times and shall be adequately lighted, ventilated and maintained in a clean and hygienic condition.
- **(iii)** ***Medical Appliances*:** In every mine, prescribed number of first-aid boxes or cupboards equipped with prescribed contents shall be provided and maintained so as to be readily accessible during all working hours. Every first-aid box or cupboard shall be kept in the charge of a responsible person who is trained in first-aid treatment and who shall always be readily available during the working hours of the mine. In every mine suitable arrangements shall be made for the conveyance to hospitals or dispensaries of injured or ill

persons. In every mine wherein more than 150 persons are employed they shall be provided and maintained a first-aid room of the prescribed size having the prescribed equipment and staff.

(iv) ***Safety*****:** If in respect of the operations of a mine it appears to the Chief Inspector or any Inspector that any matter, thing or practice is dangerous to human life or safety or is defective so as to threaten bodily injury to any person employed therein, he may give written notice to the owner, agent or manager of the mine asking him to remedy the situation within the specified time and in the specified manner. If the terms of the notice are not complied with, he may prohibit the employment in or about that mine of any person after the expiry of the specified period.

Similarly, in case of urgent and immediate danger to the life or safety of any person employed in any mine or part thereof, e.g., when there is flooding of a mine or emission of poisonous gas, the Chief Inspector or any authorised Inspector may prohibit, through a written order, the employment of any person therein until that danger is removed.

Every person whose employment is prohibited in or about a mine on grounds of unsafe working conditions, as discussed above, shall be entitled to payment of full wages for the period his employment is prohibited, unless such person is provided with alternative employment at the same wages (Section 22).

Notice of Accidents

The owner, agent or manager of the mine shall give notice of:

(a) an accident causing loss of life or serious bodily injury, or
(b) an explosion, ignition, spontaneous heating, outbreak of fire or irruption or inrush of water or other liquid matter, or
(c) an influx of inflammable or noxious gases, or
(d) a breakage of ropes, chains or other gear by which persons or materials are lowered or raised in a shaft or an incline, or
(e) an over winding of cages or other means of conveyance in any shaft while persons or materials are being lowered or raised, or
(f) a premature collapse of any part, or
(g) any other accident which may be prescribed.

To the prescribed authority in the prescribed manner, and he shall simultaneously post one copy of the notice on a special notice board in the prescribed manner at a place where it may be inspected by trade union officials, and shall ensure that the notice is kept on the board for not less than fourteen days from the date of such posting.

The owner, agent or manager of the mine shall maintain a register of accidents in the prescribed form and copies of the entries therein shall be furnished to the Chief Inspector quarterly (Section 23).

Notice of Certain Diseases

Where any person employed in a mine contacts any notified disease connected with mining operations, the owner, agent or manager of the mine, as the case may be, shall send notice thereof to the Chief Inspector and to such other authorities, in such form and within such time as may be prescribed.

Besides, if any medical practitioner attends on any such person who is suffering from any notified disease, he shall without delay send a report in writing to the Chief Inspector stating:

(a) the name and address of the patient,

(b) the disease from which the patient is or is believed to be suffering, and

(c) the name and address of the mine in which the patient is or was last employed.

Where the medical practitioner's report is confirmed by a certifying surgeon, the Chief Inspector shall pay to the medical practitioner such fee as may be prescribed, which shall be recoverable from the owner, agent or manager of the mine.

Mines Inspection

The formulation of a legally enforceable mine safety code should be accompanied by the establishment of some form of inspection service, with varying degrees of authority to secure through inspection visits that the required standards are being observed. To attain maximum effect, mine safety legislation should be drafted and administered with the guidance of personnel having a high level of technical competence and experience, and a detailed knowledge of the mining industry as well as of the more general administrative aspects of inspection services. The objectives of safety legislation can be achieved only when the code is supported and enforced by a body of specially qualified inspectors who can visit the mines regularly and fulfil three essential tasks:

(a) determine that the legal safety standards of the code are being carried out and, if not, take appropriate action to ensure compliance.

(b) assist both workers and management by supplying technical information and guidance to comply with the legal safety requirements and also to improve safety standards and work procedures generally; and

(c) notify the appropriate authority of any defects or technological changes whether may or may not be covered by the existing legal code.

International Activities

Various Conventions and Recommendations have been established by the International Labour Conference dealing with such matters as the employment of women underground (Convention No. 45, 1935), the minimum age (Recommendation No. 96, 1953), Convention No. 123, 1965, (Recommendation No. 124, 1965), conditions of employment (Recommendation No. 125, 1965), medical examination (Convention No. 124, 1965) of young persons underground, and hours of work in coal mines (Convention No. 31, 193, and Convention (Revised) No. 46, 1935). A tripartite technical conference in 1949 adopted the Model Code of Safety Regulations for Underground Work in Coal Mines for the guidance of governments and the coal mining Industry. Disastrous accidents in mines during the 1950s led the ILO to direct particular attention to the problems of mine safety and resulted in a series of publications on this subject including Accident Prevention in Mines other than Coal Mines (1954), Safety in Coal Mines (1955), and three codes of practice on Prevention of Accidents Due to Electricity Underground in Coal Mines (1959), Prevention of Accidents Due to Fires Underground in Coal Mines (1959), and Prevention of Accidents Due to Explosions Underground in Coal Mines (1974). The meetings of the ILO Tripartite Technical Committee on Mines Other than Coal Mines have put emphasis on the human factors in mine accidents, workers' co-operation in accident prevention and vocational training (1957), and on safety training for underground mine workers (1968).

Mine Safety Legislation

In drafting mine safety legislation, the most important principle is to provide a code of safety regulations which is both practicable and technically sound, and which will make an effective contribution to the objective of raising the safety of the workers to the highest attainable level. It is difficult to compile a safety code for mining because nearly every mine will have its own special conditions. Any mine safety legislation must be sufficiently flexible to cover as many special considerations as possible. Flexibility may be obtained by providing means and modifications as they become necessary through technological changes in mining methods, equipment and procedures. The principle of consultation with experienced and qualified representatives of employers, workers and other interested parties should not be overlooked when framing or amending safety legislation.

Owners of mines and quarries have a duty to make provisions to ensure that the mine or quarry is managed in accordance with the requirements of the Act and its Orders and Regulations. No mine shall be worked unless fully qualified persons are appointed. Either through the Act or by Regulations provisions have to be made at each mine for:

(a) the keeping of plans of the working of the mine;
(b) securing safe ingress and egress;
(c) securing shafts and entrances to disused workings;
(d) the construction and maintenance of roadways;
(e) safe operation of winding and rope haulage apparatus and conveyors;
(f) systematic support to roofs and sides;
(g) adequate ventilation;
(h) lighting and lamps;
(i) contraband, electricity and electrical apparatus, blasting materials and devices, fire precautions and rescue, dust precautions and external dangers to workings;
(j) duties of officials and workmen in case of danger;
(k) machinery and apparatus;
(l) surface buildings and structures;
(m) training and discipline;
(n) prohibition of heavy work by women and young persons;
(o) general welfare provisions.

Ports and Docks

Dock work nowadays involves a wide range of operations carried out on board ship, on the quarry side and in port warehouses and terminals, and frequently calls for the use of great variety of equipment. Dock workers are commonly employed in the loading and unloading of ships, the transfer of goods from a ship to a barge or to another means of transport. Ports may be natural or man-made, on the sea or on island waterways, large or small. Some docks deal with world trade, other may serve a particular factory or installation.

Accident injuries in ports and docks may be caused by environmental factors like bad weather, confined spaces, poor lighting, and so on. The hazards are numerous and they are intensified by frequent changes in the place of employment, by the diffusion of responsibility and, in many instances, by the need to work fast. The commonest accidents as a rule involves slips and falls, falling loads or objects or manual handling of mechanical equipment. In some cases, poor visibility or traffic congestion may lead to serious and costly accidents. The majority of the injuries are to the hands, face, lower limbs and feet. Inspite of introduction of sophisticated equipments, the top work remains a dangerous occupation as accidents occurring in the loading and unloading of ships and related occupations is particularly high.

Dock workers may be exposed to dangerous gases, vapours, fumes and dusts that escape from broken or inadequately packaged containers. Skin or eye contact with corrosive substances may result in chemical burns, and the handling of radioactive materials may entail a radiation hazard.

The promotion of health and safety at docks is rendered difficult by a number of special factors related to the nature of the work. In many cases there is no fixed place of employment and it is difficult to locate health and welfare amenities in a suitable place. Work is carried on during all seasons although, in some countries, dock labour is being decasualised, much of the labour force is still hired on a casual basis and is employed only intermittently; consequently, opportunities for training are often inadequate.

A number of accidents could be attributed to non-use of personal protective equipment. One of the reasons of reluctance for the use of personal protective equipment is discomfort to the wearer of the personal protective equipment. A little amount of discomfort is bound to be there with any personal protective equipment and the wearer should bear with it in the larger interest of his safety.

The type and degree of hazards during dock work vary according to the nature of the cargo and the working conditions. Therefore, the selection of suitable personal protective equipment depends on the type of hazard and the part of the body exposed.

Dock Workers (Safety, Health and Welfare) Scheme, 1961

A comprehensive Dock Workers (Safety, Health and Welfare) Scheme has been framed for all major ports and is administered by the Chief Adviser, Factories (Factory Advice Service and Labour Institutes). It is framed under the Dock Workers (Regulation of Employment) Act, 1948. Amenities provided in the port premises include provision of (a) urinals and latrines; (b) drinking water; (c) washing facilities; (d) bathing facilities; (e) canteens; (f) rest shelters; (g) call stands; (h) first-aid arrangements.

According to Regulation 91 of the Dock Workers (Safety, Health and Welfare) Regulations, 1990, a reportable accident is one which either causes loss of life to a worker or disables him from work for more than 48 hours. However, a notice is required to be sent in all cases when a worker is disabled from work for the rest of the day or shift.

Regulation 91(4) prescribe the causes of dangerous occurrence which are reportable whether death or disablement is caused or not.

Plantations

Convention No. 110 of the International Labour Organisation (ILO) defines plantation as "an agricultural undertaking regularly employing hired workers which is situated in the tropical region and which is mainly concerned with the cultivation or production for commercial purposes of coffee, tea, sugarcane, rubber, bananas, cocoa, coconuts, groundnuts, cotton, tobacco, fibres (sisal, jute and hemp), citrus, palm oil, cinchona, or pineapple. It does not include family or small scale holdings producing for local consumption and not regularly employing hired workers."

The term plantation is widely used to denote large scale agricultural units and the development of agricultural resources. The plantation employment entails considerable hard physical work, in such tasks as hand planting, axe wielding, tree climbing and trench digging. Efforts should be made to minimise arduous work by mechanisation, good work organisation, training in economical work methods, and so on. Transport should be provided to carry workers from their accommodation to the worksite. Manual lifting and carrying should be minimised. The maximum weight to be carried by one person should be limited to 50 kg.

Tools and implements should be of good quality and appropriate to the work for which they will be used. Implements should be so designed as not to cause injury or fatigue to the operator, or to the draught animals used with them. Ropes and chains are commonly used in plantation work for pulling down trees, clearing operations, skidding, loading and transportation. Workers using them should have a knowledge of their limitations and their maintenance requirements. Moreover, careless handling of electric installations and equipment can cause severe shocks, burns and quite frequently fatalities. The installation should be protected by adequate fuses.

One of the most important factors affecting the health of plantation workers is the increasing use of toxic substances in agricultural operations such as pesticides, herbicides, chemical fertilisers and various corrosive, explosive and inflammable substances. The education of workers in the specific hazards of these substances, and the way in which they can enter the body is essential for the prevention of poisoning.

Personal protective equipment should fulfil certain fundamental requirements in order to be readily accepted by the workers. In addition to accomplishing the proper safety functions, it should be comfortable and as attractive as possible. Provisions should be made at plantations to clean, sterilise and service at suitable intervals. Proper storage and regular inspections are other means of keeping personal protective equipment in order.

The proper use of personal protective equipment is vital for the safety of the workers. Therefore, workers should be thoroughly instructed in its purpose, use and limitations, and should be convinced of their operational benefits. Plantation buildings should be safe and healthy to work and live in. They should be properly designed, constructed and equipped so as to provide good working conditions and reduce, as far as possible risks of fire, collapse of structure, fall of objects and fall of persons. Particular care should be taken to ensure a high standard of cleanliness and good housekeeping. Plantation buildings should be made, whenever possible, of fire-resistant materials. Fire-resistant construction is essential for buildings stocking highly combustible materials of flammable substances.

One of the first steps that should be taken to improve safety and health organisation on a plantation is to appoint a safety officer and/or form a joint safety and health committee. Safety officers should have the duties of seeing that buildings and equipment are kept in a safe condition and that work is done in a safe manner. Joint safety and health committees are particularly valuable because they bring management and workers to participate directly in safety work. Persons applying for work on plantations should receive a pre-employment medical examination followed up by periodic examinations.

Plantations Labour Act, 1951

The Plantations Labour Act provides for the following health measures:

(i) ***Drinking Water*:** Every employer of a plantation shall make effective arrangements at convenient places for the supply of wholesome drinking water for all workers.

(ii) ***Latrines and Urinals*:** Sufficient number of latrines and urinals separately for male and female workers and conveniently accessible to them are also to be provided by the employer and maintained in a clean and sanitary condition.

(iii) ***Medical Facilities*:** The employers are required to provide and maintain the prescribed medical facilities for the plantation workers and their families. If in any plantation the requisite medical facilities have not been provided, the Chief Inspector may cause such facilities to be provided and maintained and recover the cost thereof from the defaulting employer.

(iv) ***Compensation for Accidents due to House Collapse*:** If death or injury caused to any worker or a member of his family as a result of the collapse of house provided under Section 15, and the collapse is not solely or directly attributable to the fault on the part of any occupant of the house or to natural calamity, the employer is liable to pay compensation. The amount of compensation is to be determined as provided in the Employees Compensation Act, 1923, as in force for the time being.

Offices

Many large organisations today, such as insurance, banks, governmental financial companies, and IT companies consist almost entirely of office workers. Accidental injuries are just as painful, severe and expensive when they occur to office workers as that of production workers. It is true that the risk of an occupation-related accident or injury to the office worker is lower than the risk to those employees involved in manufacturing or transportation. However, office risks often go unrecognised and unmanaged and some could eventually lead to serious injuries and property loss.

Most people would consider the average office to be a reasonably safe workplace, and do not appreciate the risks particularly that of fire. Poor standards of housekeeping is one of the principal causes of accidents in offices. Bad office planning and layout with regard to the furniture and fittings can result in numerous minor accidents. Abuse and misuse of electrical appliances and equipment can result in fires and the risk of electrocution. The commonest accident among office employees is the fall, in the office itself or on staircases and passageways. There may be bruises caused by falling objects or by doors suddenly opened. Cuts and minor wounds may be caused by paper, metal objects, and certain implements such as clipping and trimming devices.

People working on certain machines, such as punches and invoicing machines may be subject to repeated shocks or vibration. The eyes suffer from the continuous reading of documents, especially in poor light. A variety of problems including visual complaints, have appeared with the introduction of computer terminals and electronic data processing.

Office work is becoming so varied and so highly mechanised that old premises can no longer be used without substantial modifications. Office equipment should be designed in accordance with height and shape of seats, cables, and other office equipment. People who traditionally have to work in standing position should be allowed to sit down during rest intervals; and places of work should be designed accordingly.

It is exceedingly important that scrupulous cleanliness be observed if the staff are to remain in good health. Proper ventilation is very essential in offices, and in various countries there are legislative stipulations in this regard. Offices and other places where people work should be ventilated either naturally or artificially, or both.

Lack of exercise is a major problem for office employees. Exercises can be done in the office; they improve the circulation, keep limbs supple, and are good for the lungs.

There is a general impression that people who work in an office and not in a factory, do not have to bother about accidents and injuries. Also, there is a general misconception that offices are pretty safe. But the fact is that accidents can occur in offices too. However, understanding the dangers and removing the hazards can make an office a safe working place.

One reason why office safety programmes are not more widespread is that many people believe office injuries are minor. This complacency is one of the prime causes of office accidents. Employees and employers both give little thought to safety because office work is not perceived as hazardous. In order to create awareness the company should establish a formal safety programme for office employees and organise training sessions for them.

Given below are a few tips for your safety in the office:

- Do not store materials on top of cupboards.
- Do not store flammable or combustible materials under the staircase.
- Do not place anything that may obstruct your safe movement on stairs or in pathways.
- Locate telephone cables, electrical cords in such a way that you do not trip over them.
- Close filing cabinet drawers after use and have only one drawer open at any time.
- Use suitable knives and cutters for paper, board, string, etc.
- Use a pencil sharpner for sharpening pencils instead of a razor blade.
- If you have to climb, do not use office chairs or boxes. Use a suitable step ladder or step stool.
- Do not carry any load too heavy for you and make sure that you can see over any load you carry.
- Do not obstruct access to fire fighting equipments and emergency exits.
- If you smoke, use an ash tray. Do not use waste paper bins for discarding spent matches as well as cigarette butts or for knocking out a pipe.
- Do not smoke for atleast 30 minutes before finally leaving your office.
- Before you leave your workplace switch off all electrical appliances, office equipment, etc. and remove plugs from sockets, if not required.
- Do not store any combustible materials near/below electric mains. Keep safe access to electrical mains.
- Watch your steps while using stairs and use handrails, if available.
- Do not talk, read or write while walking or using stairs.

- Report all accidents to your superior and get first-aid immediately. Don't neglect small cuts or abrasions.
- Do not drop drawing pins or any other sharp material on the floor.
- Do not operate any electrical equipment if your hand is wet.
- Keep your floor dry.
- Eliminate fire hazards. Know what to do in case of fire. Know the evacuation plan.

Buildings

Building construction is engineering in action. It is performed by the skillful application of human effort to create designs and convert them into physical facilities. The traditional image of building construction is that of a series of manual activities. This is true to an extent even today. However, modern projects are progressively becoming large in volume and complex in design. The demands of economy, accuracy, quality and timely completion of the projects and use of newer building materials require mechanised operations, sophisticated plant and instrumentation and superior construction techniques. In our country, accidents are very common in building and construction industry. The Government of India enacted the Building and Other Construction Workers' Act, 1996 in order to ensure health and safety of construction workers. Under this Act, an employer shall be responsible for providing constant and adequate supervision of any building or other construction work in his establishment so as to ensure compliance with the provisions of the Act relating to safety and accident prevention. In this regard, the employer's obligation and responsibilities include:

1. the safety of building and health of workers employed therein;
2. the safe means of access to, and the safety of any working place, including the provision of suitable and sufficient scaffolding;
3. the precautions in connection with the demolition of the whole or any substantial part of a building;
4. the handling or use of explosive under the control of competent persons;
5. the erection, installation, use and maintenance of transporting equipment, such as locomotives, trucks, wagons and other vehicles;
6. the erection, installation, use and maintenance of hoists, lifting appliances and lifting gear;
7. the adequate and suitable lighting of every workplace and approach thereto;
8. the precautions to prevent inhalation of dust, fumes, gases or vapours during any grinding, cleaning, spraying or manipulation of any material;

9. the safeguarding of machinery including the fencing of every flywheel and every moving part of a prime mover and every part of transmission or other machinery;
10. the safe handling and use of plant, including tools and equipment operated by compressed air;
11. the precautions in case of fire;
12. the limits of weight to be lifted or moved by workers;
13. the safe transport of workers to or from any workplace by water and provision of means for rescue from drowning;
14. the prevention of danger to workers from live electric wires or apparatus including electrical machinery and tools and from overhead wires;
15. the keeping of safety nets, safety sheets and safety belts;
16. the precautions with regard to pile driving, concrete work, work with hot asphalt, tar or other similar things, insulation work, demolition operations, excavation, underground construction and handling materials; and
17. the provision and maintenance of medical facilities for building workers.

A brief outline on the central electricity authority (safety requirements for construction, operation and maintenance of electrical plants and electric lines) regulations, 2022:

1. Emergency management plan - availability and installation of required equipment.
2. General aspects of safety covering safety policies, safety committee(s), safety training and awareness, safety promotional activities, safety metrics and accident control measures, and hazard identification.
3. Compliance of the Boilers Act, 1923 covering inspection, certification, and operation and maintenance of boilers.
4. Certification with the environmental provisions or regulations.
5. Periodic health checkup of workers in hazardous industries.
6. Internal and external audits, including ISO 4500 safety audit.
7. Fire detection, protection and maintenance system with jockey pumps, fire hydrant and fire tenders.
8. Safety aspects of turbine generators including steam turbine protection, turbine vibrations, and lubricating oils.
9. Safety aspects of coal handling plant and associated sub-system.
10. Maintenance of electrical control and instrument systems with associated sub-system.

Transportation (Road)

In view of large potential accidents involving vehicles carrying hazardous chemicals, there is a lot of scope for improving the quality of vehicles used for transporting chemicals and also the quality of roads, highways and amenities for the crew members.

Because of lack of awareness, stiff competition and to cut the cost, the hauliers (transport contractors) pay little or no attention to the condition of their vehicles, and crew training and follow-up. Also, the crew members are very poorly paid, which prompts them to indulge in malpractices like carrying passengers in the product vehicles and pilferage.

Keeping in mind the above factors, it is absolutely essential for selection of hauliers after evaluating them on the following criteria:

1. Does the haulier have adequate repair facilities or arrangements for the same in the product vehicle routes?
2. Does the haulier maintain the vehicles in good condition?
3. Does the haulier replace worn out equipment or parts in time?
4. Are the crew members qualified and trained?
5. Do the crew members have adequate experience?
6. Are the crew members aware of the likely hazards and methods of handling emergencies involving their vehicles?
7. Does the haulier disqualify the crew members in case of any accident caused due to their negligence?
8. Does the haulier conduct the pre-employment and medical examination of the crew members and maintain records of the same?
9. Are the crew members adequately paid, including for overtime work and waiting hours?
10. Does the haulier pay *no accident bonus* to the crew members?
11. Are the vehicles and the crew insured? Does it include the third-party liability?
12. Are the crew members made aware of the dangers of indulging in alcohol and drugs?

The hauliers should select the drivers and cleaners who are literate and physically fit. The training programmes are to be conducted for all the product drivers once every six months (or more frequently, if needed) which should include defensive driving techniques; prescribed routes and prohibited routes and emergency procedures.

Tremcards in different languages are to be published for all products, by-products and even for the imported raw materials, which are required to be transported by road. Moreover, the instructions to drivers and cleaners in different languages are to be published.

The following points need to be checked each time a vehicle reports for loading: (i) valid driving licence; (ii) condition of the vehicle; (iii) statutory documents; (iv) instruments and fittings; and (v) availability of wheelchecks, tool box, emergency light, battery, torch, personal protective equipment like PVC gloves, safety goggles, and breathing apparatus, eye-wash bottle, and first-aid box.

The hauliers are also required to check their vehicles thoroughly and issue a certificate of fitness for each vehicle once a month.

In order to have all the chemical companies lay equal emphasis on all aspects of safety in road transportation, the meetings with other companies as also with ICMA (Indian Chemical Manufacturers' Association) should be conveyed. The idea was to have a common norm on haulier selection, driver training, standard of vehicle fitness and prevention of overloading. The meetings may also be held with the local police, fire brigades, hospitals, and medical practitioners to appraise them with the nature of the products, associated hazards, appropriate measures for handling leakage or spillage and fire-fighting, and also the antidotes in the line of treatment to be given in case of exposure.

Arrangements should be made for imparting special training in transportation of hazardous chemicals by road and it should be endorsed in the driving licence. Installation of long-range wireless communication instruments should be made obligatory on all vehicles transporting very hazardous chemicals.

The following basic amenities and facilities should be provided to the crew members at regular intervals on all national highways:

- Resting places with washing and dining facilities at reasonable rates.
- Parking facility
- Vehicle repair facility

The above facilities should be away from inhabited areas.

The driver's seat and the controls should be designed ergonomically to reduce fatigue. Driver fatigue is one of the causes of accidents.

Exhibit 10.1 Participative Safety Activities

In a factory establishment, the participative safety activities normally include:

1. Safety campaigns
2. Safety contests
3. Safety meetings
4. First-aid training
5. Plant fire brigades
6. Plant inspection
7. Accident investigation
8. Job safety analysis
9. Safety suggestion systems
10. Safety committees

STUDY ASSIGNMENT

Look around your normal workplace, and the room in which you are working now and identify as many potential hazards as you can. Find at least five:

1.

2.

3.

4.

5.

CASE

You were hired about eight weeks back as the first full-time safety manager for a concern manufacturing metal products and employing an average of 800 workers. While the injury records are not in very good shape, it appears that the company has been experiencing between 40 - 50 injuries per year causing loss of working days.

You have located by inspection numerous conditions or procedures in various departments that you believe are responsible to injuries. You have pointed out these to the concerned supervisors. Most of them listen to you politely, but you find on return visits that generally nothing has been done to correct the situation.

You have convinced the HR manager that the company has been experiencing more than its warranted share of injuries. He has noted some injuries resulting in handicap of the workers in some way, and two of the supervisors have suggested to him that they carefully screen-out and stop hiring workers handicapped in terms of vision and hearing, or amputees, or epileptics. He suggested you that this is one step he can undertake to reduce injuries.

→

Questions

1. What procedures do you think you might follow in an effort to get compliance with your suggestions for correction of unsafe physical conditions, or unsafe acts in the departments?
2. What do you think of the suggestions of the supervisors to promote safety? What recommendations will you make?

SUMMARY

- Proper health and safety measures are very essential at the workplace.
- The workplace in this context may include factories, mines, ports and docks, plantations, and office buildings.
- The Factories Act deals with health and safety measures and obligations of workers and employers to ensure a safe work environment.
- The mines legislation similarly contains safety and health provisions and their enforcement by the Directorate General of Mines Safety (DGMS).
- The Plantations Labour Act, 1951 provides for health measures and medical facilities for the plantation workers.
- Efforts are to be made to ensure safety to persons employed in the offices and buildings.
- A sincere joint effort by the industry, the hauliers, and the concerned authorities can drastically reduce the number of road accidents involving vehicles carrying hazardous chemicals.

Discussion Questions

1. Evaluate the following statement: "Safety in industry today is largely a matter of complying with the government rules and regulations."
2. What are the health and safety measures under the Factories Act, 1948?
3. What are the health and safety measures under the Mines Act, 1952?
4. What are the steps that can be taken for improving safety in transportation of chemicals?

11

Accident Compensation Statutes

Organised societies enact laws, establish rules and policies, and use other means to regularise the conduct of groups of individuals. Clearly, these methods alone are unreliable where a specific mission requires the cooperative effort of a number of individuals. Organised safety developed with the passing of workers' compensation laws in different countries.

This chapter covers relevant accident compensation statutes in India relating to safety and health.

The Dangerous Machines (Regulations) Act, 1983

The object of the Act is to provide for the regulation of trade and commerce in, and, production, supply, distribution and use of the product of any industry producing dangerous machines with a view to securing the welfare of labour, operating any such machine and for payment of compensation for death or bodily injury suffered by any labourer while operating any such machine, and for matters connected therewith or incidental thereto.

The term "dangerous machine" means a power thresser, and includes any such machine intended to be used in the agricultural or rural sector as the Central Government, being of opinion that it is of such a nature that any accident in the course of operation thereof is likely to cause death to the operator, dismemberment of any limb or other bodily injury, may, by notification in the Official Gazette, specify as dangerous machine.

A person who intends to commence the manufacture, or carry on business as manufacturer or dealer, of any dangerous machine, shall have to apply for issue of a licence to the Controller.

Every manufacturer of a dangerous machine shall ensure that such machine and every part thereof complies with such standards, conforming to the standards laid down therefor by the Indian Standards Institution, as may be prescribed by the Central Government.

Whenever any person operating a dangerous machine suffers death or dismemberment of any limb or any other bodily injury -

(i) by reason of any manufacturing defect in the machine whereby such death, dismemberment or injury was caused, or

(ii) by reason of the omission of the manufacturer to comply with the provisions of this Act and the rules and orders made thereunder, such manufacturer shall be liable to reimburse the person by whom compensation had to be paid under this Act to the members of the family of the person whose death was caused by such machine or, as the case may be to the person by whom such dismemberment or bodily injury was suffered.

No dangerous machine shall be operated until it has been registered in accordance with the provisions of the Act.

The manufacturer must ensure that any part of a dangerous machine conforms to prescribed standards.

If, during his employment as an operator of a dangerous machine, death or dismemberment of any limb or any other bodily injury is caused to such operator, his employer shall be liable to pay compensation:

(i) in the case of death of the operator, to his family;

(ii) in any other case, to the operator:

provided that where the operator does not have a family, the compensation shall be paid to the person or persons nominated on this behalf by the operator in writing and notified to the controller; provided further that the employer shall not be so liable:

(i) in respect of any injury which does not result in total or partial disablement of the operator for a period exceeding three days; or

(ii) in respect of any injury, not resulting in death, caused by an accident which is directly attributable to:

    (a) the operator being at the time thereof under the influence of any intoxicant or drug, or

    (b) the wilful removal by the worker of any safety guard or other device which he knew to have been provided in the machine for the purpose of securing the safety of the operator.

The amount of compensation payable shall be determined and paid in accordance with the provisions of the Employees' Compensation Act, 1923 as if the operator were a workman within the meaning of that Act.

Every employer shall take out, as soon as may be practicable after the commencement of this Act, one or more insurance policies providing for contracts of insurance whereby he is insured against any liability to make payment of compensation. Such contract of insurance may provide for the payment of annuities to the operator, or in case of his death, to the members of his family or to his nominee, if he does not have a family.

The Public Liability Insurance Act, 1991

The objective of the Act is to provide public liability insurance for providing immediate relief to the persons affected by accidents occurring during handling of hazardous substances.

The public liability Insurance Act and its amendment in 1992 envisages mandatory insurance for the purpose of providing immediate relief to the victim of accident arising out of hazardous processes and operations. This inherent risk covers not only the workman employed in a hazardous process industry but also members of the public residing in the vicinity. In addition, it includes accidents leading to death and injury to human beings and other living beings causing damages to public and private properties.

Before this Act came into force, persons had to file civil suits for damages. Apart from the heavy expenditure, the civil suits could go on for years.

This Act mainly protects the members of the weaker sections of society who by reason of their limited resources cannot afford the prolonged litigation in a court of law. To achieve this goal, it envisages for the mandatory public insurance based on the principle of "no fault" liability.

Mandatory insurance, immediate relief, corporate accountability, environmental relief fund and summary procedure are some of the salient features which made the Act comprehensive as well as impressive. The main features of the Act are as follows:

The term "accident" means an accident involving a fortuitous sudden or unintentional occurrence while handling any hazardous substance resulting in continuous, intermittent or repeated exposure to death, of or injury to, any person or damage to any property but does not include an accident by reason only of work or radio-activity.

Where death or injury to any person (other than a workman) or damage to any property has resulted from an accident, the owner shall be liable to give such relief as is specified in Schedule for such death, injury or damage.

Every owner shall take out, before he starts handling any hazardous substance, one or more insurance policies providing for contracts of insurance whereby he is insured against liability to give relief.

An application for claim for relief may be made -

(a) by the person who has sustained the injury;
(b) by the owner of the property to which the damage has been caused;

(c) where death has resulted from the accident, by all or any of the legal representatives of the deceased; or

(d) by any agent duly authorised by such person or owner of such property or all or any of the legal representatives of the deceased, as the case may be;

The Central Government may, by notification in the Official Gazette, establish a fund to be known as the Environment Relief Fund. The fund shall be utilised for the purpose of meeting the liability arising out of any claim awarded against the owner who has created the fund and to discharge the amount awarded by the Collector.

Persons authorised by the Central Government shall have a right to enter, at all reasonable times with such assistants as he considers necessary, any place, premises or vehicle, where hazardous substance is handled.

The Act provides for penalty of imprisonment or fine or with both for contravention of the provisions of the Act. The Central Government may, from time to time, constitute an Advisory Committee on the matters relating to insurance policy under this Act.

In exercise of the powers conferred under the Act, the Central Government has framed rules known as the Public Liability Insurance Rules, 1991, and notified them in the Official Gazette.

The Employees' Compensation Act, 1923

Occupational health and safety litigation prior to the introduction of the Workmen's Compensation Act, was governed by the principles of tort and in particular the law of negligence. Workmen had to file suits for damages against their employer. These took a very long time to fructify as they were treated on par with ordinary property litigation, and very few cases succeeded as it was generally very difficult for the injured employee to establish, according to the court's rigid and biased standards, negligence of the employer. The Workmen's Compensation Act cut into common law and introduced the concept of strict liability of the employer in all cases barring few exceptions.

A beginning of social security in India was made with the passing of the Workmen's Compensation Act in 1923. Prior to 1923, it was almost impossible for an injured workman to recover damages or compensation for any injury sustained by him in the ordinary course of his employment. Of course, there were rare occassions when the employer was liable under the common law for his own personal negligence. The dependants of a deceased workman could, in rare cases, claim damages under the Indian Fatal Accidents Act, 1885, if the accident was due to a wrongful act, neglect or fault of the person who caused the death. In 1921, the government formulated some proposals for the grant of compensation and circulated them for opinion. The proposals received general support. As a result, the Workmen's Compensation Act was passed in March 1923 and was put into force on July 1924. Subsequently, there were a number of amendments to the Act. The Act contains 36 sections and four Schedules.

Salient Features

- The object of the Act is to impose an obligation upon employers to pay compensation to employees for accidents arising out of and in the course of employment.
- If any accident occurs while commuting from residence to the place of work and vice versa will be deemed to have arisen out of and in the course of employment, if nexus between the circumstances, time and place in which the accident occurred, and the employment is established.
- The Act applies to railways (other than those employed in administrative work) and persons employed in factories, mines, plantations, mechanically propelled vehicles, construction work, and certain other hazardous occupations.
- A casual employee employed for a limited purpose for a limited period to carry out repairs in a residential building, not connected with employer's trade or business cannot be held to be an employee as per the provisions of the Act.
- The Act is not applicable to the persons who are covered under the Employees' State Insurance Act, 1948.
- There is no wage limit for coverage under the Act.
- The Act provides for payment of compensation to employees or their dependants, as the case may be, for industrial accidents (including prescribed occupational diseases) arising out of and in the course of employment and resulting in disablement or death.
- The employer will not be liable to pay compensation for any kind of disablement (except death) which does not continue for more than three days.
- The employer will be liable to pay compensation for death (but not in other cases) if the fatal injury is caused when the workman was under the influence of drink or drugs or wilfully disobeyed safety rules.
- The rate of compensation in case of death is an amount equal to 50 per cent of the monthly wages of the deceased employee multiplied by the relevant factor or an amount of rupees 1.2 lakh whichever is higher (relevant factors are listed in Schedule IV of the Act).
- Where permanent total disablement results from the injury, the compensation will be an amount equal to 60 per cent of the monthly wages of the injured employee multiplied by the relevant factor, or an amount of Rs. 1.4 lakh whichever is higher.

- Where permanent partial disablement results from the injury, such percentage of the compensation which would have been payable in the case of permanent total disablement as being the percentage of the loss of earning capacity caused by that injury.
- The percentage loss of earning capacity depends on the loss of limbs and varies from 1 percent to 90 percent.
- In case of temporary disablement, a half-monthly payment of the sum equivalent to 25 percent of monthly wages of the employee has to be paid.
- Half-monthly payment as compensation will be payable on the 16th day from the date of disablement.
- In cases where the temporary disablement is 28 days or more, compensation is payable from the date of disablement; in other cases it is payable after the expiry of a waiting period of 3 days.
- There is also a provision for commutation of half-monthly payments to a lump sum amount by agreement between the parties.
- Where the monthly wages of an employee exceed eight thousand rupees, monthly wages for the above purposes will be deemed to be eight thousand rupees only.
- If an employee contracts any occupational disease peculiar to that employment, that would be deemed to be an injury by accident arising out of and in the course of employment for purposes of this Act.
- The Act is administered by the respective State Governments/union territory administrations.
- The State Governments are required to appoint Commissioners for Employees' Compensation for (a) settlement of disputed claims; (b) disposal of cases of injuries involving death; and (c) revision of periodical payments.
- The compensation payable to the employee or to his dependents cannot be assigned, attached, or charged.
- The employer is required to file annual returns giving details of the compensation paid, the number of injuries, and other particulars.
- In case the compensation is not paid by the employer, the employee concerned or his dependants may claim the same by filing an application before the Commissioner.
- An appeal lies to the High Court against certain orders of the Commissioner, if a substantial question of law is involved.
- A contract or agreement whereby the workman relinquishes his right to compensation from the employer for the personal injury arising out of and in the course of employment, is null and void to the extent to which such contract or agreement purports to remove or reduces, the liability for, the payment of compensation.

The Employees' State Insurance Act, 1948

The Employees' State Insurance Act, 1948, is a pioneering measure in the field of social insurance in our country. The subject of health insurance for industrial workers was first discussed in 1927 by the Indian Legislature, when the applicability of the Conventions adopted by the International Labour Conference was considered by the Government of India. The Royal Commission on Labour, in its report (1931), stressed the need for health insurance for workers in India.

One of the earlier decisions of the Labour Ministers' Conferences between 1940 and 1942 was to invite an expert to frame a scheme of health insurance for workers. In pursuance thereof, the responsibility for preparing a detailed scheme of health insurance for industrial workers was entrusted in March 1943 to B.P. Adarkar who submitted his report in December 1944. This was considered by the Government of India and State Governments as well as other interested parties. The Adarkar Plan and various other suggestions emerged finally in the form of Workmen's State Insurance Bill 1946, which was then referred to a Select Committee in November 12, 1947. The Select Committee extended the coverage to all the employees in factories, and changed its name from Workmen's State Insurance Bill to Employees' State Insurance Bill.

The Employees' State Insurance Act came into force from 19th April 1948. The scheme framed under the Act aims at providing for certain cash benefits to employees in the event of sickness, maternity, employment injury, and medical facilities in kind, and contains provisions for certain other matters having bearing thereon.

Salient Features

- The object of the Act is to provide for certain cash benefits to employees in the event of sickness, maternity, employment injury, medical benefits, and so on.
- The Act applies in area-wise to all factories employing 10 or more persons, and also to various other establishments employing 20 or more employees.
- "Factory" means any premises including the precincts thereof whereon 10 or more persons are employed or were employed on any day of the preceding twelve months, and in any part of which a manufacturing process is being carried on or is ordinarily so carried on.
- The Act, however, does not apply to a mine or railway running shed, and specified seasonal factories.
- Persons whose remuneration (including overtime) does not exceed Rs. 21,000 a month are covered under the Act.

- The term "employee" refers to any person employed on wages in, or in connection with, the work of a factory or establishment to which the Act applies, and includes within its scope clerical, manual, technical and supervisory functions.
- The Act covers administrative staff and persons engaged in the purchase of raw materials or the distribution or sale of products and similar or related functions.
- The Act applies to employees working in the head office or branch office of factories, if such employees are doing work connected with the administration of the factories.
- Employees employed directly by the principal employer and those employed by or through a contractor under the supervision of the principal employer are covered under the Act.
- The Employees' State Insurance Scheme provides for sickness and extended sickness benefit, maternity benefit, disablement benefit, dependants' benefit, medical benefit, funeral benefit, and so on, by specifying contributory conditions, duration, and rate for each benefit.
- Employees covered under the ESI Act cannot claim compensation under the Employees' Compensation Act as there is a bar under the ESI Act.
- The Employees' State Insurance Scheme is administered by a corporate body called the Employees' State Insurance Corporation (ESIC).
- The chief executive officer of the Corporation is its director general.
- The ESIC has set up a network of regional and local offices all over the country for implementation of the Scheme.
- The main sources of finance to the Scheme are the contributions from employers and employees and share of expenses by State Governments towards the cost of medical care.
- Employees' contribution has to be calculated individually for each employee (0.75%) payable for every wage period and employers' contribution (3.25%) on the total wages paid to all the employees covered under the Scheme.
- Both the employers' and the employees' contribution are required to be paid, in cash or by cheque, into the State Bank of India or any other bank authorised by the ESI Corporation, by filling a prescribed challan in quadruplicate, within the stipulated period.
- The Scheme has laid down "contribution period" and "benefit period" for the purpose of paying contributions and deriving benefits.
- In the case of a newly employed person, the first contribution period shall commence from the date of his employment, and the corresponding first benefit period shall commence on the expiry of 9 months from the said date.

- All the benefits under the scheme are paid in cash except medical benefit, which is given in kind.
- For sickness during any period, an insured person is entitled to receive sickness cash benefits.
- An insured person suffering from any special long-term ailment – for example, tuberculosis, leprosy, mental disease – is eligible for extended sickness benefit. This is paid in addition to the usual sickness benefits.
- An insured woman is entitled to maternity benefit, which is practically equal to full wages
- Additional maternity benefit is given in case of sickness arising out of pregnancy, confinement, premature birth of a child or miscarriage.
- If a member suffers an injury in the course of his employment, he will receive free medical treatment and temporary disablement benefit in cash.
- In case of permanent total disablement, the insured person will be given a life pension, while in case of partial permanent disablement, a portion of it will be granted as life pension.
- The dependants' benefit consists of timely help to the eligible dependants of an insured person who dies as a result of an accident or an occupational disease, arising out of and in the course of employment.
- It will be available to the widow as long as she lives or until she marries; to sons and unmarried daughters up to the age of 18 without any proof of education; and to infirm or wholly dependant off-spring as long as the infirmity lasts.
- Where neither a widow nor a child is left, the dependants' benefit is payable to a dependant parent or grandparent for life.
- A fixed amount is payable as funeral benefit to the kins of the decreased.
- The kingpin of the scheme is medical benefit, which consists of free medical attendance and treatment of insured persons and their families.
- This benefit has been divided into three parts: (a) *Restricted Medical Care*: It consists of out-patient medical care at dispensaries or panel clinics. (b) *Expanded Medical Care*: This consists of consultation with specialists and supply of such medicines and drugs as may be prescribed by them. (c) *Full Medical Care*: It consists of hospitalisation facilities, services of specialists and such drugs and diet as are required for in-patients.
- An insured person and members of his family are entitled to medical care of all the above three varieties.
- The cash benefits payable under the Act are not liable to attachment or sale in execution of any decree or order of any court.
- An insured person will not be entitled to receive for the same period: (a) both sickness benefit and maternity benefit; or (b) both sickness benefit

and disablement benefit, for temporary disablement; and (c) both maternity benefit and disablement benefit for temporary disablement.

- Where a person is entitled to more than one of the benefits, he has an option to select any one of them.
- Disputes arising out of the working of the Act can be referred to the Employees' Insurance Court constituted by the State Government.
- The ESI Corporation is empowered to recover damages from the employer who fails to pay contribution due or makes delay in its payment.
- The registration of a factory/establishment with the Employees' State Insurance Corporation is a statutory responsibility of the employer.
- The employer has to afford necessary facilities to the Social Security Officer to enter the premises of the establishment and examine any person, books or document relating to employment and wages to ensure the provisions of the Act and regulations are being complied with.
- The statutory registers to be maintained up-to-date are: (a) register of employees; (b) accident book in which every accident to employees during the course of employment is recorded; and (c) inspection book.
- The Act has laid down various offences and punishments for them.

SUMMARY

- As a result of the inadequacies implicit in the common law approach in indemnifying injured workers, compensation laws were enacted in Germany in 1885, in Great Britain in 1897, and in the United States in 1902.
- The legislative enactments relating to accidents and work injuries, and also enactments providing for payment of compensations and benefits are the Dangerous Machines (Regulations) Act, 1983; the Public Liability Insurance Act, 1991; the Employees' Compensation Act, 1923; and the Employees' State Insurance Act, 1948.

DISCUSSION QUESTIONS

1. What are the statutes related to accident compensation?
2. What are the benefits under the Employees' Compensation Act, 1923, and Employees' State Insurance Act, 1948?

12

Safety Audit

Safety audit is an important tool for identifying organisational and operational safety policies, practices, and their effectiveness against accident prevention programme. It subjects each area of a company's activity to a systematic critical organisation with the objective of minimising loss. A formal report and action plan is prepared and monitored. It is a structured process of collecting independent information on the efficiency, effectiveness, and reliability of the total safety management system and draw up plans for corrective action.

Accident prevention is just an aspect of efficient operation as any other industrial activity. Therefore, it is a normal part of good management system to initiate and carry out safety audit and to ensure that operations are carried out in an efficient and profitable way. Safety programme must be audited continuously as the working environment is constantly changing because of new employees, new work processes, new material, and new or revised standards.

Objectives

The following are the objectives of safety audit:

1. To carry out a systematic critical appraisal of all potential hazards involving personnel, plant, services, and operational methods.
2. To ensure the occupational health and safety standards fully satisfy the legal requirements and those of the company's written safety policies, objectives and programmes.
3. To find out the safe working conditions and the potential hazards in various operations.
4. To evaluate work practices which may cause accidents and injuries.
5. To ensure safe maintenance of plant, equipment, fittings and fixtures.

6. To identify hazards and to take remedial and preventive action towards safety and health.
7. To find out the effectiveness of the plant's safety and loss prevention programmes.
8. To ensure that deteriorating standards of safety and occupational health are detected in time, and that the safety of personnel, operation, property, environment and public is ensured and reviewed.

Benefits

A properly developed and professionally managed audit system can result in following benefits to health and safety:

- Strengthens the organisation's safety standards and programmes.
- Improves the skills and performance of managers and employees.
- Helps to create group and self awareness.
- Identifies specific areas of safety programme deficiencies.
- Provides opportunities for group and individual motivation.
- Helps to increase management's and employee participation in the safety programmes.
- Provides timely information before any loss producing incident occurs.

Types

Before undertaking any full-fledged audit, the management should ensure that at least basic elements of a safety programme are in operation. They must analyse qualitatively, their achievements in the areas of policy statement on safety, safety set-up, hazard identification, job procedures, education and training, and record keeping. This is '*basic audit*'. After being satisfied with the result of *basic audit*, the management can take up '*self auditing*'. The *self audit* should normally be done by a management staff team with professional training in the disciplines of safety, fire, environmental monitoring and control, and occupational health. The team can be drawn from maintenance, process methods engineering, training and medical. The team should preferably be from the corporate office since they are unlikely to be biased and are directly accountable for their results.

The next stage of audit is a '*professional audit*' conducted by licensed or accredited auditors who are authorised by professional bodies. This audit will be an analysis of the safety systems at micro levels. Since this will be done by an external agency, the extent of influence of the management and bias on the part of the auditor can be expected to be minimal and will consider all the elements of the audit.

Procedure

The following procedure should be adopted for conducting the plant safety audit:

(a) Establish a plant safety audit committee.
(b) Plan the safety audit method.
(c) Schedule the plant safety audits.
(d) Conduct the audits as per plan.
(e) Report the audit results to the concerned authorities.
(f) Familiarise on the operations.
(g) Interviews with officials to get information.
(h) Verification of information by actual count, random sampling, physical measurement, and professional judgement.
(i) Cross verification at the site.
(j) Close meeting with concerned executive.
(k) Post audit work.
(l) Submission of report.

Stages

Safety audit involves the following stages:

1. Initial tour of the audit team to get acquainted with the plant layout, process, and the like.
2. Discussion with top officials to know the background of the factory, its technology, and past problems in the area of accident prevention programme.
3. Discussion with senior officials and the safety officer to critically examine the safety organization.
4. Study of materials handled, from raw material stage to final product and hazards associated therein.
5. Collection of information through questionnaire regarding operations, management policy, training, accident reports, and communicating the same to the workers and management.
6. Inspection of plant areas and various utilities and facilities to know the need for updating the information received through questionnaire.
7. Review of accident statistics and records to update the information obtained through questionnaire and personal interviews.
8. Study of the process and the in-built safety system to avoid any major accident/tragedy.
9. Study of waste/affluent and their disposal treatments system.

10. Study of on-site emergency plan.
11. Discussion with workers/supervisors regarding safety status.

Functions

The functions of a safety audit are:

1. To expose the hidden hazards due to unsafe conditions or unsafe acts under the purview of the auditor.
2. To take note of habitual unsafe acts during work, particularly, the violation of safety regulations.
3. To find out the sufficiency or otherwise of the precautions taken to contain or eliminate known hazards.
4. To take cognizance of violations of safety norms.
5. To assess the extent of effectiveness of the protective steps taken to contain the unavoidable hazards prevailing in the area.
6. To take note of inadequate safety measures in the work area.
7. To check and ensure that all safety systems are in working condition.
8. To check the condition of personal protective items available in the area.
9. To check the effectiveness of the alarm installed for safety purposes.
10. To assess the progress in implementation of all approved safety recommendations emanating from the management and sectional safety meetings.
11. To check if statutory inspection of equipment is carried out as per schedule.
12. To report on the housekeeping conditions in the area under audit.
13. To ensure that all earthing terminals to drain off static charges are maintained effectively.
14. To check the condition of maintenance tools and spark-proof / flame-proof electrical fittings.
15. To check that all fire extinguishers are in place and not in an outdated condition.

In brief, during the audit phase, the following activities are normally carried out in chronological order.

- Opening meeting with site management and knowledgeable persons who have responsibility in safety.
- Familiarisation tool of the site to gain a quick overview of the operation.
- Interviews with knowledgeable persons to assess and evaluate the safety management system of the audited site.
- Verification of safety management system by conducting document review, physical condition checks, and random employee interviews.
- Exit meeting to highlight key audit findings.

Choice of the Auditor

An auditor must be familiar with the type of work that is carried out in the section for which the audit is done. For instance, if the audit pertains to maintenance activities, the auditor should have sound maintenance experience and be aware of the hazards accompanying this type of work, including the safety aspects of the tools and machines and the facilities needed to carry out the work. For auditing a plant where process operations are in progress, the auditor must be familiar with the equipment performance data and its operations, particularly emergency operations. Above all, he should be a keen observer and should be able to obtain the co-operation of the employees in getting the correct information needed to make the audit report informative and exhaustive.

A company is subjected to close scrutiny by auditors of an accredited certification body. The process of certification includes a review of the company's quality manual, interviews with senior management, and verification visits to head office and selected sites. This certification process is also known as "quality audit". Safety audit is similar to quality audit except that in safety audit, emphasis is placed on evaluating the effectiveness and efficiency of a company's safety management system instead of the quality management system.

During an audit, the auditors will be repeatedly looking for answers to the following questions:

(a) Are arrangements being made to control hazard and risk?
(b) Have adequate standards been set?
(c) Are these standards being complied with?

The frequency of auditing of any area depends upon the number and severity of the hazards in that area. In process plants, the frequency of audits depends on the number of human errors that can cause accidents and the number of fail-safe systems installed in the plant to prevent accidents due to such errors. The audit frequency is increased when there is an unusual increase in accidents in a plant or work area.

Coverage of Audit

The subject matter for audit covers the following areas:

1. Unsafe conditions in the work area.
2. Unsafe acts.
3. Availability of protective measures to prevent injuries.
4. Progress of implementation of recommendations arising from the sectional safety committee meetings.
5. Housekeeping.
6. Condition of personal protection items provided in the work area.

7. Fire extinguishers to be checked for their expiry dates.
8. Certificates of pressure vessels, hoisting-lifting equipments and checking transport vehicles for their validity/permit licenses.
9. Implementation of the recommendations provided in the equipment inspection reports.
10. Inspection of hand and power tools.
11. Incidence of illness due to occupational ailments.
12. Environmental conditions in the neighbourhood related to safety and health of employees.
13. Checking if previous audit recommendations have been carried out.
14. Displaying conspicuously emergency telephone numbers, as shown below, so that precious time can be saved in communication during emergencies.
    (a) Fire brigade
    (b) Police
    (c) Ambulance
    (d) Hospital

The plant safety audit broadly covers the following:

1. Plant accident audit.
2. Accident investigation audit.
3. Environmental hazard control audit.
4. Plant safety inspection audit.
5. Employees safety training audit.
6. Employees safety attitude audit.
7. Plant safety observation audit.
8. Safe job methods audit.
9. Miscellaneous safety items audit.

Safety Audit Questions

Safety audit questions may be classified as under:

1. Health and safety policy
2. Safety and health organisation
3. Accident reporting, investigation and analysis
4. Safety inspections
5. Safety education and training
6. Safe operating procedures
7. Hazard identification and control
8. Plant layout
9. General working conditions
10. Personal protective equipment
11. First Aid

12. Fire protection
13. Emergency preparedness

The questionnaire for chemical plants should include the following aspects:

1. **Health and Safety Policy**
    - Is the organisation having a health and safety policy?
      If yes, please attach one copy
    - Do you have any corporate safety policy?
      If yes, please attach one copy
    - Who has signed the health and safety policy?
      Indicate his position
    - Whether it is prepared as per guidelines of the statutory provisions?
    - When was the safety policy declared and adopted?
    - How many times has it been updated till now?
    - Whether the policy is made known to all?
    - When was it last updated?
    - Does it find a place in the annual report?
2. **Safety Organisation**

    (A) ***Safety Department***
    - Does the factory have a safety department?
      If yes, furnish the following information:
      (i) Head of the safety department
        (a) Name
        (b) Designation
        (c) Qualification
        (d) Experience
        (e) Status
      (ii) Strength of the safety department including safety officers and staff
    - Does the head of safety department / safety officer report to the chief executive?
    - How often are the safety officers retrained in the latest techniques of total safety management?
    - What additional duties the safety officer is required to perform?

    (B) ***Safety Committee(s)***
    - Does the factory have a safety committee(s)?
      If yes, give details of their types, structures, and terms of reference
    - How are the members of safety committee(s) selected/elected/nominated?

- How often are the meeting of safety committee(s) held?
- What are the matters and problems discussed in the meeting?
- How are the recommendations of the committee(s) implemented?
- Are the minutes of the committee(s) meeting circulated among the members?
- Are the minutes forwarded to the trade union(s) and chief executive/occupier?
- How do the management and trade union(s) play their active roles in supporting and accepting the committee(s) recommendation?
- How are the committee(s) members appraised of the latest developments in safety, health, and environment?

(C) *Safety Budget*

- What is the annual safety budget?
- What is the percentage of this budget in relation to the total turnover of the company?
- How much budget has been utilised till date?
- Is the safety budget adequate?
- How is the safety budget arrived at?

3. Accident Reporting, Investigation and Analysis

- Whether the accident data for the last three years for reportable and non-reportable accident available?
- Are all near-miss incidents and accidents reported and investigated?
- For how many years are the investigation report retained?
- Who maintains the accident statistics data?
- How is the management appraised of these data?
- Is the accident statistics effectively utilised? If yes, how?
- What type of injuries have occurred during the last three years?

4. Safety Inspection

- What types of safety inspections are carried out and what are their frequency?
- Is there any system of internal inspection?
- Who does the inspections?
- Is a check-list prepared for these inspections?
  Specify item-wise, for example, housekeeping, fire protection, etc.
- To whom are the recommendations submitted?

5. **Safety Education and Training**
   - Is there any training department?
   - What infrastructural facilities with audio-visual support are available for training?
   - Do the programmes cover the plant safety rules, hazard communication and any other special safety rules or procedures unique to the plant or specific departments?
   - Whether the training programmes are conducted in the local language?
   - Whether visits to safety institutions/organisations are arranged?
   - Are all the employees trained? What is the frequency of such training?
   - Do the training programmes cover safety and health aspects, and if so, how much in terms of number of sessions/hours?
   - Do the trained supervisors train their own employees in safety and health aspects?
   - Is the retraining performed whenever new hazards/process changes are followed/introduced?
   - What is the manner of participation of the senior management personnel in safety and health programmes?
   - How many employees have been trained in safety and health in the last five years?
   - How do you ensure that the training is put to use by the employees trained in safety and health?
   - What is the training plan for the next two years? Give details.
   - What documentation system has been established regarding safety and health training?
6. **Safety Communication/Motivation/Promotion**
   - Are public address system available in all plant areas?
   - Are public address system provided with uninterrupted power supply?
   - Whether public address system is checked periodically for its proper functioning?
   - What is the means of communicating emergency in the plants?
   - Does the factory have safety suggestion schemes? Give details.
   - Does your factory participate in a national awards/suggestion schemes?
   - Did your factory get safety awards during the last five years?
   - Are safety contests organised in the factory? Give details.
   - What are the news publications of your organisation? Do they include information on safety and health subjects?

- Is the literature on safety and health made available to the employees?
- How is the issue of safety and health publicised in your factory?
  - (i) Bulletin boards
  - (ii) Newsletter
  - (iii) Others (specify)
- Does the organisation celebrate safety day/week or organise safety exhibition?
- When was the last safety day/week celebrated?

7. First Aid/Occupational Health Centre

- How many first aid boxes are provided? Are they adequate? Give location details.
- Is there any first aid / ambulance room?
- Are qualified / trained first aid personnel available in each shift?
- Whether occupational safety and health centre is provided or not?
- Are the medical attendants/doctors available in each shift?
- Is ambulance van available in each shift?
- Is there any liaison with the nearest hospital(s)? Give details.

8. General Working Conditions

(A) ***Housekeeping***

- Are all the passages, floors and the stairways in good condition?
- Do you have the system to deal with the spillage?
- Do you have sufficient disposable bins clearly marked? Are they suitably located?
- Do you have adequate localised extraction and scrubbling facilities for dust, fumes, and gases?
- Whether walkways are clearly marked and free from obstruction?
- Do you have any inter-departmental competition for good housekeeping?
- Whether your organisation has good housekeeping practices and standards?
- Are there any working conditions which make the floors slippery? If so, what measures are taken to make them safe?

(B) ***Noise***

- Are there any machine/processes generating noise? Specify.
- Was any noise study conducted?
- Which are the areas having high level noise?

- Have engineering and administrative controls been implemented to reduce noise exposure below the permissible limits?
- Whether the workers are made aware of the ill-effects of high noise?
- Whether any personal protective equipment along with ear muffs / plugs are provided and used?

(C) ***Ventilation***

- Whether natural ventilation is adequate or not?
- Whether dust/fumes/hot air is generated in the process? Give details.
- Is there any exhaust dilution, ventilation system in any section of the plant?
- Whether any ventilation study has been carried out in the section(s) to check the record?
- Are periodic preventive maintenance of ventilation system carried out and record is maintained?
- Is the work environment assessed and monitored?
- Whether personal protective equipment are given to workers exposed to dust/fumes and gases?

(D) ***Illumination***

- Was any study carried out for the assessment of illumination level?
- Is there any system of periodical cleaning and replacing the lighting fittings / lamps in order to ensure that they give the intended illumination levels?
- Are the workers subject to periodic optometric test and records maintained? Give details.

9. Hazard Identification and Control

- Are all the hazardous areas identified?
- What are the types of hazards? Physical hazards (noise, heat, etc.) and chemical hazards (fire, explosion, toxic release, etc).
- What steps have been taken to prevent these hazards? Give details.
- Are there any safety inter-locks alarms and trip systems? Give details.
- Are these tested periodically? How often? Specify?
- Are there audit or HAZOP or any other studies carried out and the recommendations implemented? Give details.

10. Personal Protective Equipment (PPE)

- Whether a list of required PPE for each area/operation been developed and the required PPE is made available to the workers?
- Is the safety department and the workers consulted in the selection of PPE?
- Have the workers been trained in proper use of PPE?
- What is the system of replacement/issue of PPE?
- Are the contractor's workers provided with the required PPE? Who is responsible for this? Give details.
- Does the PPE conform to any standard? Give details.
- Who is/are responsible for inspection and maintenance of PPE?

11. Material Handling

- Are there adequate storage facilities available?
- Are these areas clearly defined?
- Do you have adequate equipment for handling materials?
- Do the workers know the hazards associated with manual material handling?
- Are the workers trained in manual handling?
- Do workers follow safe procedures for storage of materials?
- Whether contract workers are trained in safe material handling?

12. Safe Operating Procedures

- Are written safe operating procedures available for all operations?
- Whether the written safe operating procedures displayed or made available and explained in the local language to the workers?
- Whether the safe operating procedures are prepared jointly by the plant and safety department(s)?
- What system is used to ensure that the existing safe operating procedures are updated? Give details.
- Have the workers been informed of the consequences of failure to observe the safe operating procedures?
- Are contract workers educated and trained to observe safety at workplace?

13. Waste Disposal System

- Is identification done for various types of wastes? Give details.
- What are their disposal modes?
- What is the system/measures adopted for controlling air/water/land pollution?
- What is the system of effluent treatment? Whether it is approved by the competent authority?
- How are the treated effluents used?

14. Plant Layout, Vessels, and Equipment

- What is the system of classification of hazardous zones in the plant for electrical installations? Specify.
- Whether periodic inspection and preventive maintenance of electrical installations is done by a qualified person and record is maintained?
- Give the details of the piping and vessels which are operated at a pressure greater than the atmospheric pressure.
- What standards/codes of practice are adopted for design, fabrication, operation and maintenance of the pressure vessels? Give details.
- How are the pressure vessels tested? Give details.
- Give details of safety devices available for these pressure vessels.
- Whether the log book for pressure vessel and the pressure plant is maintained?
- What is the system for effecting any change in the existing plant, equipment or process? Whether it is approved by the appropriate competent authority?

15. Lifting Machines, Tackles, and Vehicular Traffic

- Whether all the lifting machines are marked with their SWL?
- Are all the examinations and tests documented in the prescribed form?
- Are all the examinations and tests carried out and certified by competent person(s)? Give details.
- Are adequate lifting tackles provided at all the necessary places? Give details.
- Are the trained operators engaged for operating the equipment? Give details.
- What is the system of training for such operators?
- Are all the lifting machines and tackles maintained in good conditions? Give details.
- Are all the mobile equipments in good condition?
- Are trained drivers engaged for fork-lift trucks?
- What systems do you adopt to assess the standard of driving? poor/fair/satisfactory/good
- Are the hazards associated with transportation within the plant identified and safety measures taken? Give details.

16. Tank Storage Vessels and Gas Cylinders

- Give storage vessels designation.
- Give the names of storage materials in each of them.
- What are the vessel sizes (capacity in tonnes)?

- What is the material used in construction of each vessel and what standards followed in designing/fabricating the vessel?
- What are the operating pressure and temperature?
- Where are the vessels located?
- Indicate whether vessels are above ground / underground?
- Is any of the tanks storing flammable material?
- Whether electrical installations are flame proof or not?
- Are these storage vessels bunded/dyked?
- How are the vessels isolated in the event of a mishap?
- Are the vessels provided with alarms for high level, high temperature and high pressure?
- Are standby empty tanks provided for emptying in case of emergencies?
- What are the provisions made for fire fighting/tackling emergency situations around the storage vessels?
- What are the various gas cylinders used in the plant? Give details.
- What are the storage facilities of gas cylinders?
- What are the measures taken for combating any emergency in the cylinders storage area?
- Are valid licenses available for storing all these cylinders?
- Whether integrity test certificates are obtained from the suppliers of the cylinders?

17. Transportation

- What potentially hazardous materials are transported to or from the site (including wastes)?
- What modes of transport are used?
  Road/Rail/Pipelines
- Does the company employ licensed vehicle of its own/outside sources?
- Are the loading/unloading procedures in accordance with safety measures?
- Do all truck and tanker drivers get training in handling emergencies during transport?
- What hazardous materials are transported by rail?
- Does the company have a direct siding on-site?
- Are tankers or other wagons used in transportation?
- What materials are transported to and from the site by pipeline(s)?
- Are the pipelines underground or overground?
- Are the pipelines extended in the public domain?
- Are the pipelines dedicated for each type of chemicals?
- Are the pipelines fitted with safety equipment such as leak detectors, automatic shutoff valves, etc.?

- What is the frequency and method of testing of the pipeline?
- Is there a written procedure for tackling leakages in pipeline?

18. Fire Protection

- Give the location, number and type of portable fire extinguishers.
- Are the fire fighting systems and equipments approved, tested, and maintained as per relevant standards.
- What is the inspection and maintenance schedule of the above extinguishers? Who performs these functions?
- What areas of the plant are covered by the fire hydrants?
- What is the capacity of dedicated water reservoir for supply to the hydrants? What is the source of water?
- Are all personnel conversant with the fire prevention and protection measures? Give details.
- Do you have fixed or automatic fire fighting installation(s) in any section of your plant?
- Are the fire alarms adequate and free from obstruction?
- Do you have the fire department? If yes, give details?
- What is the system for conducting mock drills? Give details.
- Do you have any mutual aid scheme with any of your neighbouring industry or any local organisation(s)?
- Are there any safe containers for the movement of small quantities of hazardous chemicals? Give details.
- Are all self-closing fire doors in good condition and free from obstructions?
- How many major and minor incidents/fires were there in the factory during the last five years? Give department/plant-wise details.
- Have all the fires/incidents been investigated and corrective actions taken? Give break up.

19. Emergency Preparedness

- Is there an on-site emergency plan for your factory? Attach a copy of the plan.
- What is the frequency of conducting mock drills or on-site emergency plan?
- How many emergency control centre(s) and assembly points are there? Where are they located?
- Whether emergency team or the key personnel identified?
- Are suitable and adequate protective and rescue equipments available?
- How is the emergency rescue team trained to use the above equipments?

- How is the emergency communications with local bodies and other organisations ensured? Give details.
- Whether any alternate power source been identified? Give details.
- Whether there is any medical emergency response system? Give details.
- Are you a member of any Mutual Aid Scheme of your area? If so, give details.
- How many emergency alarm system(s) is/are available? Give details.

SUMMARY

- The objective of the safety audit is to (a) determine the effectiveness of the company's policies and programmes; (b) examine and assess critically all potential hazards involving personnel, plant and machinery; and (c) ensure that the company is complying with all statutory requirements and also the management's laid down safety standards and promotional programmes.
- Perception inventory survey is the primary performance tool and complimentary to safety audits.
- Audits assess what are in place; perception surveys assess how effective they are as perceived by the employees.
- The quality audit system may be classified as: (a) internal audits, (b) external audits, and (c) regulatory audits.
- Safety audit is a useful tool in predicting the accident potential so that action can be taken well in advance to prevent the occurrence of loss producing events which may be in the form of accident or property damage.
- Safety audit, like financial audit, should be part of the management practice to ensure that critical business operations are carried out in a profitable, efficient and safe way.
- Safety audit needs to be undertaken by a team of experts in different disciplines of science and engineering with a coordinator, and involvement of the factory personnel at different stages of audit is also essential.
- The audit is carried out against the checklist which covers among other activities:(a) the personal attitudes, (b) communication, (c) training/education, (d) safety of the workplace, (e) safety procedures, (f) occupational health and safety, and (g) environmental controls.

→

- A detailed audit report along with required forms and findings and action plan containing various observations and recommendations is to be prepared and submitted to the authorities concerned within reasonable period.

DISCUSSION QUESTIONS

1. How a properly developed and professionally managed audit can act as an important tool for improvement of safety performance in an organization?
2. Give a detailed account of the areas to be covered and questions to be asked in a health and safety audit.

13

Stress at Work

Stress comes from the Latin word *Stringere* (to pull tight). When the word was first used, it had the meaning of having difficulty or suffering hardship or affliction, only later did stress imply strain or pressure. Hans Selye first used the term stress in the 1930s. Stress can be defined simply as the rate of wear and tear of the body systems caused by life.

Stress is the body's reaction to any demand made by it. It is a psychological and physiological state that results when certain features of an individual's environment, including noise, pressures, job promotions, monotony, or the general climate, impinge on the person. Both positive and negative occurrences can give rise to stress. In stress situations individuals generally go through three stages of alarm, resistance and exhaustion.

In recent decades, stress at work has become increasingly common. Rapid advances in technology and escalating social and economic changes have all contributed to occupational stress.

Stress may be due to any demand made on a person from the environment. The demand may relate to situations or other people within that environment. The resulting stress may be positive or negative depending on the type of demand and the response of the person.

Stress is a necessary stimulus to an active life and all areas of human experience involve some stress and challenge. In the vast majority of instances, people master and grow from the demands placed upon them. However, in some instances the demands prove too great and the person reacts with a negative stress response.

The causes or sources of stress are commonly called stressors, with the resultant outcome being the stress response.

Stress is a common aspect of the work experience. It is expressed most frequently as job dissatisfaction, but it finds expression also in more intense and aroused affective states–anger, frustration, hostility, and irritation. More passive, but perhaps no less negative, are such responses as boredom and tedium, burnout, fatigue, helplessness, hopelessness, lack of vigour, and depressed mood.

Stress is not only generated by work or the workplace. It is not possible, however, to separate the harmful effects of stress into "work" and "non work" related components. It is the combined result which affects the performance of the individual at work. Additionally, most persons spend a significant amount of their time involved in work and associated activities. Sometimes these include difficult tasks or work in difficult situations. It is not surprising, therefore, that work creates or contributes to stress responses in individuals. 'At risk' people include workers, employers and the self-employed.

Work stressors which cause significant harm to the individual can result from a series of exposures, such as ongoing organisational conflict, or from exposure to a single highly stressful incident, such as an armed robbery or serious workplace accident. It is acknowledged that risks resulting from poor health and safety practices can lead to negative stress responses. Such risks can be reduced or eliminated at the workplace through effective management of personnel at work.

Causes of Stress

Environmental Stressors

It includes extremes of temperature, inadequate lighting and ventilation, presence of dust, fumes, vapours and gases, noise and vibration, and poor amenities.

Occupational Stressors

Typical stressful conditions at work include:

(i) too heavy or too light a workload;
(ii) a job which is very difficult or very easy;
(iii) working overtime;
(iv) conflicting job demands;
(v) too much or too little responsibility;
(vi) poor human relationships;
(vii) incompetent superiors/supervision;
(viii) lack of participation in decision making;
(ix) reduced career prospects, threat of redundancy or premature retirement; and
(x) deficiencies in interpersonal relations and skills;

Social Stressors

It is associated with family, marital relationships, bereavement due to death of a family member, indebtedness, alcoholism and smoking, trouble with the neighbours, and so on.

Exhibit 13.1 A Menu of the Potential Sources of Stress

Organisational stressors

- Work overload or underload – repetitive work, unrewarding work
- Poor job design – ergonomically, environmentally, lacking autonomy or variety
- Role ambiguity or role conflict
- Poor quality leadership or supervision
- Lack of participation in decision-making
- Poor quality relationships – horizontally, vertically or externally
- Responsibility for people or for achieving targets and objectives
- Organisational change
- Organisational climate
- Being subjected to bullying or harassment
- Pay, poor working conditions and job insecurity

Personal stressors

- Unhealthy eating, sleeping or exercise
- General health and its abuse
- Family and social relationships
- Significant life events
- Conflicting personal and organisational demands

Wider environmental stressors

- The general economic situation
- Political uncertainty
- Social change and threats to personal values and standards
- Concern with the natural environment
- The pace of technological change
- Changing male and female role perceptions

The signs and symptoms of physiological stress include headaches, inability to sleep, fatigue, over-eating, constipation, back pain, allergies, nervousness, nightmares; alcohol abuse, indigestion, dermatitis, menstrual distress, nausea, irritability, asthma attacks, loss of appetite, peptic ulcers, heart palpitations, minor accidents and so on. The common psychological effects of stress are anxiety, depression, anger and so on.

Costs of Stress

The cost of stress in the workplace is high in both human and financial terms. The direct and indirect effects of stress for an individual can result in psychological responses such as depression, anger, anxiety, fatigue and insomnia. Stress can also lead to physical complaints such as back, neck or shoulder pain and it can interfere with learning, memory, concentration, decision making and general work performance. Additionally, research indicates stress can negatively affect an individual's immune responses therefore leaving the person more susceptible to disease.

It must be emphasised, however, the presence of any of the above signs and symptoms does not necessarily mean a person is exhibiting a stress response. The whole situation must be considered.

As a result of the stress responses of workers, workplace efficiency and profits can be diminished by:

- poor morale
- poor staff relations with resultant conflict and communication problems
- poor public image
- decreased job interest and motivation
- increased accident rate
- increased incidence of disease and pain reporting
- increased absenteeism
- lost productivity
- increased wages / overtime payments
- increased staff turnover
- increased workers' compensation premiums.

Handling of Stress

Effective management of stress at work makes good business sense. This effective management means reducing negative sources of stress and controlling negative effects. Besides the cost aspects of workplace stress, employers have specific responsibilities under workplace health and safety legislation. Under this legislation, employers are required to ensure the health and safety of all workers at work. While it may not be practicable to remove all forms of negative stress from work and the workplace, employers should take reasonable steps to manage existing and foreseeable workplace stressors.

Prevention of serious occupational stress can be accomplished by early identification of potential stressors and the people who may be affected. By applying risk management procedures, an employer can play a proactive role in management of stress at work. Risk management is a logical and systematic approach which involves:

(a) Identifying potential stressors
(b) Assessing the effect of stressors
(c) Developing control measures to minimise stressors
(d) Reviewing control measures.

Identifying potential stressors

As workers are most likely to know the work situations which they find stressful, effective consultation and communication between managers/ supervisors and workers should be considered a priority for identifying and managing stressors at work. In addition to consultation, an employer can use several other methods to identify stressors at the workplace, or outside stressors which are impacting at the workplace. Organisational culture, the type of work and work processes will determine the method selected. A combination of methods may give the most complete results. Methods to help in the identification of potential stressors include:

- investigation of incident, accident, injury and disease reports. This may indicate, for example, the involvement of worker fatigue in a high proportion of incidents occurring during the night shift
- identification of organisational aspects such as the work environment, work processes and conflict levels, which have the potential to result in stress problems
- absenteeism rates - high rates may be related to occupational stress
- workers' compensation claims for stress related injuries or diseases
- observed uncharacteristic behaviour of workers which may suggest a stress response, for example, anger, depression or sloppy work
- consultation with specialist practitioners, industry associations, unions and government bodies who may be able to provide additional assistance.

Employers should assess the nature and extent of any identified potential stressors and their effects at the workplace. They should consider the following:

- work groups at increased risk, for example, counter staff
- number of workers likely to be affected
- possible cumulative effects
- frequency and duration of exposure

Developing control measures to minimise stressors

At a broader level, however, employers should consider the roles of organisation of work and training in stress minimisation. Some factors of work organisation are governed by industrial awards and agreements.

However, other factors relating to the organisation of work which can contribute to effective minimisation of stress include:

- encouraging workers to use and develop their own resources appropriate to the job content and its control;
- maximising opportunities for workers to influence work situations, methods and pace;
- addressing the social environment of the workplace so that it allows for human contact and encourages co-operation;
- ensuring the workplace layout and environment fosters effective work outcomes, for example, placing people in quiet areas if high levels of worker concentration are required.

Training

Effective training programmes can lower the potential for worker stress. If workers have an overview of the organisation in which they work, know what is expected of them and have received training for their job, they are more likely to do their job well and feel confident in how they go about their work. On the other hand, a lack of effective training and supervision can result in workers making mistakes, being confused about their duties and displaying a poor understanding of workplace issues. These outcomes have the potential to increase worker frustration and stress, and as a result, negatively affect workplace effectiveness.

In some occupations such as policing and the helping professions, potential stressors such as dealing with violence, death or other people's problems are part of the job. Where these stressors cannot be reduced or eliminated, it is extremely important for management to be supportive. This support may include adopting a team approach, providing professional counseling if needed, and ensuring communication channels between supervisory and other staff remain open. Additionally, stress management training should be made available to workers who have increased exposure to stress at work.

A training programme in stress management could include:

- development of self assessment skills so that a worker is able to recognise the signs of stress;
- the importance of stress modifiers in a person's daily life. Examples of modifiers are regular exercise, sound nutrition, social support and ensuring a balance between work and leisure time;

- information that drug dependence (including tobacco, alcohol, caffeine and prescription and other drugs) is an unhealthy and ineffective method of coping with stress;
- teaching communication and other skills in such areas as assertion, time management and dealing with volatile or difficult situations;

- teaching coping skills such as relaxation techniques; and encouraging alternative views, for example, seeing potential stressors such as conflict and change, as challenges rather than difficulties.

Reviewing control measures

Having implemented procedures to manage identified stressors and people at risk, it is important that the effects of these procedures are monitored. This will include review of those indicators used in the identification and assessment steps. If stress problems remain or become worse, or other difficulties develop, urgent reassessment should be carried out. Where changes have resulted in improved outcome, periodic reviews of control measures should still be undertaken. Changes in the workplace environment and personnel over time can make existing procedures for stress management inappropriate.

Sources of Stress and Control Measures

The following discussion on sources of stress at work is not meant to be comprehensive. The stressors and control measures mentioned are examples only. Employers will need to consider the stressors that occur in their workplaces and respond accordingly.

1. ***Conflict with supervisors and co-workers***

   Workplace conflict can be a significant source of stress at work.

   Because of the amount of time people spend in the workplace and in the company of work colleagues, it is inevitable that conflict among workers and between workers and managers/supervisors will arise from time to time. While much of this conflict may be of minor nuisance value only, some conflict may escalate to serious levels and create significant stress for the people involved.

   Management procedures which could reduce the potential for such conflict include:

   - supervisory and managerial practices which display to workers a concern for their rights, health and well-being.
   - recognition that conflict is a legitimate concern of workers and managers
   - whenever possible, facilitating dispute resolution
   - use of consultation between employer and workers, for example, consultative committees.

2. ***Abuse, threats or harassment***

Abuse, threats or harassment are forms of violence which can occur at the workplace. Persons dealing with the public or those working in institutional settings are at risk of these forms of abuse. This abuse can escalate to physical violence. Whether the abuse is of a minor nature or more serious, the end result is that staff can become stressed.

Risk reduction procedures to deal with abuse, threats and harassment from clients or members of the public could include:

- addressing the potential for violence at work in staff training programmes which incorporate communication and conflict resolution skills;
- ensuring workplace efficiency and effectiveness is maintained so that client and public frustration can be minimised;
- implementing clear guidelines to deal with a potential or real violent encounter;
- setting up a system for alerting co-workers that urgent help is required;
- using job rotation so staff can have a break from client or public contact.

3. ***Critical incidents***

Critical incidents in the workplace include such events as armed robbery, physical assault, an accident which results in injury or death to a worker or other person, or other traumatic episodes. After such an incident, victims or others involved, including witnesses to the event, are at risk of developing post traumatic stress. This form of stress response can have severe repercussions for a person and for an organisation. If appropriate help is available immediately to people involved in critical incidents, negative stress responses are usually effectively diminished.

To minimise the risk of critical incidents and their effect on workers, control measures could include:

- use of suitable procedures;
- ensuring 'at risk' workplaces such as banks and service stations are designed with security in mind, for example, installing protective barriers, security lighting, cameras and alarms;
- education in appropriate selfcare and peer support, and in the common psychological reactions following a critical incident; and
- using a counseling or support service for people involved in critical incidents to reduce the impact of the event.

4. ***Process work***

Research has shown that work stress can be related to repetitive tasks, piece-rated payment and machine pacing of work. The stressors in these examples may arise from monotony, pressure to complete tasks quickly, and perceived lack of control a worker has over job-related decisions. Where process work is organised so that it offers little scope for skill utilisation or the exercise of initiative, it can lead to boredom, loss of concentration, anxiety and frustration. Research also suggests that increased pain reporting, for example, lower back pain, can indicate stress resulting from this type of work.

Management procedures for consideration include:

- using a team approach or job rotation to help relieve monotony;
- ensuring, where possible, workers have some control over machine rate;
- ensuring scheduled breaks are taken. In particular, meal breaks should not be taken at the work station.

5. ***Poor organisational support***

The perception by workers of management as unsupportive or insensitive to workers' needs can contribute to stress levels in an organisation. Poor communication or lack of stated policy on issues affecting personal needs are of particular importance. Organisational support procedures include:

- advising workers of organisational changes which affect them or their work;
- providing career options;
- offering effective information and instruction;
- responding to complaints/suggestions;
- supporting staff in difficulty, for example, at times of illness or death of a family member;
- regular feedback on performance;
- clear definition of duties and expectations;
- effective consultation and communication between employers and workers;
- use of timely procedures to ensure unacceptable behaviour by any one worker is dealt with. A lack of action may contribute to workplace tension and stress responses in other staff.

6. ***Disciplinary action***

A stress response to disciplinary action is to be anticipated and can affect both the giver and the receiver. It can however be minimised by:

- use of disciplinary procedures developed in consultation with workers;
- ensuring that the person taking disciplinary action has the communication skills to minimise any stress response in the receiver;
- disclosure of full facts and reasons for the discipline;
- conducting disciplinary discussions in private.

7. ***Forced redeployment, relocation or termination***

   As with disciplinary action, forced redeployment, relocation or termination can create stress for the individuals involved. To minimise any stress response, affected staff should be informed of decisions as early as possible. Staff should be encouraged to discuss their concerns, and where possible, employer support should be offered. For example, redeployment may require a consideration of career development issues and further training. Relocation or termination can have devastating effects on workers and their families. An employer could offer, for instance, assistance in finding suitable accommodation for a relocated worker. In the case of termination/redundancy, support in seeking alternative employment may be appropriate.

8. ***Role conflict***

   Role conflict can occur when workers lack the information needed to perform required tasks effectively, or when performance demands outstrip the time available to do the job. It can also occur if workers perceive work practices to be unnecessarily rigid. The outcome of these forms of role conflict can lead to a stress response with associated confusion, low productivity, anxiety and anger. Control measures which could reduce the likelihood of role conflict include:

   - information on the business objectives and expectations, training and appropriate supervisory support;
   - encouraging effective communication between staff and management;
   - recognition that workers may have, from time to time, demands on their time which affect their productivity at work, for example, study commitments or a sick child;
   - where possible, utilising workers' individual approaches to their jobs.

9. ***Shiftwork***

   An estimated 20 to 25 percent of workers undertake shiftwork in a wide range of industries. Shiftwork usually affects sleeping time and quality which can lead to fatigue and domestic and social difficulties.

These and other problems associated with shiftwork can create stress for workers and their families and contribute to workplace conflict and accidents. Management procedures which could reduce this form of organisational stress include:

- rearranging duties to reduce the need for shiftwork;
- consulting with shiftworkers or their representatives when designing or changing rosters;
- informing workers of the possible disruptions to personal and family life;
- if practical, incorporating tasks which involve activity or interaction with other team members, to help maintain alertness;
- incorporating safety checks into work practices to help guard against accidents resulting from decreased alertness;
- ensuring effective security arrangements are in place, particularly for night and weekend workers.

10. ***Physical work environment***

A work environment which exposes workers to adverse physical conditions such as noise, cold, heat and poorly designed equipment can increase stress levels in staff. The possible physical outcomes of deafness, hypothermia, heat stress or musculoskeletal injury, mental stress can be aggravated by uncomfortable environmental conditions. Productivity losses can also be associated with inadequate working environments. It is important, therefore, for employers to ensure that the working environment is as comfortable as possible and also designed specifically for the tasks to be undertaken.

Stress-Related Problems

In psychological stress responses such as anxiety and depression, stress can play a role in the development of ulcers, headaches, high blood pressure and heart disease. Research also indicates stress from low job satisfaction and poor relations with work colleagues and supervisors can lead to increased reporting of musculoskeletal disorders such as back, neck and arm pain.

These stress related symptoms can lead to absence from work. This absence may be in the form of sick leave, recreation or long service leave, leave on superannuation payments or on workers' compensation. Research shows a stress related claim for workers' compensation is often preceded by higher than normal levels of work absence.

A worker may be entitled to workers' compensation where employment has been the major significant factor causing the illness or aggravation of pre-existing illness. However, some illness may relate to events in the workplace for which compensation is not payable for example -

- action taken to transfer, demote, discipline, redeploy, retrench or dismiss the worker
- a decision not to promote, reclassify or transfer a worker, or give leave of absence, or provide a benefit in connection with the worker's employment
- the worker's expectation or perception of reasonable management action being taken against the worker
- circumstances in which a reasonable person, in the same employment as the worker, would not have been expected to sustain an injury.

The permanent loss, or lengthy absence, of skilled workers because of stress can create significant costs to the organisation in the selection and training of replacement staff and in such areas as workers' compensation. In the event of a stress related work absence, it is important for employers to focus on effective management of return to work programmes.

1. ***Rehabilitation***

   The aim of rehabilitation is to achieve a rapid and safe return of a worker to suitable employment. Research shows the longer a worker is absent from the workplace, the less likely is a return to work. Effective rehabilitation can help prevent these lengthy absences. Worker benefits from effective rehabilitation include faster recovery and reduce suffering, maintenance of social contacts, return to productive work and job, and financial security. These outcomes also benefit the employer. Additionally, employers reap benefit from less downtime and lost productivity, reduced training costs and reduced claims costs.

   The rehabilitation is most effective when it is set up in consultation with the worker, supervisor and the treating doctor. If appropriate, a union representative and an occupational health professional also should be consulted. On-going liaison between the employer and all parties will encourage the exchange of useful information and help ensure a safe and speedy return to work. Written consent from the worker is necessary before exchange of any confidential information. It is essential that the worker is involved in the decision-making process to encourage commitment to any decisions.

   The treatment of occupational stress depends on the causes of the stress and the person's response to it. Likewise, rehabilitation must include a consideration of the work and social environment and the person's coping abilities.

   The rehabilitation process must be continually reviewed to ensure rehabilitation intervention is not needlessly prolonged. In this regard, a rehabilitation plan should be initiated at the first opportunity. This plan should outline the rehabilitation objectives and the steps required to achieve the objectives. The rehabilitation counselors can assist employers to establish individual rehabilitation plans for workers.

2. ***Workplace rehabilitation***

   Various forms of rehabilitation may be required during recovery from an occupational stress-related condition. However, the most important issue for employers and workers involves workplace-based rehabilitation. This form of rehabilitation programme is performed entirely or in part at the worksite using normal or selected duties in a therapeutic manner. It involves the use of regular work tasks that are time limited, specifically controlled and upgraded according to the recovery rate of the injured worker, and as approved by the treating doctor.

   Steps in the development of a rehabilitation programme at a particular workplace could include:

   - consulting and informing workers
   - developing rehabilitation policy and procedures
   - identifying alternative duties
   - training a co-ordinator or contact person.

   Rehabilitation for occupational stress may be managed in the following ways

   - appropriate support, such as counseling, can be given to workers to enable them to continue their usual tasks
   - permanent or temporary modification of work systems and resources to accommodate the affected worker and to remove stress causes. The assistance of rehabilitation professionals may be required to establish suitable modifications
   - selected duties based on matching tasks to recovering abilities. This mode of return to work is by nature temporary and likely to lead to a return of injured workers to their own jobs.

3. ***Early intervention***

   Early intervention by an employer can significantly reduce the amount of time lost from work due to injury or illness. Early assessment, decision making and referral are significant factors in controlling lost work time due to occupational stress. Early intervention also can promote worker reaction to occupational stress.

   Early intervention requires answering such questions as:

- Can the worker with a stress related condition continue working on normal, modified or new duties as part of an on-site rehabilitation programme?
- Is the condition a more complex or of severe nature where absence from work is required?
- Is outside help required for example, from a doctor, psychologist or occupational therapist?

Stress Reduction

Psychotherapy can be very effective with people who are suffering very severely from job-related stress. Certain personality traits seem to make people more resistant to stress, and this may be because people who have them are more prone to using coping strategies that are likely to be effective. So, whether any particular tactic to cope with stress will do the trick or not depends on several factors, such as the demands and constraints of the context in which it is being used and the skill with which it is applied.

***Yoga*:** Yoga can be defined as the art of harnessing one's instinctual energies and transforming them so that they can be placed at the service of a higher consciousness. Yoga is not just a glorified form of physical exercise or relaxation therapy. At its deepest level, yogic practice can bring about the refinement of every dimension of the individual, from the most subtle to the most gross. The appeal of yoga lies in its emphasis on a holistic purification of the individual. Yoga gives equal importance to the ethical principles of conduct and physical conditioning, breath control, concentration, and meditative absorption. Since the body is the grossest part of ourselves, *assanas* give us the doorway to enter the inner realms of our system. The *assanas* refine the body, the soul's abode, so that it serves the human being efficiently.

Vipassana is one of India's most ancient techniques of meditation. Around 2,500 years ago it was re-discovered by Gautam Buddha and taught by him as a universal remedy for universal ills. At first the technique flourished in India but gradually it became corrupted, lost its efficacy, and disappeared from its land of its origin. Fortunately, the neighbouring country of Burma preserved the original form of *Vipassana* through the millenia. Since 1969 it has been re-introduced in India, where it is now striking deep roots.

Vipassana is a pali word meaning insight, seeing things as they really are. It is not a blind faith or philosophy, and has nothing to do with sectarian religion. Instead *Vipassana* is a practical method that can be applied by anyone of average intelligence. Its goal is to purify the mind, to eliminate the tensions and negativities that make us miserable.

Various *asanas* (yogic postures) and *pranayama* (breathing techniques) help to overcome stress and exercise both the internal as well as external organs. In brief, common *asanas* are:

1. *Padmasana* or Lotus pose which is ideal for practicing meditation and breathing exercises.
2. *Paschimotasana* is good for the back and hips and keeps them supple.
3. *Bhujangasana* is ideal for the back, shoulders and internal organs of the stomach.
4. *Vrikshasana* or tree pose helps one with his balance and develops concentration.
5. *Shavasana* is the ultimate form of relaxation of the mind and body and lowers blood pressure and is a must in today's stressful time.

The following factors are to be borne in mind in any stress reduction programme:

- Identify management practices that cause stress and try to change them.
- Try to keep physically fit in order to stay mentally healthy as well.
- Use the Indian concept of actions divorced from desire to de-stress the self.
- Develop the idea of an executive group retreat as a tool to release stress.

Hints for reducing stress on the job as suggested by Marc G. Singer:

1. Rate your tasks in order of importance and complete the most important ones first.
2. Whenever you can, delegate.
3. Set realistic deadlines and establish a schedule. Then, stick to it.
4. Plan ahead for future events.
5. Stop and take breathers during the work day. Everyone fatigues sooner or later, and fatigue usually results in lowered productivity.
6. Don't procrastinate on making decisions. Worrying about them is sometimes more damaging than the negative consequences of making a bad decision.
7. Determine whether or not you control the outcome. It serves no purpose to worry about something you can't control anyway.
8. Eat sensibly; avoid indulging in alcohol, tobacco, and drugs; get adequate rest; and exercise regularly.

The extent of dysfunctional stress could be evaluated by performing a stress audit, which helps identify the symptoms and causes of stress. To perform an audit, raise the following questions:

1. Do any individuals demonstrate cardiovascular, gastrointestinal, allergy-respiratory, or emotional distress irregularities?

2. Is job satisfaction low, or job tension, turnover, absenteeism, strikes, or accident proneness high?
3. Does the organisation's design contribute to the symptoms?
4. Do interpersonal relations contribute to the symptoms?
5. Do career-development variables contribute to the symptoms?
6. What effects do personality, sociocultural influences, and the nonwork environment have on the relationship between stressors and stress?

Alcoholism

According to a definiton given by an Expert Sub-Committee of the World Health Organisation (WHO), "Alcoholics are those excessive drinkers whose dependence upon alcohol has attained such a degree that it shows a noticeable mental disturbance or an interference with their bodily and mental health, their inter-personal relations and their smooth social and economic functioning. They therefore require treatment."

Due to the different unfortunate consequences resulting from an excessive consumption of alcohol and in order to avoid any confusion in the terminology, the norm today is to speak of "alcohol-related problems" or "alcohol-related disabilities" rather than "alcoholism".

Alcoholism is an age old social, physical, and environmental problem, but today it is generally considered a psychiatric one. Actually, mental problems may not be the underlying cause of alcoholism since non alcoholics often have the same apparent mental problems as alcoholics. Nevertheless, alcoholism creates "lost weekends" increases marital flare-ups, destroys physical health, brings financial crisis, and causes anti social behaviour. Numerous statistical investigations have shown that alcohol is at the origin of offences and accidents of all kinds. The chronic alcoholic may subconsciously feel that work itself is one his problems, and that all these problems interfere with his drinking, which is his only solace. Whenever drinking consistently interferes with the job, the physician is obliged to call it alcoholism.

Industry has recognised the disastrous effects of alcohol on its workers over the years. It has found that excessive drinkers have a higher rate of sickness, inefficiency, absenteeism, and a higher accident proneness. General opinion is coming to accept that action must be taken within industry itself to reduce the problem of alcoholism. This action is justified by the fact that alcoholism gets detected and treated at a sufficiently early stage, and is usually curable.

The incidence of alcoholism among supervisors and executives is another serious problem, but it is difficult to assess because of inconsistent policies and data in this respect. Even without data, a diagnosis may still be made by observing certain general characteristics of alcoholism such as, emotional distress and inappropriate changes in appearance, ability, coordination, use of authority, attitudes, and adaptability.

The three important predisposing factors to alcoholism are hereditary sensitivity, personality disorders, and the widespread practice of social drinking. These three combined factors make for a potential alcoholic once the habit begins. Early identification of alcoholism is important, and the earliest cause for action is a poor work performance. Supervisors who suspect that an employee's work difficulties may be due to problem drinking should discuss the matter with the employee. Personal rehabilitation plan, if any, can be obtained from doctors, counseling personnel, psychiatrists, governmental agencies, or alcoholics anonymous. There is no one "best" way to handle an employee with a drinking problem. The approach will vary according to the individual. In any case, the supervisor should not exact or demand any promise the employee might make to stop drinking. Moreover, the problem should be kept confidential. There is nothing to gain and much to lose when someone's personal problems are disclosed.

Although alcoholism, or problem drinking is a problem with medical, social, and economic ramifications; experts view it as a disease with identifiable causes and a cure. Prevention should be at three levels: education, improvement of working conditions, and catering for workers' food and drink requirements. Worker education must be conducted with skill and persistence and the assistance of both management and unions is invoked. It should take place on an individual basis or be directed at small groups of workers who run a high risk of alcoholism. Educational action should be undertaken by the industrial medical officer, the industrial nurse, the safety officer, and related social welfare agencies. The improvement of working conditions can also contribute to the reduction of alcoholism. Attention to the workers' nutritional requirements and dietary intake is an important measure in preventing alcoholism in the industry.

Alcohol problems at work cause extensive economic loss through inefficiency, absenteeism and accidents. An alcohol education programme should provide information and advice on sensible drinking, and recommend safe limits and inappropriate drinking.

First the man takes a drink, then the drink takes the man, says a Japanese proverb. It takes about 10 years for alcoholism to develop fully, but in the early stages it is difficult to distinguish problem drinking from social drinking. If people recognise the danger signs early enough, alcohol dependence can be controlled. Here is how the disease advances:

- Occasional drinking in company.
- Drinking more than others in the group.

- Relief drinking to feel better about problems.
- Finding excuses to drinking.
- Having drinks secretly or lying to cover up drinking.
- Hiding or protecting liquor supply so you are never without it.
- Personality changes even when not drunk.
- Inability to do a job responsibly.
- Negligence in eating and self career.
- Complete surrender to addiction.

Drug Abuse

The current explosion of non-medical drug usage has become a true personnel problem. A much greater number of people of all ages are becoming involved in the abuse of drugs. A new awareness of this hazard has resulted in both governmental and private groups working toward the proper solution of the continuing problem. Nevertheless, non-medical drug abuse is a growing cause of concern and will require expert management.

According to experts, when a drug-dependant person misses his normal dose, he feels a general uneasiness. This may not include a return of the original problem, yet there is a feeling of deprivation, irritability, unsteadiness and anxiety. In case of stronger drugs, the withdrawal symptoms could include a flu-like feeling, muscular aches, sweating, nausea and palpitations.

Smoking

The use of tobacco by workers and the general population continues to increase world-wide. Smoking is thought to be an important factor in the development of many disorders of clinical and public health importance. Smoking in the workplace may lead to contamination of cigarettes and other tobacco products, as many substances present in tobacco are also present in the working environment. Therefore, smoking can increase exposure to those chemical agents. For instance, cold dust and cigarette smoking produce an additive effect in obstructive airway disease. Similarly, in cotton workers cigarette smoking will increase the prevalence of the early stages of byssinosis. There is possible relation between smoking and lung cancer. Asbestos workers with smoking habits have been found to show a much higher lung cancer risk than asbestos workers who were non-smoking.

CASE 1

Alcoholism

Many companies have adopted the modern view that alcoholism is a form of sickness. When an employee is afflicted with the problem, they hold off from termination of service, and instead offer an opportunity to obtain medical and other professional help.

In company X, the supervisor Patil was convinced that Sitaram was an alcoholic. The supervisor tipped off the personnel department that ran the company's alcoholics aid programme.

Sitaram was directed to enroll in a three-week rehabilitation programme conducted by a local hospital. Sitaram wailed "you people did me a wrong. I am not a boozer".

However, Sitaram's protests went unheared. Fearing that he would be subjected to disciplinary action, he reluctantly agreed to enter the hospital. Sitaram remained two weeks at the institution. He steadfastly denied that he was an alcoholic. As a result, hospital discharged him early because he was uncooperative.

When Sitaram went back to work, he was discharged from service after a domestic enquiry. Sitaram took the help of his union which initiated a move for withdrawal of termination order. The union and management agreed to refer the dispute to arbitration.

At a session before an arbitrator, management stated its side:

- We extend a helping hand to every employee who suffers from alcoholism and willingness to do anything to that end.
- But Sitaram resists any effort to help him. That is why the hospital threw him out.
- He is a confined alcoholic, and we do not want him to continue in our employment.

Sitaram responded

Sure I take a sip now and again. But this does not mean, I am an alcoholic.

I was forced to enter the hospital under the threat of dismissal. When I did not permit those staff at the institution to handle me the way they did with their really sick patients, they became annoyed and tossed me out.

→

Questions

1. Under the circumstances, should the employee be taken back or remain under termination? Why?
2. Why should an organisation be concerned about alcohol and drug usage by its employees?
3. Critically assess the role of a hospital in handling the cases of alcoholism and drug abuse, keeping in view the present case.

CASE 2

Managing Stress in Healthcare

'Stress' has received a great deal of coverage in recent years in the media and has been the subject of considerable psychological research. Over the past few years, occupational psychologists have conducted research aimed at identifying work stressors; that is external, relatively permanent features of the work environment or the work itself. This has led to the development of models postulating a wide variety of job characteristics that are regarded as stressful. Individual differences in personality, coping mechanisms are important in understanding stress. It is possible to identify those features of the work environment which are likely to be a source of strain for many workers.

Health and clinical psychologists have recognised the importance of minimising patients' stress in health care settings. Here psychologists' interventions have primarily focused on a variety of interventions to help the individual to cope with stressful medical procedures. An individual is often subject to a wide range of diverse stresses. A hospital patient suffering acutely may cause irritation or restlessness among the fellow patients.

It is clear that work stress is not a simple issue. For example, the heavy workload that junior doctors experience may well be a source of stress, but the extent to which it leads to strain will depend on a number of intervening variables. The extent of resources to deal with the work load, the amount of autonomy the doctor has, and the amount of social support will all be important. There are a large range of widely diverse professional groups within the health service who are subject to one type of stress or the other. A number of studies have suggested that health workers experience more stress than comparable samples of non-health workers. Many health professionals are likely to suffer from organisational stressors that are intrinsic to the nature of the job; for example providing terminal care, counseling bereaved parents and

→

relatives or dealing with disturbed and violent patients. An additional problem for health service workers is that their stress may have a direct effect on the recipients of the service, such that communication with patients may deteriorate. This is confirmed by a study which found that nurses reporting higher levels of subjective stress were rated by their supervisors as poorer on human relations aspects of their work performance, such as interpersonal effectiveness and tolerance. Obviously, stress is an important issue in maintaining a high standard of patient care.

A study by Parkes (1982) compared stressors and strains of student nurses during assignment to medical and surgical wards. She found that medical wards caused higher levels of depression and lower levels of job satisfaction among the students. Surgical wards offered work which was rated higher on a number of key characteristics such as control on job discretion and opportunity for use of skills and acquisition of new skills. Work on medical wards was more emotionally demanding with lower levels of social support.

Junior doctors have also been the focus of a number of studies. Suicide rates, alcoholism and depression have been found to be high in the medical profession and particularly among junior doctors. The long hours of work are a particular problem, but the responsibilities of dealing with death and dying and the possibility of making mistakes are among a number of other important work stressors.

It should, however, also be pointed out that there are a number of positive aspects about many jobs within the health service which might also be expected to lead to higher levels of satisfaction. The work is valued by society and some 'vitamins' are likely to be present (for example, variety, skill utilization, and intellectual challenge) which are likely to compensate for other potential stressors. Nevertheless, the numerous studies indicate that work stress in the health service is a significant problem. The emphasis in psychological research so far has been to view stress as an individual problem to be treated by individual stress management interventions. There is, however, a need for greater focus on organisational factors and environmental factors, and approaches to organisational change which may minimise stress for both staff and patients.

Questions

1. How can the stress be minimised amongst healthcare professionals?
2. What are the main stress-related problems in nursing administration and practice?

SUMMARY

- Stress is a mental or emotional tension or strain resulting from adverse or demanding circumstances.
- Hans Selye views stress as a non-specific strain on the body caused by irregularities in normal body function.
- Stress can affect all aspects of life, including emotions, behaviours, thinking ability, and physical health.
- Long-term health and safety performance related issues may be identified and addressed through stress audit in an organisation.
- In terms of human and economic costs, it is important to address stress issues in the workplace, and stressors of most relevance should be identified.
- Employers need to consider the frequency and extent of stressors, and their effects in the workplace.
- Procedures for reducing stress and stress responses should be implemented effectively.
- Affected workers can benefit from rehabilitation programmes for stress related conditions which impact on work performance.
- Benefits can be maximised if it is recognised that both employer and worker can contribute to the minimisation of stress at work

DISCUSSION QUESTIONS

1. What is stress? What are the causes of executive stress?
2. Suggest stress reduction measures at different levels of organisation.

14

Environmental Pollution and Protection

Pollution is not a new phenomena. It is as old as civilisation itself. It was conceived the day Adam dawned on the planet earth. With the advancement of civilisation, it also matured and expanded in proportionate dimensions. It is the bye-product of development and infact a price for progress. Dhanwantri, an outstanding physician and philosopher of India stressed the importance of good air and water and emphasised the need and necessity of proper sanitation.

Industrial revolution of 19th century lead to environmental disaster. The environmental problem assumed colossal importance with the transition from feudal system to industrial capitalism. Man's capacity to become master of his surroundings and his quest to enhance the quality of life has caused incalculable harm to human beings and environment. Today, the entire world is worried about the deteriorating and depleting conditions of environment. The protection and improvement of environment has become a major issue which affects the well being of the people and economic development. Atmosphere being a common heritage of entire mankind, it is the duty of all to preserve it. Kurt Waldheim, the then Secretary-General of United Nations addressing a Conference in 1972, observed that pollution of environment is a problem "no nation, no continent, no hemisphere, no race, no system can handle alone". He further observed that "the quality of our atmosphere can be nothing else but the bye-product of the behaviour of the nations". The Charter of Economic Rights proclaims that "the protection, preservation and enhancement of environment for the present and the future generations is the responsibility of all States".

Pollution is the deliberate or accidental contamination of the environment with human waste. It has been defined as "matter in the wrong place", or "anything released into the environment which degrades it". It is the release of substances or energy into the environment by human beings in quantities that damage either their health or resources.

Pollution has become the first enemy of mankind. Today, the whole mankind is more afraid of pollution rather than the nuclear holocaust.

Air Pollution

We can live without food and water for days together but for only five minutes without air. On an average a man needs 30 lbs of air everyday. So air pollution is indeed of great immediate concern than any other aspect of pollution. The dry air has concentration of certain gases, which are naturally present in the atmosphere. Whenever the balance of the natural composition of air is disturbed and have an adverse effect on man and environment can be termed as pollution of air. Thus, air pollution is the accumulation of any substance in the air in sufficient concentration to effect the man, animal, vegetation or other materials. Pollutants may occur as solid particles, liquid droplets, gases, or in various combinations of these forms. Pollutants may be directly emitted into the atmosphere from the identifiable sources and include solids, aerosole, such as carbon particles, metallic dust fluorides, cigarettes and industrial or vehicular smoke, pollens and sulphates, course particles, nitrogen compounds. Pollutants may also be produced by interaction among two or more primary pollutants or by reaction with normal atmospheric constituents, with or without the aid of sunlight.

The effects of air pollutants on man are varied. When carbon monoxide is inhaled it displaces the oxygen in the blood and reduces the amount of oxygen carried to the body tissues. The gas is believed to impose an extra burden on those already suffering from anaemia, diseases of the heart and blood vessels, chronic lung conditions and overactive thyroid. Sulphur oxide can cause temporary and permanent injury to the respiratory system, irritating the upper respiratory tract, lung tissues.

Many other pollutants are a growing public health worry even though they may not constitute immediate and direct threats. Air pollution has inflicted wide-spread damage on plant life, buildings and materials. Air pollutants also damage man's durable products; steel corrods 2-4 times faster in urban industrial centres than in rural areas where much less sulphur bearing coal and oil are burnt. Air pollution disasters have caused a colosal loss of human life, animal and the plant life. The devastating example is the "Bhopal Tragedy" in which thousands of people died. Many more were disabled. The animals that perished in this disaster have not been counted at all.

Pollution control consists of the legal, institutional, scientific and technological arrangements established to avoid or mitigate such excesses in the environment. It can be accomplished by containing pollutants at the source, by devising new technologies of manufacture that eliminate or reduce pollutants, and by recycling materials and commodities through reprocessing and resource recovery.

Water Pollution

Water is of such critical importance to life and the environment that it would be difficult to think of life on any planet without it. Water is plentiful, but pure water is scarce. Of all the types of pollution that affect the health of people, water pollution is by far the most serious. According to Central Board for Prevention and Control of Water Pollution, the fresh water that is so essential to our lives is only a small portion of the earth's total water supply; it is only about two percent of the total. Almost 85 per cent of the rain falls directly go into the sea and never reaches the land. The small remainder precipitates on land. It is this water that fills the lakes, wells, underground supplies, and keeps the river flowing. Humanity is left with only one tea-spoon full of sweet water for every five litres of total water.

Water is thus a scarce commodity and is getting scarcer everyday as communities, industries and agriculturalists discharge their filth, muck and harmful wastes into the nearest sink. In view of the importance of water, several Acts have been passed by the Government of India and State Governments.

Water pollution has been defined to mean "such contamination of water or such alteration of the physical, chemical or biological properties of water or such discharge of any sewage or of trade effluent or of any other liquid, gaseous or solid substance into water (whether directly or indirectly) as may or likely to create a nuisance or render such water harmful or injurious to public health or safety, or to domestic, commercial, industrial, agricultural or other legitimate uses or to the life and health of animals or plants or of aquatic organisms".

John R. Holum defined water pollution as "The addition to water of an excess of material (or heat) that is harmful to humans, animals or desirable aquatic life, or otherwise causes significant departures from the normal activities of various lining communities in or near bodies of water". The National Water Commission (U.S.) stated that "Water is polluted if it is not of sufficiently high quality to be suitable for the highest uses people wish to make of it at present or in the future".

In India there are many sources of water pollution. Firstly, community wastes of human settlements give rise to water pollution. Most of these wastes are discharged untreated into the water courses. Secondly, industrial effluent directly entering into a stream or through a municipal sewer or through a discharge on land meant for irrigation causes water pollution. Thirdly, anything stored on earth, such as, any raw material, solid refuse of a mine or quarry on any land may cause pollution by rain washing it into a stream. Fourthly, the use of fertilisers also causes water pollution when unused nitrogenous fertiliser is drained out of soil into lakes and rivers. It causes overgrowth of green plants resulting in pollution of water. The use of pesticides in agriculture may cause pollution due to rain water washing it into stream. Fifthly, water pollution may also be caused by air pollutant.

Even ground water is not free from pollution. It is being polluted due to dump trade or sewage effluent into under-ground strata. But ground water may be polluted due to seepage or percolation from the surface. Human wastes may cause pollution by seepage from improperly constructed or improperly placed septic tanks, or leaking sewer lines. Industrial wastes, including highly poisonous chemicals, may be introduced to the ground water either by intention or accident. Even the dumping and covering of vegetable materials in garbage result in their decomposition and they carrying down of decomposition products, including carbon dioxide gas, by percolating water into underlying ground-water bodies from the article.

Noise Pollution

Man is born with noise and dies with that; thus, noise is an integral part of human life and a natural product of human life and a natural product of human environment. It is connected with man's life from cradle to grave, even before and after it. We, in India, have a traditional liking for noise. All our happy and sad moments of life are expressed through noise. It may be in the form of bursting of crackers, playing of music or recitation of religious scriptures. Unconscious of the immediate and ultimate ill effects of this traditional attitude, life goes on without any tangible protest from the public, public servants or the persons occupying the chairs of this democratic Republic. In fact there is a silent compromise by us to take noise as a normal part of our life without realising that we are playing with not only the health of the present society but with the health of the posterity.

The environmental action, started with the Stockholm Conference (1971), at national and International level, was not much concerned with the noise pollution to begin with. It was only after scientific research exposed the most dangerous effects of noise on man's health that a serious thought was given by international community to control this dangerous monster. The growth of cities and development and concentration of big and small industries around them changed the amount and composition of our society, and created the most acute problem of our century–the problem of noise pollution. In one degree or the other this problem is faced by all the cities, though our villages cannot be said to be totally out of the grip of this problem as technological developments have also made inroads into the agricultural sector which is by and large a profession confined to our villages.

Commercial outlook towards life and nature has further added to the intensity and extent of the problem of noise pollution. Population explosion has further aggravated the problem of noise pollution. Every year man increases pressure on land by ever increasing population. This has exerted pressure on all available resources besides the significant rise of noise level. Increase in population has resulted into massive increase in human activity in all spheres connected directly or indirectly with the human beings. It may be in the form of increased traffic of automobiles, railways, aircrafts, or new construction of buildings, industries, roads and railway lines etc. The ultimate result of all this is, that noise has gained such an intensity that it grates on everyone's nerves.

Noise may have any one of the following effects on man. It may change a man's psychological state, may cause heart attack, may cause chronic effects as hypertension or ulcers and damage one's hearing. In addition to health hazards, noise costs heavily in terms of money to the country which may not possibly be estimated correctly.

Although there is no single universally accepted criterion of what constitutes noise pollution or excessive noise, it is widely accepted that excessive noise has an adverse effect on human health. Noise has been defined as an excessive, offensive, persistent or stoutling sound. In American Jurisprudence noise has been defined as unwanted sound that produces unwanted effects. It is generally considered to be a form of pollution and has begun to be recognised as a major evil. Similarly, a large number of industrial psychologists have defined that term noise in different terms. In the words of Harrell, "Noise is an unwanted sound which increases fatigue and under some industrial conditions it causes deafness". Blum defines noise as "a distractor and therefore, interfering with efficiency". According to J. Tiffin, "Noise is a sound which is disagreeable for the individual and which disturbs the normal way of an individual".

***Sources of Noise*:** Sources of noise are numerous but may be broadly classified as: (a) industrial, and (b) non-industrial. The industrial may include noises from various industries operating in cities, like transportation, vehicular movements such as car, motor, truck, train, tempo, motorcycle, aircrafts, rocket, defence equipments, explosions, and so on. Among the non-industrial sources, important ones are the street noise due to hawkers, use of loudspeakers, thunder, demonstrations, and so on. However, the list is not exhaustive as the number is on the increase with the industrial and technological advancement.

***Effects of Noise*:** Noise has many ill effects on living as well as non-living things. Its general effect on human beings is that it causes disturbances in sleep which lead to other side effects. It has auditory effects like loss of hearing. Broadly speaking it has (a) psychological, and (b) physiological effects. Many behavioural changes are recorded as a result of exposure to high decibel noise in human beings as well as in animals. Certain symptoms can be observed outrightly. The undesired sound may cause annoyance. Interruptions in speech communications may impair performance, lead to errors and lower output and efficiency. Noise can cause tension in muscles, nervous irritability and strain. No doubt, the noise reaction varies to a large extent in different individuals. Noise also produces physiological effects on human body. Prolonged chronic noise can also produce stomach ulsers as it may reduce the flow of gastric juice and change its acidity. It may lead to abortions and other congenital defects in unborn children.

Many countries have enacted specific legislation to control noise pollution. In England, there is the Noise Abatement Act, 1960, of which Section 2 provides that loudspeakers shall not be operated (a) between the hours of nine in the evening and eight in the following morning for any purpose; (b) at any other time for purpose of advertising and entertainment, trade or business. There are exceptions, of course, prescribed in the Act.

The United States Noise Pollution and Abatement Act, 1970 is an important legislation for regulating control and abatement of noise. Under this law, the environmental protection agency, acting through the office of Noise Abatement and Control, holds public meetings in selected cities to compile information on the noise pollution. In some U.S. States, environmental rights have been embodied in their constitution.

In India, it is important to know that there is no law which exclusively deals with problems of noise and its control. Certain States have made legal provisions regulating and restricting the use of loudspeakers. What is warranted for is a uniform law for controlling noise pollution so that the society is free from this hazard. A national environmental policy and ecological plan must be promoted by law so that man and nature love and live in harmony. There is no escape from pollutant. It reaches us through, the air we breathe, the water we drink, the food we eat and the sounds we hear. In view of the alarming proportions of noise and its impact, what is called for, is 'noise control' through technology, and by other remedial measures. Only a healthy environment can provide a healthy body and a healthy mind.

***Remedies*:** Some of the concerted efforts that can be made to control noise pollution can be divided into the following categories:

(a) ***Administrative Remedies***: Noise pollution is one of the fields in which administrative process seems to enjoy advantage over judicial mechanism in its control and abatement. Since administration is not only the strong arm of the State with all the coercive powers of the State but also dominates in policy and planning of all the development programmes of the State. As a matter of rule legislature confers powers upon administrative agencies for the implementation and enforcement of pollution control laws. Further, the administrative agencies play a vital role in giving effect to any law or legal process and the success or failure of any legislative attempt largely depends upon the sincerity and commitment to the cause of the administrative wing of the State.

(b) ***Judicial Remedies***: Courts can play a very constructive role by their pragmatic judicial approach towards the problems dealing with noise pollution. Firstly, this is possible by exercising the discretion of judicial review in favour of administrative orders or regulations; and secondly, judiciary must reflect its concern for environmental protection by providing effective remedies to persons who approach the court in any case of environmental pollution.

(c) ***Legislative Remedies***: Law is an instrument of social change and the legislature is the authority in control of law-making process. All other authorities like administrative agencies, courts, and other social institutions concerned with environmental protection derive their powers from the law passed by the legislature and operate within the limitation prescribed by law. Thus, legislative action is the most effective approach to governmental and judicial action against noise pollution.

(d) ***Public Co-operation***: Law remains in suspended animation unless the public becomes conscious of their rights and availability of various remedies for their enforcement. This fact becomes more conspicuous in case of noise pollution because our masses are still ignorant of the unhealthy effects of noise pollution. People's involvement and active co-operation is essential for the success of any or all the programmes directed towards securing pollution free environment.

(e) ***International Co-operation***: International co-operation can also help in a great way to deal with the problems of noise pollution. National action needs to be supplemented by international measures and co-operation. Take for example motor vehicles, the single most significant source of noise, require international technological co-operation for reduction in noise emission from the vehicles. International assistance can be secured through exchange of appropriate technology, research programmes, legal and other methods, and learning from the success or failure of others.

Pollution is not something new. With the growth of civilisation, it also grew in various directions. Awareness about the need for the protection of environment is increasing the world over day by day. A number of conferences, seminars and workshops are being organised every year. Various steps have also been taken both at national and international level. But these steps and measures adopted are not sufficient to meet the menace of pollution. The accelerated exploitation of natural resources and industrialisation, unchecked use of chemicals and pesticides lead to more pollution. Even natural calamities like typhoon, earthquake also affect environment. Destruction of environment will result in depletion of life on this earth. The task is himalayan, but the resources are scarce. So we must pool all our resources, knowledge and means to check pollution of water, air, destruction of forests and other natural resources to ensure health and life for us and for our future generations.

Environmental Policies

Environmental policy can be defined as the sum of objectives and measures designed to regulate society's interaction with the environment as a natural system; it comprises aspects of environmental conservatism, restoration, and management. Since society's long-term existence ultimately depends on the natural environment and its resources, environmental policy in principle involves all societal rules governing the use of nature by human beings.

Practice, however, does not conform to such a broad definition. Generally, only selected parts of the relations between environment and society becomes the subject of environmental policy. So far, specific policies concerning air quality, water quality, noise abatement, and waste disposal have been formulated. This conventional type of environmental policy was based on the simple recognition that a number of waste products resulting from production and consumption processes are detrimental to human health. Only in recent years it has been recognised that both the efficiency of the ecological and economic systems and, ultimately, the sustainability and the acceptability of the social system are threatened.

Political reaction to this realisation brought about changes during the late 1960s and the 1970s to reduce the amount and degree of harmful waste products. Contrary to original assumptions, however, discrete policy areas that control the use of resources and the corresponding technical systems– especially those policy areas governing industry, infrastructure, technology, raw materials, energy and agriculture - proved to be vital factors in the relationships between environment and society. Of course, conventional media-specific environmental regulation is useful and still necessary. However, it carries the inherent risk that the measures needed may not be coordinated, that problems may be spatially shifted and or displaced from one environmental medium to another (problem shifting; problem displacement).

A discussion on the relationship between environment and society should thus not be solely confined to the questions of conventional environmental policies as described above. A broader, more comprehensive perspective on environmental problems is needed in which the political response of the various social factors, government, industry, environmentalists, and of society as a whole, to the issue of the use of nature and of harmful waste products should become the focus of discussion.

After a decade or so of applied environmental policy, the basic lines of response to environmental problems can be pointed out. Of primary importance is the question of political response or implementation of policy, namely, how rapidly and in what way or style does such response occur and how much freedom do the various societal actors have in actively selecting existing options to deal with environmental problems?

The actual or alleged tendency to remain more or less curative, re-active stages of response has been viewed by many critics of environmental policy as a structural deficiency of industrial society. Whether, and to what extent, processes of ecological adaptation and structural change take place depends partly on the knowledge acquired about environmental damage itself, on the availability of low emission technologies, and on the relative strength of economic and political interest groups at the time when environmental damages are assessed and the costs of response are to be distributed. Successful environmental policy, then, results in part from improved information and technology; it is, however, also determined by structural economic and social change. It may be assumed that as knowledge about the state of the environment, its trends, and its determinants increases, responses can be expected that go beyond mere reaction. Environmental policy would then progress from its curative phase (react-and-cure strategy) to prevent one (anticipate-and-prevent strategy). It would evolve from a policy of *ex post* environmental protection into an integrated *ex ante* environmental policy, devoted to an ecologically sound development of economic and technological systems.

The concept of preventive environmental policy could, in short, be described by several basic patterns of goals and means, such as:

(a) prevention of the spread of all harmful emissions that exceed the assimilative capacities of ecosystems through more and improved recycling, introduction of low emission technologies, and pre-emptive substitution of environmentally harmful products and production processes.

(b) conservation of non-renewable resources, encouragement of the use of highly efficient renewable resources, and drastic reduction of combustion processes, and

(c) active management of the natural environment allowing for greater participation of hitherto under represented public interests in all relevant planning procedures, and aiming for a strong institutionalisation of the prevention principle throughout the society.

By consequence, the successful application of preventive environmental policy requires structural changes of the existing political and administrative system because it must cover a scope broader than that of conventional, media-specific environmental policy. It may be expected that preventive environmental policy is going to encounter political opposition that will attempt to restrict the scope of the policy, to deinstitutionalise parts of it, and ultimately to force it back into the limited realm of regulating immissions (effects) instead of emissions (causes). If such a tendency prevails, the largely re-active, curative environmental policy of the past would be preprogrammed for the future.

Healthy environment is the most essential prerequisite of human life. But to-day man in his unquenching thirst for comforts, is so arrogantly misusing his environment that he is quite close to suffocating himself. There prevails environmental degradation deterimental to the existence of living species. All components of environment, viz., air, water, soil, and noise, are so sadly affected by the pollutants that the whole environment has been polluted. Pollution occurs when environmental changes create or are likely to create nuisance or hazards to public health, safety or welfare or when they are harmful to domestic, industrial, agricultural, recreational or other legitimate uses of environmental components or livestock, wild-life, fish aquatic life and other biological species. All aspects of pollution or environment are directly or indirectly related to human health and well-being. Pollution thus has a great effect on living organism and in case appropriate steps are not taken to control environmental degradation, the life will become miserable. In fact, pollution has become a great menace to life and all the countries, whether developing or developed suffer from environmental crisis. The environment to-day is under severe threat from the pressure exerted by the growth of human and animal populations, poverty and the misuse/unplanned use of natural resources.

The impact of pollution is so alarming that according to World Health Organisation's estimate as much as 80 per cent of the world's diseases are traceable to pollution. Jaundice, typhoid, cholera and other gastro-enteric diseases are attributed to water pollution. Added to it, air pollution causes innumerable health hazards. Noise pollution or unwanted sound is another form of environmental hazard. Noise pollution is dangerous for human life. It affects the brain and prolonged exposure to noise may cause blood vessels to contract, sometimes resulting into hypertension. Thus, it can be said that pollution of whatever kind it may, causes serious damage to man, animals, birds, aquatic life, crops and vegetation.

The environmental pollution can be attributed to several factors like urbanisation, industrialisation, automation, unsound planning and lack of general awareness in the masses. Urbanisation, though a global phenomenon caused due to industrialisation, migration of labour from rural to urban areas and other pull and push factors, has brought in its wake series of problems like congestion, insanitation, slums and degraded environment. In most of our cities congestion and over-crowding exist to such an extent that constitutes a menace to public health and living. Over-crowding signifies congestion and shortage of houses. The production of chemical fertilisers used for the increase in production, and of pesticides for the prevention of decay in agricultural production too, sometimes affect the environment. The gases discharged by the industrial units pollute the atmosphere and affect the health of human beings. One of the chief reasons for the present environment crisis is that great amount of materials have been extracted from the earth, converted into new forms, and discharged into the environment.

Certain industries give rise to a considerable amount of dust which not only pollutes the environment but is associated with various forms of ill health. Constant inhalation causes irritation of nasal passages and produces lung's diseases. The emission of black smoke from chimneys of factories is another source of nuisance to the residents in their immediate vicinity.

Transportation is another source of pollution today. The smoke released by the driving vehicles spread pollution which vitiates the whole of atmosphere. Since the mid of this century transportation through road has increased tremendously.

Control over sewerage and industrial discharge into rivers and disposal of waste is so fragile that almost no pollution control is possible. The sewage is generally discharged into rivers. The pollution of rivers affects the surroundings. In India, the sewage of all the big cities is being discharged into the rivers and subsequently it leads to water pollution. The defects in the sewerage systems too, add to pollution. Solid waste collection and disposal methods add to the over-all problems of pollution. The garbage from households and their disposal at each point spreads pollution throughout the city.

Lack of general awareness and poverty among the masses is also responsible for environmental crisis. Majority of the people being poor are always concerned with bread earning and pass the lifetime. They do not get time to think of pollution eradication nor are aware of its effects.

Unsound planning, too, add to the environmental crisis. Lack of planning has caused uneven growth of cities and towns. The haphazard expansion stand in the way of having modern municipal amenities like sewerage, proper drainage system, water supply, roads and pavements, cause an atmosphere injurious to health.

Exhibit 14.1 Environmental Policy of Tata Steel

Tata Steel reaffirms its commitment to minimise the adverse impact of its operations on the environment. Towards this end, it shall endeavour to:

- Set around environmental objectives and targets, and integrate a process of review, as essential elements of corporate management.
- Install, maintain and operate facilities to comply with applicable environmental laws, statutes and other regulations.
- Conserve natural resources and energy by constantly seeking to reduce consumption and wastage.
- Minimise process waste, and promote the recovery and recycling of materials.
- Phase out pollution-prone processes and install state-of-the-art technology for pollution prevention, and the continual improvement, in environmental performance.
- Develop and rehabilitate waste dumps through afforestation and landscaping.
- Develop an environmentally aware workforce.

Environmental Programmes

Though provisions have been made in the Constitution of India for environmental protection and a number of acts have been enacted, environmental protection did not receive any significant attention until the early seventies. The fourth plan gave explicit recognition to the fact that planning for harmonious development is possible on the basis of comprehensive appraisal of environmental issues and it was necessary, therefore, to introduce the environmental dimension into the planning and development process. To advise the Government on environmental problems, a National Committee on Environmental Planning and Co-ordination (NCEPC) was constituted on February 1972. On the recommendation of this committee, the Government set up a separate department of environment on 1st November 1980. The Union Government also asked the State and Union Territory Governments to establish similar departments to deal with environmental considerations. Among other things, the department is responsible for monitoring and control of air and water pollution. The department is assisted by several other agencies.

Several Acts have been promulgated for the prevention and control of air and water pollution. These include mainly: (a) the Water (Prevention and Control of Pollution) Act, 1974; (b) the Water (Prevention and Control of Pollution) Amendment Act, 1978; (c) the Water (Prevention and Control of Pollution) Cess Act, 1977; (d) the Air (Prevention and Control of Pollution) Act, 1981; and (e) the Environmental Protection Act, 1986. The Central Pollution Control Board is responsible to administer these Acts with the help of the State Pollution Control Boards.

The Government has taken steps to promote environmental research and sanctioned several research projects to the universities, research institutes and non-governmental agencies. To boost the programme of environmental education, training and awareness, several workshops have been organised in different parts of the country. Just like the Centre, several steps have been taken at the state and local level for environmental protection and prevention of pollution. Despite all these efforts, no phenomenal success has been achieved and people both in urban and rural areas are living in an atmosphere of dirt, pollution and degradation.

The following measures need to be considered in improving environmental protection:

1. Unfortunately, the municipal organisations today which are required to enforce existing provisions against pollution are themselves the major pollutors. Their own out-dated methods of solid waste management, the crude dumping grounds, inadequate drainage system, inadequate administrative machinery, are some of the glaring examples of their ineffectiveness in this field. The municipal institution should take these problems as their own and adopt the modern techniques to manage the sewage, discharge of contaminated water, and other contaminants.
2. The financial inadequacy too, is responsible for this crisis. No doubt, the Central Government has made lofty plans for eradication of pollution but the implementation lies with the state and local authorities which lack the fiscal power and resources to meet environmental challenge. Therefore, it is essential that all implementation agencies should be given adequate financial help and autonomy to meet the menace of environmental pollution.
3. The local authorities lack sufficient power and authority to control the agencies responsible for environmental crisis. Particularly, they feel helpless with regard to many big and medium industries. The local bodies and the State Governments should be entrusted with sufficient powers to punish those who violate the law and cause pollution.
4. Our policies with regard to environment management are good but their implementation is poor for want of trust and dedicated

personnel. Therefore, steps should be taken to impart right type of training to the concerned people in the process of environmental management.

5. Like other programmes there is lack of people's participation. The people generally do not take interest and consider it to be the responsibility of the government to work for the protection of the environment. As far as possible, voluntary agencies should also be involved in the environmental protection schemes.

To sum up, pollution problem in the country has assumed gigantic dimensions and if necessary steps are not taken to meet this challenge and contain this crisis, there will be further environmental deterioration having cancerous effect on human life and other species. There is a need to have a national thinking without bias and partial considerations.

Pollution Control

To achieve environmental harmony, environmental problems are being assigned top priority by almost all the countries irrespective of their level of development. However, policy instruments used to achieve the objectives vary with the level of development. The differentials in the policy approach have to suit the variations in magnitude and sources of environmental problems.

Environmental concern in India rests not so much on asethetic and health grounds as on survival and livelihood issues. Though asthetic and particularly health considerations are important, it is the survival and livelihood issues which lend even greater urgency to environmental policy than in the developed countries. Apart from the concrete legislative and institutional measures, the need for environmental protection was explicitly incorporated into the Constitution by the Constitution (42nd Amendment) Act of 1976. The Amendment also sought to evolve a national policy and institutional framework for implementing the policy by including 'forests' and 'protection of wild animals and birds' in the concurrent list where both the central and state governments had a say; these categories were earlier in the exclusive domain of the states. Article 253 of the Constitution was added to empower the national parliament to make laws for the whole country including subjects in the State List.

To serve as a guideline to the government, Article 48A was added to the Directive Principles of State Policy in 1976 which states: 'The State shall endeavour to protect and improve the environment and safeguard the forests and wildlife of the country'. Article 51A(g) also imposed a similar responsibility on every citizen.

India's environmental concern with regard to pollution can be traced back to 1853 when the Shore Nuisance (Bombay and Kolaba) Act was passed. The Indian Penal Code Act of 1860 was a more comprehensive legislation containing several provisions for pollution control. The Motor Vehicles Act of 1939 empowered State Governments to regulate emission of smoke and other nuisance. However, the approach to environmental regulation was piecemeal and was based on the tort law. The action against the pollutors was taken only by courts on the basis of proper representation by the affected people. The penalty provided was not deterrent enough, and the course of legal action took a long time. Environment related legislation was not accorded priority, and no special institutions were created to monitor such legislation and prevent adverse impact on the environment.

This situation continued after Independence upto 1972. There was, however, legislative activity between 1947 and 1972 which related to pollution control among other things. The prominent steps were: the Factories Act, 1948; the Industries (Development and Regulation) Act, 1951; the River Boards Act, 1956; the Atomic Energy Act, 1962; the Insecticides Act, 1968; the Merchant Shipping (Amendment) Act, 1970; and the Radiation Protection Rules, 1971. These Acts often dealt only incidentally with pollution and proved ineffective in handling it. A breakthrough came in 1972 in the wake of the Stockholm Conference. The preparatory committee for this conference which identified areas of environmental concern became the precursor of the National Committee on Environmental Planning and Coordination (NCEPC) which was set up in 1972. It had the responsibility of reviewing policies and programmes for the environment. It also undertook the appraisal of several, if not all, development projects from the environmental angle and made suggestions for their modification to avoid adverse impact.

The early 1970s witnessed a milestone in environmental action in the form of the enactment of the Water (Prevention and Control of Pollution) Act of 1974. Though there were earlier several sporadic and piecemeal attempts to control water pollution as noted above, the 1974 Act was the first serious attempt at controlling it. Under this law, Boards for Prevention and Control of Pollution of Water were established both at the Central and State levels with the necessary technical competence and legal power to monitor the implementation of the law. Without waiting for the affected people to launch legal action, the Boards could on their own initiate proceedings against those individuals and firms who infringe the law. The Boards could lay down standards for discharge of effluent and take action to implement them. Factories or persons are required to obtain consent to discharge effluents which can be rejected if they do not maintain the standards prescribed. The Boards can also advise on appropriate sites for locating new industry. The Boards are autonomous in nature, headed by the Chairman. To meet the expenses of the Boards, the Water Cess Act of 1977 was passed requiring industries to pay a cess on their water consumption.

The next important step was the setting up of the Department of Environment (DOE) in the Central Government in 1980. The DOE coordinates all environmental programmes, reviews and clears development projects from an environmental angle, and directs policy and administrative responsibilities concerned with environmental laws. By 1985, the Department was transformed into a full-fledged Ministry of Environment and Forests (MEF). Subsequently, it was replaced by a new body called National Committee on Environmental Planning (NCEP) in April 1981. Some of the important functions of NCEP had been taken over by the Department of Science and Technology and then by the new DOE.

It was not until 1981 that steps were taken to control air pollution. This was done through a separate Act–the Air (Prevention and Control of Pollution) Act, 1981. To enable an integrated approach to pollution control, the Water Pollution Control Boards were authorised to deal with air pollution besides water pollution (Hence they are now just called Central/State Pollution Control Boards). The States and State level boards are required to prescribe and enforce emission standards for industries as well as automobiles after consulting the Central Board and noting ambient air quality standards. The approach of the boards were judicial. The Acts were administered through criminal prosecutions initiated by the Boards and through applications to magistrates for issuing instructions to restrain pollution. This took a long time and often the pollutors got away with mild penalties which did not prove a deterrent.

The Bhopal Gas Tragedy which took place after the midnight of December 3, 1984, proved to be a turning point in the evolution of environment policy in India. Forty tonnes of highly toxic Methyl Iso Cyanate (MIC) gas stored in the pesticide plant of Union Carbide located in Bhopal city leaked into the atmosphere killing around 3,500 people and seriously injuring about 200,000, a good proportion of them permanently. This led to a further spate of legislative activity and the tightening up of environmental law and implementation.

The Government of India has taken recourse to various legislative enactments for prevention and control of pollution as shown in Table 14.1.

Table 14.1 Environmental Legislation in India.

* The Water (Prevention and Control of Pollution) Act, 1974.
* The Water (Prevention and Control of Pollution) Rules, 1975.
* The Water (Prevention and Control of Pollution) Cess Rules, 1978.
* The Air (Prevention and Control of Pollution) Act, 1981.
* The Air (Prevention and Control of Pollution) Rules, 1982-83.
* The Environment (Protection) Act, 1986.
* The Environment (Protection) Rules, 1986.
* The Hazardous Wastes (Management and Handling) Rules, 1989.
* The Manufacture, Storage and Import of Hazardous Chemical Rules, 1989.

Table 14.1 ***Contd….***

* The Manufacture, Use, Import, Export and Storage of Hazardous Micro-Organisms, Genetically Engineered Micro-Organisms or Cells Rules, 1989.
* The Public Liability Insurance Act, 1991
* The Public Liability Insurance Rules, 1991
* The Environmental (Protection) Rules 1992 and 1993 - "Environmental Statement".
* The Environmental (Protection) Rules, 1993 - "Environmental Standards".
* The Environmental (Protection) Rules 1994 - "Environmental Clearance".

As the list of environmental legislation is very comprehensive, three important enactments have been selected for discussion:

The Water (Prevention and Control of Pollution) Act, 1974

The Act was passed with the following objects:

(a) to provide for the prevention and control of water pollution and maintaining or restoring wholesomeness of water (in the streams or wells or sewer or on land);

(b) to establish Central and State Boards with a view to carrying out the above purposes;

(c) to confer and assign powers and functions to the Boards.

To achieve these objectives, the Act requires that no person shall knowingly cause or permit any poisonous, noxious or polluting matter determined in accordance with such standards as may be laid down by a board to enter (whether directly or indirectly) in any stream or well. Further, it imposes a prohibition on a person to discharge sewage or trade effluent into a stream without consent of the Board.

Under the Act, "pollution" means such contamination of water or such alteration of the physical, chemical or biological properties of water or such discharge of any sewage or trade effluent or of any other liquid, gaseous or solid substance into water (whether directly or indirectly) as may, or is likely to create a nuisance or render such water harmful or injurious to public health or safety, or to domestic commercial, industrial, agricultural or other legitimate uses, or to the life and health of animals or plants or of aquatic organisms.

The Government may constitute Central Board, State Board or a Joint Board for prevention and control of water pollution. The main function of the Central Board shall be to promote cleanliness of streams and wells in different areas of the states. The Central Government and State Governments grant financial contribution to their respective boards to discharge their various functions under the Act.

The Act provides penalties for failure to comply with the provisions of the Act. In exercise of the powers, the Central Government has framed rules under the Act known as the Water (Prevention and Control of Pollution) Rules, 1975.

The Air (Prevention and Control of Pollution) Act, 1981

The National Committee on Air Pollution appointed by the Central Government submitted a draft legislation. In pursuance of the above recommendations the Air (Prevention and Control of Pollution) Bill, 1978 was conceived. The Air (Prevention & Control of Pollution) Bill, 1978 was introduced in the Lok Sabha on April 17, 1978. On November 5, 1980 it was again introduced in the Lok Sabha as Bill No. 187 of 1980. The Bill No. 14 of 1981 received the assent of the President on March 29, 1981 and was published in the Gazette of India of March 30, 1981. Under Section 1 of the Air Act the Central Government is to bring the Act into force on such date as it may by notification in the Official Gazette appoint. Accordingly the Central Government notified on May 15, 1981 that the Act shall come into force from May 16, 1981. A number of lacunaes and loopholes have been noticed in the implementation of the Act. This led to the amendment of the Act in the year 1986.

The Act has defined the term "air pollutant". It means "any solid, liquid or gaseous substance (including noise) present in the atmosphere in such concentration as may be or tend to be injurious to human beings or other living creatures or plants or property or environment". "Air pollution" means "the presence in the atmosphere of any air pollutant". This will not only cover the industrial air pollution rather it will include all kinds of operations like burning of rice and wheat straw in the fields causing air pollution.

Under the Act, any person discharging any water whether polluted or not has to obtain the consent of the Board. The Act provides for the constitution of Central Board, State Board and Joint Board.

The Central Board for the prevention and control of water pollution constituted under Section 3 of the Water (Prevention and Control of Pollution) Act, 1974, shall exercise the powers and perform the functions of the Central Board for the prevention and control of air pollution under this Act. In any State in which Water (Prevention and Control of Pollution) Act, 1974 is in force and the state government has constituted for that State a State Board, such State Board shall be deemed to be the State Board for the prevention and control of air pollution.

The term "Joint Board" means a Board for the prevention and control of water pollution constituted under Section 13 of the Water (Prevention and Control of Pollution) Act, 1974.

The functions of the Central Board are:

(a) to advise the Central Government on any matter concerning the improvement of the quality of air and the prevention, control or abatement of air pollution;

(b) to plan and cause to be executed a nationwide programme for the prevention, control or abatement of air pollution;
(c) to co-ordinate the activities of the State Boards and resolve disputes among them;
(d) to provide technical assistance and guidance to the State Boards, carry out and sponsor investigations and research relating to problems of air pollution and prevention, control or abatement of air pollution;
(e) to plan and organise the training of persons engaged or to be engaged in programmes for the prevention, control or abatement of air pollution on such terms and conditions as the Central Board may specify;
(f) to organise through mass media a comprehensive programme regarding the prevention, control or abatement of air pollution;
(g) to collect, compile and publish technical and statistical data relating to air pollution and the measures devised for its effective prevention, control or abatement of air pollution;
(h) to lay down standards for the quality of air;
(i) to collect and disseminate information in respect of matters relating to air pollution;
(j) to perform such other functions as may be prescribed.

The functions of State Boards are:
(a) to plan a comprehensive programme for the prevention, control or abatement of air pollution and to secure the execution thereof;
(b) to advise the State Government on any matter concerning the prevention, control or abatement of air pollution;
(c) to collect and disseminate information relating to air pollution;
(d) to collaborate with the Central Board in organising the training of persons engaged or to be engaged in programmes relating to prevention, control or abatement of air pollution and to organise mass education programmes relating thereto;
(e) to inspect, at all reasonable times, any control equipment, industrial plant or manufacturing process and to give, by order, such directions to such persons as it may consider necessary to take steps for the prevention, control or abatement of air pollution;
(f) to inspect air pollution control areas at such intervals as it may think necessary, access the quality of air therein and take steps for the prevention, control or abatement of air pollution in such areas;
(g) to lay down, in consultation with the Central Board and having regard to the standards for the quality of air laid down by the Central Board, standard for emission of air pollutants into the atmosphere from industrial plants and automobiles or for the

discharge of any air pollutant into the atmosphere from any other source whatsoever not being a ship or an aircraft;

(h) to advise the State Government with respect to the suitability of any premises or location for carrying on any industry which is likely to cause air pollution;

(i) to perform such other functions as may be prescribed or as may, from time to time, be entrusted to it by the Central Board or the State Government;

(j) to do such other things and to perform such other acts as it may think necessary for the proper discharge of its functions and generally for the purpose of carrying into effect the purposes of this Act.

The Central and State Government has power:

(a) to declare air pollution control areas;

(b) to give instructions for ensuring standards for emission from automobiles;

(c) to restrict on use of certain industrial plants;

(d) to furnish information to State Board and other agencies or authorities.

A State Board has power to obtain information regarding the types of air pollutants emitted into the atmosphere and the level of emission of such pollutants from the occupier or any other person carrying on any industry or operating any control equipment or industrial plant. A State Board or an officer empowered by it has the power to take samples of air or emission from any chimney or any other outlet, for the purpose of analysis.

The State Government may, by notification in the Official Gazette establish one or more State Air Laboratories.

The Central Government may extend financial contribution to the State Board to enable it to perform its functions under the Act.

The Central Government has to submit an annual report giving full account of its activities under the Act during the previous financial year.

In exercise of the powers conferred under the Act, the Central Government has framed the Air (Prevention and Control of Pollution) Rules, in consultation with the Central Board.

The Environment (Protection) Act, 1986

The aftermath of Bhopal gas tragedy saw the passing of the Environment (Protection) Act, 1986. This Act empowers the Central Government to take all necessary measures to protect and improve the quality of environment and to prevent, control and abate environmental pollution. The Act assigned the Central Government to deal with environmental problems of the country so that an integrated and holistic policy can be implemented with regard to the environment. Unlike the earlier enactments, the scope of Environment (Protection) Act is broad, covering water, air, land and the inter-relationships that exist among water, air, land and human beings and other living creatures. The general powers of the Central Government to take measures to protect and improve environment include all or any of the following matters, namely:

1. co-ordination of action by the State Governments, officers and other authorities under this Act, or the rules made thereunder;
2. planning and execution of a nation-wide programme for the prevention, control and abetment of environmental pollution;
3. laying down standards for the quality of environment in its various aspects;
4. laying down standards for emission or discharge of environmental pollutants from various sources whatsoever;
5. restriction of areas in which any industries, operations or processes shall be carried out subject to certain safeguards;
6. laying down procedures and safeguards for the prevention of accidents which may cause environmental pollution and remedial measures for such accidents;
7. laying down procedures and safeguards for the handling of hazardous substances;
8. examination of such manufacturing processes, materials and substances as are likely to cause environmental pollution;
9. carrying out and sponsoring investigations and research relating to problems of environmental pollution;
10. inspection of any premises, plant, equipment, machinery, manufacturing or other processes, materials or substances and giving directions to authorities, officers and persons concerned;
11. establishment or recognition of environmental laboratories and institutes and to ensure that they carry out statutory functions entrusted to them;
12. collection and dissemination of information in respect of matters relating to environmental pollution;
13. preparation of manuals, codes or guides relating to the prevention, control and abatement of environmental pollution;

The Central Government may appoint suitable officers and entrust them powers and functions as it may deem fit.

Other matters regarding prevention, control, and abatement of environmental pollution includes:

(a) no persons carrying on any industry, operation or process shall discharge or emit or permit to be discharged or emitted any environmental pollutants in excess of such standards as may be prescribed;

(b) no person shall handle or cause to be handled any hazardous substance except in accordance with such procedure and after complying with such safeguards as may be prescribed;

(c) persons responsible for the discharge of any environmental pollutant in excess of the prescribed standards shall have to furnish information to authorities and agencies;

(d) any person empowered by the Central Government shall have the right to enter, at all reasonable times with such assistants as he considers necessary;

(e) the Central Government or any officer empowered by it shall have power to take samples of air, water, soil or other substance from any factory, premises or any other place for the purpose of analysis;

(f) the Central Government may establish one or more environmental laboratories and assign them such functions as it deems fit.

(g) the Act provides for penalties for contravention of the provisions of the Act and the Rules.

(h) the Central Government may take into consideration the following factors while prohibiting or restricting the location of industries and carrying on of processes and operations in different areas -

(i) standards for quality of environment in its various aspects laid down for an area are:

  (i) the maximum allowable limits of concentration of various environmental pollutants (including noise) for an area,
  (ii) the likely emission or discharge of environmental pollutants from an industry, process or operation proposed to be prohibited or restricted,
  (iii) the topographic and climatic features of an area,
  (iv) the biological diversity of the area which, in the opinion of the Central Government needs to be preserved,
  (v) environmentally compatible land use,
  (vi) net adverse environmental impact likely to be caused by industry, process or operation proposed to be prohibited or restricted,
  (vii) proximity to a protected area under the Ancient Monuments and Archaeological Sites and Remains Act, 1958 or a sanctuary, national park, game reserve or closed area notified as such under the Wild Life (Protection) Act, 1972,
  (viii) proximity to human settlements.

The Ministry of Environment and Forests has issued a notification in the year 1994 that expansion or modernisation of any activity or new projects shall not be undertaken in any part of India unless it has been accorded environmental clearance by the Central Government.

It is not necessary for the MEF to perform all the functions assigned to it. There are institutions like the Pollution Control Boards (PCBs) which are empowered to act on its behalf, with overall control vested with the MEF. The necessary amendments were made to Water and Air Pollution Prevention and Control Acts to empower PCBs in this regard. The Air Pollution (Amendment) Act, 1988 sought to bring noise pollution also under control, with procedures laid down to monitor and control it.

Since the Bhopal gas tragedy involved the handling of risk from a hazardous substance, it inevitably led to a further tightening up of law in this regard. The important enactments passed are the Hazardous Waste (Management and Handling) Rules, 1989; the Manufacture, Storage and Import of Hazardous Chemical Rules, 1989; and the Public Liability Insurance Act, 1991. The last is to provide immediate relief to persons affected by accidents occurring while handling any hazardous substance. Steps are also being initiated to introduce environmental audit of local municipal bodies, statutory bodies and public limited companies to evaluate the effect of their policies, operations and activities on the environment. Persons causing pollution of air and water beyond the standards prescribed are punishable with imprisonment and fine.

After a brief review of the environmental laws, the Tiwari Committee, in its report (1981, page 19-24) noted some major shortcomings which may be summarised as follows:

1. Many of these laws are outdated.
2. They lack statements of explicit policy objectives.
3. They are mutually inconsistent.
4. They lack adequate provisions for helping the implementing machinery.
5. There is no procedure for reviewing the efficacy of the laws.

Inspite of all the stringent provisions, the process of taking action was lax before the mid-1980s as the Boards had little scope for direct administrative action. They had to approach a magistrate to enjoin the pollutor. The amendments to the Air Act in 1987 and to the Water Act in 1988 not only strengthened penal provisions of these laws but also empowered the enforcement agencies to close down polluting industries and to stop their electricity and water supply. The shift from judicial to administrative enforcement of environmental laws is meant for better compliance. However, the Boards still have no powers to exact fines, order imprisonment or compel compliance with their detections, and have to depend on the courts of law. With the emergence of public interest litigation (whereby even citizens not directly affected by pollution can launch litigation in public interest), the role of the higher judiciary has undergone a radical and creative transformation. In the process, the Supreme Court has interpreted the right to life and personal liberty to include the right to a wholesome environment. The Supreme Court embarked even on necessary administrative action such as appointing expert committees, and based on their findings, giving direction for closure of polluting industries. As environmental awareness and responsibility increases in the public mind as well as industry, as has indeed been taking place, the future prospects are more optimistic.

Environment Policy of a Company

- Our company shall strive to provide a safe and healthy working environment and comply, in the conduct of its business affairs, with all regulations regarding the preservation of the environment of the territory it operates in.
- Our company shall be committed to prevent the wasteful use of natural resources and minimise any hazardous impact of the development, production, use and disposal of any of its products and services on the ecological environment.

SUMMARY

- The term *environment* is derived from the French word *environ* which means 'to encircle or to surround'.
- Environment is a mutually interacting system of living and non-living components.
- Types of environmental pollution include air pollution, water pollution, soil pollution, solid waste, and noise pollution.
- It has to be ensured that the industrial sector pays utmost attention in adopting proper pollution control measures, in addition to conservation of material usage and waste recycling methods.
- The industries must create a separate cell with qualified personnel for operation and maintenance of the effluent treatment plant.
- Pollution, be that air, water or noise, is a menace to the society; all the developed and developing countries are facing this hazard.
- With the industrialisation of the country problem of pollution comes in.
- This is not the problem of one country, it is the problem of all countries.

→

- If it is allowed to go unabated there will be serious health hazard to the human community.
- Of course, no country can prosper without industrialization; but at the same time, it should be the endeavour of all to control the pollution.
- Pollution cannot altogether be eliminated; it can only be minimised.
- The three important legal enactments regarding environmental pollution and protection are: Air (Prevention and Control of Pollution) Act, 1981; Water (Prevention and Control of Pollution) Act, 1974; Environment (Protection) Act, 1986.
- Adequate environment protection measures especially for waste water, chemicals, and dust suppression systems, shall be provided in line with Air and Water Act and provisions of appropriate Pollution Control Board.
- The most effective control of hazardous waste is the reduction or elimination of the waste.
- There should be procedures to control the generation, transportation, storage and treatment or disposal of hazardous waste.

DISCUSSION QUESTIONS

1. What is pollution? What are the common sources of pollution?
2. Pollution control laws have not proved very effective. What do you think are the reasons?

15

Disaster Management

Disaster is an undesirable occurrence of events of such magnitude and nature that adversely affect production, cause loss of human lives and property as well as damage to the environment. Industrial installations are vulnerable to various kinds of natural and man-made disasters. Events such as bomb explosions have reinforced the critical need for preplanning the emergencies whether they be man-made or natural in origin. Effective preplanning and emergency procedures can save lives and business property.

Disasters can include:

- man-made disasters such as civil war, nuclear attack, and terrorist attacks
- criminal activities such as sabotage, riots, and extortion
- natural disasters such as tornadoes, hurricanes, earthquakes, blizzards, floods, dust storms, and heavy thunder storms
- accidents such as fire, explosion, hazardous material spills, vehicle crashes, and major mechanical or structural failures.

Communications can easily become disrupted during an emergency due to lack of power, storm damage, fire damage, and so on. It is impossible to forecast the time and nature of disaster, which might strike an undertaking. However, an effective disaster management plan helps to minimise the loss in terms of human lives, plant assets, and environmental damage and helps resume work as soon as possible.

Objectives of Disaster Planning

Disaster arrives without any warning inspite of all precautions and preventive measures taken. However, an efficient control/response plan can minimise the losses in terms of property, human lives and damage to environment.

The main objective of the plan is to minimise the effect of the disaster. The plan is developed in such a way to make best possible use of the resources at the command of the undertaking as well as resources available outside like local fire services, police, civil administration, and hospitals and the neighbouring industries. Advance meticulous planning minimises chaos and confusion, which normally occurs in such a situation and reduce the response time of disaster management organisation.

The objectives of disaster management plan are:

- To contain and control the incident.
- To rescue the victims and arrange for their treatment in quickest possible time.
- To safeguard persons and evacuate them to safer places.
- To identify persons affected or dead.
- To give immediate warning signals to the employees as well as the people in the surrounding area in case such a situation arises.
- To inform relatives of the casualties.
- To provide authoritative information to news media and others.
- To preserve the affected area as well as the equipment as evidence for inquiry/investigation.
- To restore normal working condition at the earliest.
- To investigate the cause of the accident so that similar happenings do not occur.

Emergency Planning

Major emergency can be defined as an accident that has potential to cause serious injury or loss of life. It may cause extensive property damage and adversely affect the environment as a whole. The emergency may be caused by factors like failure of safety system, operational error, vehicle crash, sabotage, earthquake, and so on. Irrespective of the cause, the emergency will generally manifest itself in one of the three basic forms i.e. fire, explosion, or release of toxic substances.

Risk of an accident can be reduced to a great extent by good design, operation, maintenance, and inspection of the plant. Absolute safety is not achievable and hence emergency planning is an essential part of major hazard control programme.

Emergency planning has a vital role to play in the mitigation of the consequences of any incident at a site storing and/or processing hazardous substances. It encompasses the principles of identification, assessment and control, and mitigation.

Emergency planning is based on certain factors, namely, (a) recognising that accidents are possible; (b) assessment of possible consequences of the credible accidents; and (c) preparation of both on-site and off-site emergency procedures to be implemented in the event of an emergency.

The objectives of an emergency plan are:

- to contain and control incidents;
- to safeguard employees and others who might be affected;
- to minimise damage to property or the environment;
- to rescue and treat the casualty;
- to identify casualties, notify their relatives, and render necessary help to them;
- to rehabilitate those who are affected;
- to provide right information to media and government agencies; and
- to preserve necessary records/equipments for subsequent enquiry.

Effective emergency planning at any location depends on the identification of the range and scale of possible accidents, the estimation of their consequences, the construction of a flexible plan to deal with those consequences, and the proper rehearsal of the plan by those who carry the responsibilities. It requires close co-operation between the company, the emergency services, and the local community.

The emergency plans must be capable of dealing with the largest incidents that can reasonably be foreseen, but detailed planning should concentrate on those events that are most probable. They must also have sufficient flexibility built-in so that the response is tailored to the severity of the incident. This will prevent unnecessary calls on the external emergency services such as the police, fire and ambulance services, if the matter can be dealt with completely using the company's own resources.

A wide range of events may have to be considered in an emergency plan. All the events identified should be considered when preparing the on-site emergency plan. Those events which are capable of giving rise to consequences off-site should be identified by the company and information on their nature, extent and likely effects should be passed to the relevant local authority. These events should form the basis for the off-site emergency plan.

Emergency plans need to be tested when first devised and thereafter to be rehearsed at suitable intervals. Individual personnel with duties under the plans should be qualified and experienced in their day-to-day operations. Some duties, however, such as firefighting for the works fire team, are not routine and special training will be needed. In addition, key personnel will need training in their emergency roles, both individually and as a team.

In order to control the emergency effectively, we should have emergency procedures, and all concerned people must be trained about their duties and responsibilities in emergency situations.

Emergency procedure for meeting emergencies should include the following:

- Proper control, co-ordination and communication by the concerned individuals.
- Specific duties of key personnel.
- Instructions to the personnel not engaged in responding to emergency.
- Site evacuation plan during toxic release.
- Emergency response plan and resources mobilisation in controlling fire.
- Outside assistance in handling the emergency situations.
- Maintenance of a list of doctors, ambulances, and chemists at all important locations.

As a first step in preparing an emergency plan, companies should carry out an assessment of the activities at their premises to ensure that all that is reasonably practical has been done to avoid or reduce danger.

Identification and assessment of hazards in emergency planning forms the basis for both on-site and off-site plans. This requires systematic study of the plant to identify emergencies that can occur. It is desirable to have the hazards identification at the design stage of a new project, and also an operating plant so that necessary safety features can easily be incorporated. Techniques like Preliminary Hazard Analysis (PHA) and Hazard and Operability Study (HAZOP) can be used. The overall assessment of major hazard assessment provides a basis for preparing both on-site and off-site plans.

On-Site and Off-Site Emergency Planning: The management should prepare an on-site plan based on residual risks. This plan must be specific to the needs of the plants i.e. size, complexity, nature and quantum of hazard. Each on-site plan will therefore be different from the other. Reliable and prompt alarm and communication system is extremely important in controlling a major emergency. Actual alarm system will depend on the size and specific needs of the plant.

Certain individuals are to be nominated to take specific responsibility during emergency. A responsible person has to be nominated to take full charge at the site of the incident, assess the situation, decide and initiate all the efforts at the site.

Also, a major emergency may affect areas outside the works and hence off-site emergency plan is an essential part of a major hazard control system. A large number of government bodies are involved in formulation and operation of an off-site plan. On the whole, the local administration is responsible for off-site plan; and the works administration is responsible for providing basis for the plan. An off-site plan must clearly identify an emergency co-ordinating officer as overall incharge of all off-site actions.

The basic aspects of an off-site plan are:

(a) a clear organisation structure giving details of a command structure, warning procedure, and emergency control structure;
(b) details of safety organisations like the fire services, police, hospitals, factory inspector, voluntary organisations, and emergency services of industry which can assist during emergency;
(c) plans for evacuation, safe routes, medical treatment and rehabilitation;
(d) system to inform public and media;
(e) educating public whereby people who are liable to be affected can take actions as per plan.

Close co-operation of all the agencies concerned is most essential for the formulation and effective implementation of any plan.

Emergency planning is often assigned to the safety and health professionals, although ultimate responsibility resides with top management. In developing an emergency plan, organisations must:

1. identify and evaluate potential disasters;
2. assess the potential harm to people, property and environment;
3. estimate warning time needed to mobilise the plan;
4. determine what changes must be made in company operations; and
5. consider what power supplies and utilities may be required to handle the emergency.

Basic emergency planning usually includes establishing a chain of command, an alarm system, medical treatment plans, communication system, shutdown and evacuation procedures.

Once the initial background work has been done, the next step is to develop a working emergency management plan. This plan should be created not only within the facility but in cooperation with other facilities in the area. To administer the emergency plan, someone within the company should be appointed emergency planning director or coordinator. The organisation can develop a manual or handbook to cover many of the emergency steps that might be needed for implementation.

Disaster Preventive and Pre-emptive Measures

Risk analysis forms an integral part of disaster management plan and any realistic disaster management plan can only be made after proper risk analysis, study of the activities, and the facilities provided. Correct assessment and evaluation of the potential hazards, advance meticulous planning for prevention and control, training of personnel, mock drills, and liaison with outside services available can minimise losses to the plant assets, rapidly contain the damage effects, and effectively rehabilitate the damaged areas.

After identification and assessment of disaster potential, the next step in disaster management plan is to formulate and practice the preventive and pre-emptive measures. Proper preventive and pre-emptive measures can reduce the disaster potential to minimum. Such measures shall be taken from the design stage itself. Preventive measures include:

- procurement and use of material and machinery in construction work as per relevant code/specification;
- judicious layout of the equipment with proper operating space;
- use of instrumentation for proper operation, safety with automatic alarm, and shutdown system, if the operating parameters suddenly go out of control;
- installation of proper safety, fire detection and fire fighting systems; and
- inspection of the materials by some reputed agency.

The next step in pre-emptive measure is to formulate a detailed disaster management plan, both on-site as well as off-site, where actions to be taken before, during, and post disaster period should be clearly mentioned.

Responsibility of preparation of off-site disaster management plan lies on district administration. A district hazard management committee should be formed to take care of off-site emergency. In case of both on-site and off-site disaster management, the most important activities which are to be taken are:

Awareness: Creation of awareness amongst employees/public living in the vicinity of the undertaking and the role they have to play in case of any disaster occurring is very important. The employees of the undertaking as well as the public shall be educated as to what should be done by them to take care of themselves as well as to protect the properties in case of emergency. For employees regular training programme on safety, fire-fighting, and disaster control shall be conducted. Public awareness should be created through public address system, film shows and periodical distribution of leaflets in the local language.

Mock Drill: Mock drill is very important to know the strengths and weaknesses of the disaster control response team. Efficient fire fighting arrangement shall be provided effectively in the factory premises. Fire hydrants, portable fire fighting and safety equipment shall also be provided at vulnerable locations. Mock drill shall be carried out for fire as well as for other emergencies.

Communication: Proper and timely communication to the response team members helps to minimise the effect of any disaster by taking suitable and timely action. The undertaking shall have internal as well as external telephones for communicating with its different sections as well as outside. In addition public address system shall be provided in different places for informing fire services department as well as others about the location of hazard. Sirens shall be provided at different places of the undertaking to inform the employees and the public outside about emergency/disaster.

Communicating in a crisis is different from normal. In a serious crisis, all affected people act differently, like: (a) taking in the information; (b) process information; and (c) act on the information.

Key to successful communication:

- Be the first source of information
- Express sympathy early
- Show competence and expertise
- Remain honest and open
- Commit and remain dedicated

Entry of Personnel: Entry of personnel shall be restricted for preventing sabotage or any untoward incident due to inadvertence. This will also eliminate any terrorist activity.

Formation of Disaster Management Cell: Disaster management cell should be formed both for on-site as well as off-site emergency condition. A detailed on-site emergency disaster management plan shall be formulated by the undertaking. Similar planning and formation of off-site disaster management cell is necessary to be formulated by district administration with police, factory administration, and nearby hospitals.

***Mutual Aid Scheme**:* Although an organisation may have enough facility to control hazards envisaged, it may so happen that outside help may be necessary to survive and recover from a disaster or major accident. For this, facilities existing outside such as fire tenders with trained personnel in near-by fire stations, hospitals, and nursing homes need to be called wherever such situation arises. Mutual aid plans with other firms can help to pool the knowledge, resources, equipment, and personnel of many organisations.

Emergency Medical and Transportation Services: Emergency medical services are a vital part of any emergency management plan. They should be headed by a doctor, if available, or by the health and safety professional.

An emergency transportation unit includes ambulance service, transporting emergency service crew and supplies, and moving employees to and from work as necessary.

Security: Security and protection is a high-priority concern for management. As a result, top management must ensure that security staff are properly trained not only in their basic duties but also handling of such hazards as fires, chemical spills, personal injuries, and so on.

Protection from Personal Attacks : One major responsibility of the organisation is to protect employees from personal attack like muggings, rapes, and robberies. The management may enlist the advice and aid of local law enforcement agencies, and train workers in the basics of personal protection.

Disaster Control Plan

Both on-site as well as off-site disaster management response plan can be sub-divided into: (a) equipment plan, (b) organisational plan, and (c) action plan.

During an emergency easy access to the required equipment and facilities are of paramount importance. Equipment plan needs arrangement of sufficient and proper appliances needed to combat any disaster after careful study of requirements including alarm and communication system as well as provision of vehicles for communication and relief measures.

Efficient and adequate measures shall be made available within the premises to combat any emergency inside the premises. Efficient fixed fire fighting arrangement and portable fire fighting shall also be provided inside the factory premises. An emergency control centre shall be provided at a safe place from where designated emergency director/coordinator shall function for on-site emergency. The emergency control centre shall be provided with adequate personal protective equipment, alarm and communication network (siren, telephone, mobile phone, public address system), route map, fire hydrant, monitor layout, and copy of detailed disaster management plan.

The disaster management manual should contain:

(a) the names, telephone numbers (including mobile) of the response team members and their responsibilities;

(b) the names and telephone numbers of key personnel from outside agencies, district authorities, mutual aid schemes, fire station, state hospitals and doctors;

(c) map of the factory and surrounding areas, evacuation routes, fire hydrant network, first-aid, and other important information.

Emergency shelter places shall be chosen sufficiently away from likely affected site. Employees who are not in the emergency management team shall be asked to take shelter in those places to keep themselves away from fire explosion and release of any toxic gases. More than one emergency shelter shall be designated for the purpose so that a proper shelter point can be chosen out of them depending on wind direction and other factors.

The names, residential addresses and telephone numbers of key personnel shall be clearly written in disaster management manual and to be widely circulated. It is to be noted that first few minutes after start of the incident is most vital in prevention of its escalation. The personnel available in the plant round the clock play an important role.

The action plan for tackling disaster should include the following:

- Action to be taken in pre-disaster period.
- Action to be taken during disaster period.
- Action to be taken in post disaster period.

Some of the pre-disaster measures in case of on-site include:

- Ensure implementation of disaster planning.
- Ensure that all people drafted for emergency undergo regular training.
- Ensure all team members in different disciplines are prepared for tackling emergency/disaster.
- Ensure mock drills are performed to assess the strengths and weaknesses of the response team.
- Ensure awareness among employees through regular training.
- Ensure good liaison with all agencies and industries in the neighbourhood for getting help if situation arises.
- Ensure adequate stock of safety appliances and personal protective equipments in good working condition.
- Ensure awareness amongst public in the neighbouring areas.

Action during disaster include the duties and responsibilities of the key personnel which are as follows:

- To assess the magnitude of the disaster and decide whether a major emergency condition exists or likely to be developed.
- To exercise direct operational control for combating emergency.
- To assess whether evacuation of the employees is necessary in the affected as well as other non-affected areas.
- To review continuously the situation in consultation with the chief emergency co-ordinator.
- To keep liaison with senior officials of police, fire brigade, factory inspectorate, and other agencies.

- To ensure that the persons affected are getting proper relief and rehabilitation care.
- To release authoritative information to press and public through the coordinator.
- To conduct inquiry about the incident and restore normalcy as early as possible.

The major follow-up questions to be asked after a disaster gets over should include:

- How effectively was the emergency assessed?
- How effectively was the information passed on to other personnel?
- How effective was the disaster handling by the person-in-charge?
- Were shut down procedures followed in a timely and effective manner?
- If a response team was used, did they perform efficiently?
- Was outside assistance called in time and directed effectively to the areas of need?
- Did the alarm system work as designed?
- Did the evacuation scheme work as planned?
- Were shut down procedures followed in a timely and effective manner?
- Was there adequate emergency equipment available?
- Did the emergency equipment function properly?
- Was the media handled properly?

Development of a Contingency Plan

- Specification of individual authority for initiating the contingency plan.
- Liaison with external authorities.
- The appointment of an emergency team co-ordinator and emergency control centre.
- Notification to local authorities.
- Call-out of competent persons.
- Immediate action on site.
- Evacuation procedure.
- Access to records.
- External communication arrangements.
- Public relations matters.
- Catering and temporary shelter arrangements.
- Contingency repair and reinstatement arrangements.
- Training of key personnel and other employees.

In Case of Fire or Other Emergency

- Avoid panic and confusion.
- Find out the safest way out of the building.
- Know the location of exits.
- Know the location of nearby fire extinguishers.
- Know how to report a fire or other emergency.
- Follow exit instructions.
- Walk to your assigned exit.
- Leave your workplace calmly and without running.
- Close doors behind you if you are the last person.
- Use the stairs if you are above the ground floor– lifts can be dangerous in fires;
- Think of anyone who may need your help or any visitor for whom you are responsible.
- Do not go back to the building until you are told that it is safe to do so.

In case of Bomb Threat

Any bomb threat, received by telephone, letter, or other means of communication, should be treated seriously; and a formal company procedure should be written to cover such incidents, including the responsibilities of designated individuals for ensuring evacuation, dealing with the civil authorities, and agreeing for re-entry to buildings.

Mutual Aid and Response Group (MARG)

A mutual aid and response group is functioning in Thane District with active support of the Directorate of Industrial Safety and Health, Thane, Government of Maharashtra. District of Thane is known for a heavy concentration of chemical industries and important industrial clusters. The response group comprises of members drawn from corporate body and company level representatives.

Aims and objectives of MARG are as follows:

The units should offer help to other unit/s on request or call in event of fire, explosion or release of toxic gases in that unit. The emergency response shall be without lapse of time and should be aimed at

- Preventing danger to life, health and property.
- Minimising the damage of environment and social well-being.
- Creating community awareness.

Mutual help may be in areas like:

- Fire fighting equipment

- Trained/technical expertise
- Transport and communication
- Personal protective equipment
- Medical and health
- Security
- Any other item mutually agreed.

MARG aims at interacting with local agencies like collectorate, police, fire brigade, local council, health and first aid volunteers and others, and co-operate and co-ordinate with district authorities in implementation of off-site disaster management plan.

The scheme envisages efforts in the direction of taking general public living in the vicinity of such industrial clusters, in confidence and imparting safety knowledge and awareness to them by developing an out-reach programme without bringing fear/panic.

The scheme aims at making organised efforts in imparting adequate training in safety, health and environment and emergency measures to persons likely to be involved in such activities.

Bhopal Gas Disaster

The entire country is shocked and dismayed at the serious accident in the plant of Union Carbide at Bhopal resulting in leakage of a highly poisonous gas in the environment. In the biggest ever environmental disaster in the world, it is reported that about 2,000 people - majority being women and children - died and a few thousands were crippled as a result of exposure of this gas.

The tragedy began with the discharge of the deadly poisonous gas from the pesticide plant of the company located in the outskirts of Bhopal in the early hours of 3rd December, 1984. The plant manufactures a particular brand of pesticide known by the generic name Carbaryl and the trade name Sevin. In the manufacture of this product, both phosgene and methyl isocyanate (MIC) along with other substances as intermediaries are utilised. The MIC is stored in tanks before its conversion to the pesticide. Due to increase in temperature and pressure, the liquid MIC gushed out as a gas through the safety valve and passed through the stack into the outside environment leading to the calamity that befell the population of Bhopal.

The accident occurred at the time when the plant was completely shut down and as such there was no production; only usual maintenance was carried out during this period. For the storage of MIC, three tanks which are interconnected were utilised. Generally, one tank is always kept empty and the other two tanks filled with MIC. On the day of the accident, the operating staff noted a high pressure build-up in the one of the tanks filled with MIC. It was reported that by 12.55 a.m. the pressure in the tank went up to almost 55 lbs, per sq. inch and as such the safety valve was blown off. The poisonous gas from the tank thus gushed out and passed through the scrubber. Since the scrubber was designed for neutralisation of a very small quantity of the gas that might leak out, the whole quantity of the gas passed unneutralised through the pipeline into the outside atmosphere. The lethal methyl isocyanate gas enveloped an area of about 40 sq.kms. on the wintry midnight.

Many people dropped dead on the road while fleeing. As the time ticked away more and more victims of the gas poisoning were moved to the hospitals. It is estimated that virtually one-fifth of the eight lakh population in the city was affected by the deadly gas. Thousands of survivors are reported to be suffering from the dangerous effects of the gas. Besides human beings, a large number of animals such as cattle, dogs, cats, birds succumbed to the effects of the poisonous gas. There are reports of adverse effects on vegetation too.

In the wake of the leakage of this poisonous gas from the plant of Union Carbide the question arose how to neutralise the remaining MIC in the other tank. It was decided by the government that the ideal way would be to convert the remaining MIC into the finished product i.e. Carbaryl. The operation of neutralisation of the gas or rather conversion of this gas into carbaryl was carried out under the direction of the Director General, CSIR. All precautions such as covering the factory with tarpulin, spraying of water over the factory and the surrounding areas with the help of helicopters was taken. The government instituted a judicial enquiry and the investigation of the accident was entrusted to the CBI. Preliminary investigation as reported in the press reveals that the leakage of MIC was primarily due to a build-up of high pressure inside the tank, code numbered 610, which contained about 45 tonnes of MIC leading to the rapture of the valves and the escape of the gas. The build-up of a high pressure inside the tank was possibly due to an ingress of certain quantity of water into the tank. It is known that water reacts violently with MIC resulting in increase of pressure.

Two major safety lapses were possibly responsible for the mishap. One was disconnecting the pipeline running from the vent gas scrubber to the flare tower and the second was the failure to keep the tower flame burning.

Since MIC is primarily used in the chemical industry as an intermediate, not much work has been done on the toxicity of this chemical. It is, however, known that the substance is absorbed into the body by inhalation, ingestion, and through the skin. Absorption of this chemical into the body leads to causation of nausea and vomitting, chest pain, difficulty in breathing, coughing, redness and pain in the eyes as well as blurred vision.

The cause of the death is primarily due to injury of the lungs and pulmonary oedema. The long-term effects of MIC are not known.

The post-mortem of the victims of Bhopal gas tragedy have raised some pertinent questions regarding the toxic effects of the gas to which the people were exposed. Some contend that the adverse effects resemble those as to be expected from the exposure to phosgene. The after effects of exposure to this gas can only be found out if a concerted follow-up study of the affected persons could be carried out. This will throw light to show whether this chemical is a systematic poison, neurotoxic, carcinogenic, mutagenic or teratogenic.

It is reported that the safety performance of the plant of Union Carbide at Bhopal was not up to the mark. There had been a number of accidents in this plant prior to this disaster due to non-compliance with the provisions of the Acts. It has been reported that the computerised safety devices which are in existence in the other plants of the multinational in abroad are not provided in this particular plant. The answers to different queries raised will provide information as to the causation of such a serious accident.

The Bhopal gas tragedy reminds all of us to be more vigilant while dealing with toxic chemicals. It calls for more research and training in the field of occupational safety and health with a view to providing better working conditions in the industry.

The management of Union Carbide India has served a notice on April 11, 1984 under the provisions of the Industrial Disputes Act, 1984, to close its Bhopal plant. According to V.P. Gokhale, Managing Director "After the unfortunate accident on the intervening night of December 2/3, 1984, on instructions received from the authorities of the State Government, Union Carbide India had closed its Bhopal plant operations. The licence under the Factories Act, which expired on December 31, 1984, has not been renewed by the authorities. The management of the company had refrained from taking any precipitate action in regard to the closure of the plant because it was felt that alternatives to plant closure should be fully considered and explored. In compliance with the State Government instructions, the factory has not been operating effective from December 3, 1984, but the company has continued to pay full wages to all its workmen and had taken no further steps despite the press report to the effect that the company will not be allowed to restart its manufacturing operations at the Bhopal plant. During recent meetings with the government authorities, it was made abundantly clear to us that permission to restart the factory will not be given. The position concerning the operation of the plant is now quite clear. Management sincerely regrets this most unfortunate development."

The Union of India has filed a law suit in the United States' southern district court of New York on April 8, 1985, against Union Carbide Corporation Ltd. (UCCL) holding it 'absolutely liable' for the worst chemical disaster in Bhopal. It has sued the American firm on seven counts: multinational enterprise liability, absolute liability, strict liability, negligence, breach of warranty, misrepresentation and punitive damages. The decision to lodge the suit came after the collapse of negotiations which had been under way for settling the matter out of court. The demands of the plaintiffs are to award compensation of an amount appropriate under the facts and the law i.e., fully, fairly and adequately to all persons having claims and also to award punitive damages in an amount sufficient to deter the Union Carbide and any other multinational corporation from wilful, malicious and wanton disregard of the rights and safety of the citizens of those countries in which they do business.

The case was unique in more senses than one. Apart from being potentially the biggest compensation claim in legal history pertaining to the world's worst accident, it was first time that a government was taking such a step against a multinational corporation. The case will make a legal history if it is heard to the end.

The Madhya Pradesh Government, which refused to renew the factory's licence after the leak, permitted the Carbide to close down the plant. The plant has been under the control of the Central Bureau of Investigation (CBI) since the release of toxic pesticide. The management's three months notice for the closure of the plant as required under the Industrial Disputes Act, expired on July 10, 1985. The plant was formally closed down on July 11, amid protests from its employees demanding alternative jobs or adequate compensation. The employees, who have been agitating inside the factory premises since June 14, put up on its outer wall effigies of top officials of the Union Carbide including that of the multinational's chairman Warren Anderson. The management issued notices in local newspapers announcing that the compensation and dues would be mailed to the hourly workers, who had refused to accept the compensation.

In a sudden turn of events, the Supreme Court on Feb. 15, 1989 awarded \$470 million (Rs. 715 crores) for those killed and maimed by the leakage of the lethal MIC gas from the multinational Union Carbide Corporation (UCC) plant in Bhopal. In a consensus order, which has been agreed to by the UCC and the Indian government, a five-judge constitution bench presided over by the chief justice, R.S. Pathak, directed the UCC to pay the amount to the Indian government on or before March 31, 1989. Following this agreement, all legal cases arising out of the disaster and pending before different courts stand transferred to the Supreme Court. The Supreme Court had to intervene because of a legal battle between the Union of India and the UCC regarding the Madhya Pradesh High Court order of April 4, 1988, reducing the interim compensation from Rs. 350 crores to Rs. 250 crores. The protracted legal battle which appeared to many legal luminaries as an unending one, finally came to a halt. In his opening remarks, justice Pathak observed that the case was "preeminently fit for an overall settlement in view of the acute suffering of the victims." He also noted that $470 million was a just, fair, equitable and reasonable sum" to settle the compensation claims of the gas victims, many of whom have been incapacitated for life.

In the last 38 years, the Bhopal gas disaster case has seen several twists and turns. In 1985, India sued UCIL for $3.3 billion in the United States. But in 1989, the Centre and UCIL struck an out-of-court deal for $470 million as compensation, which was described by victims as grossly inadequate. In 2010, curative pleas were filed by the CBI (later junked) and the Centre seeking enhancement of damages. It said in 1989 compensation was based on a preliminary assessment of 3,000 deaths, 20,000 people suffering from severe injuries, and 50,000 from minor ones. Since then, it said the death figures rose to 5,295 with 35,000 people suffering from serious injuries and 5,27,000 from minor ones. In October 2022 the Centre assured a 5-judge bench of the Supreme Court that it would push ahead with its 12 year old curative petition seeking compensation over and above the $470 million paid by UCIL. The Supreme Court agreed for a detailed hearing.

In the course of hearing the curative petition, the Supreme Court on January 12, 2023 strongly repelled a suggestion that the $470 million settlement in the 1980 Bhopal gas tragedy case took place due to the court's anxiety, stressing it did not thrust the settlement upon parties and that it was the union government and the chemical company which mutually agreed to the amount towards "all past, present and future claims".

SUMMARY

- Disaster management is emerging globally as a full-fledged discipline.
- Accidents involving highly flammable and toxic substances either through plant failure or human error, can result in fire and explosion or toxic release leading to extensive damage to plant and equipment, environment and loss of life.
- People from diverse backgrounds such as workers, engineers, social scientists, or medical professionals are in the domain of disaster response.

→

- No industrial, commercial, mercantile, or public sector organisation is immune from disaster.
- Emergencies can arise at any time and from many causes, but the potential loss is the same – injury and damage to people, the environment, and property.
- Advance planning for emergencies is the best way to minimise this potential loss from natural or human caused disasters and accidents.
- A written emergency management plan is intended to provide preplanned response to those unexpected disastrous events that can strike any organisation.
- Emergency planning must provide for the safety of employees and the public, protect property and the environment, and establish methods to restore operations to normal as soon as possible.
- The management should be prepared to control and curtail the loss to the industry to the minimum possible extent by careful emergency planning and training of their personnel.
- Successful handling of emergencies will depend on appropriate action and decisions on the spot.
- It is not possible to predict every action which may be required to be taken on the spot.
- Some of the major actions that may be taken during emergencies include: (a) setting in motion the containment procedures or plant-shut down procedures; (b) evacuating the personnel from affected areas, and (c) communicating safety precautions to individuals concerned including contractors, visitors and security personnel.
- Informing relatives about the casualties through a designated public relations officer and arranging for the authoritative release of information to all concerned.
- All employees should know the details of the emergency plan and must receive initial training in emergency procedures.
- The medical organisation for getting proper response in emergency (industrial major accidents) has mainly three components: (a) organisation at the site, (b) transportation of victims, and (c) organisation in the hospital.

DISCUSSION QUESTIONS

1. What are the objectives of disaster management plan?
2. What are the preventive and pre-emptive measures in a disaster management plan?

Appendices

Appendix A
Relevant Labour Laws

(Salient Features)

As labour is in the Concurrent list, both the Centre as well as the States have the power to make laws. All labour laws are welfare-oriented, based on the principles of social justice, natural justice, equity and so forth and they aim to protect the interests of labour, management, and the government. Labour laws are administered by: (a) State Governments only; (b) Central Government only; (c) both by the Central and State Governments; and (d) neither by the Central or State Governments. The important labour enactments in India may be broadly classified into: (a) legislation on working conditions; (b) legislation on wages; (c) legislation on social security; (d) legislation on employment and training; (e) legislation on labour welfare funds; and (f) legislation on industrial relations.

The Factories Act 1948

- The Factories Act 1948 is a most comprehensive piece of legislation and replaced all the earlier Acts.
- The object of the Act is to ensure suitable health, safety and welfare measures, and provide working hours, leave and other benefits to the workers employed in the factories.
- The term "factory" means: (a) any premises in which 10 or more workers are employed and a manufacturing process is carried on with the aid of power; and (b) any premises in which 20 or more workers are employed and a manufacturing process is carried on without the aid of power.
- Factory does not include a mine, a mobile unit belonging to the armed forces of the union, a railway running shed, and a hotel, restaurant, or eating place.
- The term "worker" means a person employed directly or through any agency including a contractor.

- The term "occupier" of a factory means the person who has ultimate control over the affairs of the factory.
- In the case of a company which owns a factory, it is only one of the directors of the company who can be notified as the "occupier" of the factory for the purpose of the Act and the company cannot nominate any other employee to be the "occupier".
- Under the Act, the occupier of a factory has to apply to the Chief Inspector of Factories for registration and to get the plans of a factory approved.
- Every factory must: (a) maintain cleanliness and be free from effluvia; (b) dispose of wastes and effluents; (c) maintain adequate ventilation and reasonable temperature; (d) prevent inhalation of dust and fumes; (e) ensure proper standards of humidification; (f) avoid overcrowding; (g) provide suitable and sufficient lighting; and (h) arrange wholesome drinking water; provide sufficient latrines, urinals, and spittoons [health measures].
- Every factory must: (a) fence dangerous parts of the machines; (b) prohibit employment of women and children near cotton openers; (c) protect workers from repairing machinery in motion; (d) maintain hoists and lifts, floors, steps and staircases in good condition; (e) protect workers from injury to eye; (f) protect workers from dangerous fumes, inflammable dust, gas, and such other materials; (g) maintain accurate and up-to-date health/medical records of the workers, if the process is hazardous; and (h) employ safety officers wherein 1,000 or more workers are ordinarily employed [safety measures].
- Every factory must: (a) provide and maintain separate and adequate washing facilities, and facilities for storing of clothes; (b) arrange sitting facilities wherever necessary; (c) provide and maintain first-aid boxes or cupboards at least one for every 150 workers, and ambulance rooms in case of 500 or more workers; (d) provide canteens if there are 250 or more workers; and adequate restrooms/shelters, and lunch rooms, if there are 150 or more workers; (e) provide and maintain crèche facilities if there are 30 or more women workers; and (f) employ a welfare officer if 500 or more workers are employed [welfare measures].
- The main restrictive provisions of the Act about the working hours of adults are: (a) a worker cannot be employed for more than 48 hours in a week; (b) a worker cannot be employed for more than 9 hours in a day; (c) a worker must be given an interval of at least half an hour after five hours' work; (d) the total periods of work inclusive of rest interval must not spread over more than 101/2 hours in a day; (e) a

worker must be given a holiday for a whole day in every week; and (f) if a worker is deprived of any of the weekly holidays, he shall be given compensatory holidays.

- The claim for overtime wages under the Act is maintainable only for work done beyond 9 hours in a day or 48 hours in a week and not otherwise.
- The Act prohibits the employment of children below the age of 14 years in factory establishments and provides for their working hours.
- A woman worker cannot be employed except between the hours of 6.00 am and 7.00 pm.
- The Act provides for leave with wages and payment of overtime at double the ordinary rate of wages and a weekly day of rest.
- The registers to be maintained under the Act are: (a) register of adult workers; (b) register of leave with wages; (c) accident register with forms; (d) muster roll and wages register; (e) inspection book; (f) register of compensatory holidays and overtime; and (g) muster roll for exempted workers.
- The manager of the factory is required to send to the appropriate authorities a notice of certain accidents resulting in death or bodily injury which prevent the injured person from working for a period of 48 hours.
- The State Government may exempt any workshop or workplace where a manufacturing process is carried on and also any factory in the case of grave emergency from all or any of the provisions of the Act for such period and subject to such conditions, as it thinks fit.
- The responsibility for administration of the Act rests with the State Governments who administer it through their own factory inspectorates.
- The Act lays down various offences and the penalties thereof.

The Mines Act 1952

- The Act is intended to amend and consolidate the law relating to the regulation of labour and safety in mines.
- The Act has defined the term "mines" and its application.
- The Act aims at providing for safe as well as proper working conditions in mines, health and safety, and certain other welfare amenities to the workers employed therein.
- The Act has laid down certain provisions regarding hours of work and stipulates limitations of employment for persons employed in the mines.

- The Central Government is the administrative authority under the Act having the power to make rules and regulations for carrying out the purposes of the Act.
- The owner, agent or manager of the mine is required to send a notice of accidents and diseases to the authorities concerned and shall maintain a register of accidents in the prescribed form.
- The Act provides for the appointment of a Chief Inspector of Mines and inspectors by the Central Government.
- The Director General of Mines Safety is entrusted with the responsibility of enforcing the Act in all mines other than coal mines.

The Building and Other Construction Workers' (Regulation of Employment and Conditions of Service) Act 1996

- The object of the Act is to regulate the employment and conditions of service of building and other construction workers and to provide for their safety, health and welfare.
- The Act provides for the constitution of a corporate body namely the Building and other Construction Workers' Welfare Board by the Central Government or the State Government as the case may be.
- The Act has defined the terms "building or other construction work" and also "building worker".
- Every employer shall make an application to the registering officer for the registration of the establishment to which the Act is applicable along with the prescribed fees.
- The registering officer shall issue a certificate of registration to the employer to enable him/her to employ building workers.
- An employer shall be responsible for payment of wages to each building worker employed by him or his contractor and such wages shall be paid on or before the prescribed date.
- Employees employed through contractors on building operations are entitled to get bonus under the Payment of Bonus Act, 1965.
- An employer is liable to pay compensation in case of death or disablement of the building worker in accordance with the provisions of Employees' Compensation Act, 1923.
- The Act provides for certain welfare amenities like temporary accommodation at or near worksites, crèches, canteens, first-aid, and washing facilities.
- The Act specifies the appointment of inspecting staff and their responsibility for effective enforcement of the provisions of the Act.
- The State Governments are responsible for enforcement of the Act in their respective jurisdictions.

The Contract Labour (Regulation and Abolition) Act 1970

- The object of the Act is to provide for the abolition of contract labour wherever possible and for the regulation of the conditions of contract labour in establishments or employments where the abolition of contract labour system is not considered feasible for the time being.
- The Act applies to: (a) every establishment in which 20 or more workmen are employed or were employed on any day of the preceding twelve months as contract labour; and (b) every contractor who employs or who employed 20 or more workmen on any day of the preceding twelve months.
- The Act will not apply to an establishment in which work only of an intermittent or casual nature is performed and such work does not exceed the total of 120 days in a year, and in case of seasonal establishments 60 days in a year.
- A contractor is one who hires and supplies contract labour to an establishment or undertaking to produce a given result.
- The Act provides for registration and licensing by every principal employer and contractor respectively.
- The principal employer has to ensure that the contractor is licensed under: (a) Contract Labour (Regulation and Abolition) Act, 1970, and (b) Inter-State Migrant Workmen (Regulation of Employment and Conditions of Service) Act, 1979.
- The appropriate government may, after consultation with the Central Board or as the case may be a State Board, prohibit the employment of contract labour in any process, operation or other work in any establishment as may be notified by it in the Official Gazette having regard to the criteria laid down in the Act.
- The contractor is required to provide certain welfare amenities to the contract labour such as a canteen (employing 100 or more workers), restrooms, sufficient supply of wholesome drinking water, sufficient number of latrines and urinals, washing facilities, and first-aid boxes.
- It is the obligation of the principal employer to ensure due compliance of all labour laws applicable to contract labour, namely, the Payment of Wages Act, Minimum Wages Act, Inter-State Migrant Workmen Act, Employees' Compensation Act or Employees' State Insurance Act, and Employees' Provident Funds and Miscellaneous Provisions Act.
- The principal employer has to ensure due compliance of statutory wages and certify the payments of contract labour engaged by the contractor.

- If a contractor fails to pay wages within the prescribed period, then the principal employers are liable to make the payments and deduct the amount from the payments to be made to the contractors.
- It is the ultimate obligation of the principal employer to provide the statutory amenities to the contract labour employed in his establishment, if the contractor fails to do so.
- The principal employer is entitled, however, under the Act, to recover the payment made by him from the contractor by deducting any amount payable to the contractor under the contract or as a debt payable by the contractor.
- It is the duty of every principal employer/contractor to maintain records giving particulars of contract labour employed, the nature of work performed by them, and the rates of wages paid to them.
- Every contractor/principal employer is required to maintain in respect of the employees employed by him, registers such as: (a) muster roll; (b) register of wages; (c) register of deductions; (d) register of overtime; (e) register of fines; (f) register of advances; and (g) wage slips.
- The contractor is required to send half-yearly returns to the licensing authority.
- The principal employer of a registered establishment is required to send to the Registering Officer concerned an annual return in the prescribed form.
- It is also the duty of the contractor to exhibit in the premises of the establishment notices containing particulars about the hours of work, nature of duty, etc. in accordance with the rules framed under the Act.
- It is well settled that non-registration of the establishment by the principal employer and non-possession of license by the contractor would not result in regularisation of the contract workers.
- There is no provision under the Act for automatic absorption of contract labour, on issuing a notification by the government prohibiting employment of contract labour in any process, operation or other work in any establishment.
- The principal employer has to ensure that the agreement with the contractor is not a sham contract.
- If the provisions of any law, agreement, contract or service or standing orders are less favourable than those of the Contract Labour Act, they will be superseded by the provisions of the Contract Labour Act; but if such provisions are more favourable than those of the Contract Labour Act, the former will prevail over the latter.

- The Act stipulates various offences and punishments, and provides for the appointment of inspectors with certain powers.

The Inter-State Migrant Workmen (Regulation of Employment And Conditions of Service) Act 1979

- The Act is intended to safeguard the interests of the workmen who are recruited by contractors from one state for service in an establishment situated in another state, and to guard against the exploitation of such workmen by the contractors.
- The object of the Act is to regulate the employment of inter-state workmen and to provide for their conditions of service and for matters connected therewith.
- The Act applies to every establishment in which five or more inter-state migrant workmen are employed.
- It also applies to every contractor who employs or who employed five or more inter-state migrant workmen on any day of the preceding twelve months.
- The Act provides for the appointment of: (a) registering officers for registration of establishments employing inter-state migrant workmen, and (b) licensing officers for grant of licences to contractors employing such workmen.
- The duties and obligations of contractors are laid down under the Act.
- The contractor is required to provide stipulated welfare facilities to such workmen and comply with various other statutory requirements.

The Maternity Benefit Act, 1961

- The object of the Act is to provide for maternity leave and payment of certain monetary benefits to female employees during the period when they are out of employment because of their pregnancy.
- The Act applies in the first instance to mines, factories, plantations, and to every establishment wherein persons are employed for the exhibition of equestrian, acrobatic and other performances and to shops and establishments employing 10 or more persons.
- The Act does not apply to any factory or other establishment to which the provisions of the Employees' State Insurance Act apply.
- There is no wage limit for coverage under the Act.
- To be entitled to maternity leave, a female employee must have actually worked for not less than 80 days (only working days are to be calculated) in the twelve months immediately preceding the day of her expected delivery.
- The period for which a woman can get maternity benefit is 26 weeks.

- The maximum period entitled to maternity benefit by a woman having two or more surviving children shall be 12 weeks of which not more than 6 weeks shall precede the date of her expected delivery.
- A woman who legally adopts a child below the age of three months or a commissioning mother (biological mother) shall be entitled to maternity benefit for a period of 12 weeks from the date the child is handed over to the adopting mother or the commissioning mother, as the case may be.
- In case of miscarriage, a female employee is entitled to six weeks' leave with pay from the day of miscarriage.
- For illness arising out of pregnancy, delivery, premature birth of a child or miscarriage, a woman employee can take extra leave up to a maximum period of one month. She has, of course, to get a certificate from a doctor in the prescribed form.
- In case of tubectomy operation, a woman shall on production of proof, be entitled to leave with wages at the rate of maternity benefit for a period of two weeks immediately following the day of her tubectomy operation.
- Every woman entitled to maternity benefit is also entitled to a medical bonus if the employer fails to provide pre-natal or post-natal care free of charge.
- The amount of maternity benefit for the period preceding the date of her delivery shall be paid in advance to the woman on production of proof that the woman is pregnant and the amount due for the subsequent period shall be paid within 48 hours of production of proof of delivery.
- The computation of maternity benefit in case of a female worker engaged on a daily wage basis or casual basis has to be made for all the days including Sundays and rest days.
- Every woman who returns to duty after delivery of a child shall, in addition to the interval for rest allowed to her, be allowed in the course of her daily work two breaks of prescribed duration, for nursing the child until the child attains the age of fifteen months.
- A female employee can ask for light work for one month preceding the six weeks prior to her delivery or during these eight weeks if, for any reason, she does not avail of her leave.
- Under the Act, not only regular female employees but also those engaged on casual basis are entitled to maternity benefit.
- In case where the nature of work assigned to a woman is of such nature that she may work from home, the employer may allow her to do the assigned work after availing of the maternity benefits for such

period and on such conditions as the employer and the woman may mutually agree.

- If a female employee entitled to maternity benefit dies before receiving such benefit, the employer shall pay such benefit to the person nominated by her and in case there is no such nominee, to her legal representative.
- When a woman absents herself from work in accordance with the provisions of the Act, it shall be unlawful for her employer to discharge or dismiss her on account of such absence.
- A female employee can be deprived of maternity benefit if: (a) after going on maternity leave, she works in any other establishment during the period she is supposed to be on leave; and (b) during the period of her pregnancy, she is dismissed for any prescribed gross misconduct.
- Every establishment having 50 or more employees shall have the facility of creche within such distance as may be prescribed, either separately or along with common facilities.
- The employer shall allow four visits a day to the creche by the woman, which shall also include the interval for rest allowed to her.
- Every establishment shall intimate in writing and electronically to every woman at the time of her initial appointment regarding every benefit available under the Act.
- The employer is required to exhibit the abstract of the provisions of the Act and the rules made thereunder.
- It is the obligation of the employer to prepare and maintain the prescribed registers, records, and muster rolls and submit the prescribed returns.
- The appropriate government can exempt any establishment from the operation of all or any of the provisions of the Act or any rule made under the Act, if the benefits provided by the establishment are not less favourable than those provided in the Act.
- The Central Government is responsible for administration of the Act in mines and in the circus industry, while the State Governments are responsible for factories, plantations, and other establishments.
- The Act provides for penalties for contravention of the provisions of the Act.

The Minimum Wages Act 1948

- The object of the Act is to extend the concept of social justice to the employees employed in certain scheduled employments, and ensure protection to workers in which they are vulnerable to exploitation on account of lack of organisation and bargaining power.

- The Act is applicable only to the scheduled employments listed under the Act, regardless of the number of workers actually employed.
- The Act covers employment in agriculture, including horticulture, dairy farming, and raising of bees and poultry.
- The Act provides for statutory fixation and revision of minimum rates of wages by the Central or State Governments within a specified period.
- The contract labour employed under the Contract Labour (Regulation and Abolition) Act including daily-rated employees, are covered under the Act.
- While fixing or revising minimum wages, the appropriate government may follow either the committee method or notification method.
- After considering the advice of the said committee or representations received, the appropriate government will, by notification in the Gazette, fix or revise the minimum rates of wages.
- Once the minimum wages are notified and become effective, the employer must pay to every employee engaged in a scheduled employment under him wages at a rate not less than the minimum rate of wages fixed by such notification for that class of employees.
- The minimum rates of wages may be fixed: (a) for different employments; (b) for different classes in the same employment; (c) for adolescents, children, and apprentices; and (d) for different localities.
- The minimum wages payable under the Act are to be paid in cash. But it also provides for authorisation of payment in kind where the appropriate government considers it necessary.
- The minimum rate fixed may consist of the basic rate of wages and cost of living allowance and the cash value of concessions in respect of supply of essential commodities at concessional rates.
- The rates of wages may be: (a) a time rate, (b) piece rate, (c) a guaranteed time rate, and (d) an overtime rate.
- The Act empowers the State Government to constitute advisory boards to co-ordinate the work of different committees and sub-committees and advise the government on the fixation of minimum wages.
- The Central Government is also empowered to constitute a Central Advisory Board to advise Central and State Governments, and to co-ordinate the work of the advisory boards.
- The appropriate government may appoint inspectors for the purposes of the Act.

- The appropriate government may appoint authorities to hear and decide claims arising out of payment of wages less than the minimum rates of wages and other incidental matters.
- The Act prohibits civil courts from entertaining any suit for recovery of minimum wages payable under the Act.
- Under the Act, the employer is required to maintain: (a) register of wages; (b) register of overtime payment; (c) muster roll; (d) register of fines; and (e) register of deductions.
- Any contract or agreement whereby an employee relinquishes or reduces his right under the Act shall be null and void.
- The Act is administered both by the Central and State Governments.
- The Act lays down various offences and penalties for violation of the provisions of the Act.

Appendix B
Health Safety and Environment

Every organisation is unique in terms of its culture and business challenges, but sustainable profitable growth is common thread cutting across all sectors of industries and business. To ensure sustainable profitable growth, the organisations need to have policies and programmes for business continuity by protecting the most valuable asset of the organisation, namely employees. The health and safety of the employees and protection of the environment forms the central theme of the related policies and programmes. The key concepts of health, safety and environment management systems are universal in nature and hence can be applied and implemented in organisations having variety in business challenges and nature of operations. At the same time, one should also understand that one side does not fit all and hence the health safety and environment management procedures and programmes should be designed tailor made for each organisation based on key risks, business challenges and organisational culture. The skills and maturity of the leadership team of the organisation plays an important role in facilitating these programmes and procedures.

All activities within an organisation are the results of human act and in turn human behaviour. That is why it is believed that all industrial incidents/accidents can be prevented. To do so we need to direct our efforts towards changing human behaviour and creating a culture that is self-sustainable and capable enough to prevent injuries/incidents.

An HSE culture can be defined as a culture where there is considerable informed endeavour to reduce risks to the man, machine, material and the environment to a level that is as low as is reasonably practicable. Influencing HSE culture is more about leadership than management. Without support from senior management, safety behaviour changes are less likely to be sustained. In order to develop a strong HSE culture, that culture must start from the top senior management and extend down to the least experienced employee in any organisation. Every employee needs to understand that HSE must be the first focus of his or her job. The final responsibility for HSE is not the responsibility of the HSE department only, but of each employee engaged with the organisation. After all, at the end of every day, all the employees should return home safely.

DuPont had carried out a study to determine how organisation culture is related to safety performance. DuPont measures culture by considering a Broadley Curve which shows the progressive maturity of the culture. The goal is to move from a reactive safety culture to dependent, independent, and finally interdependent safety culture. Some organisations show excellent HSE performance while others in the same business had poor performance. There is a strong relationship between the culture of an organisation and the development of a systems approach. The systems approach is the key to ensure sustainable profitable growth.

The three core components of a strong and world class HSE management system are: (a) HSE leadership, (b) HSE organisation, and (c) HSE processes.

HSE Leadership

The HSE culture does not appear from the sky nor does it spontaneously intertwine/ingrain itself into the fabric of a company. It is paramount that the way in which a company does business, and how the senior management is involved in the nurturing of its sustainability. The senior managers, starting with the highest executives of the company should: (a) be in tune with the elements of HSE culture, (b) be true believers of the 'creed' of zero harm, (c) demonstrate HSE value, and (d) be knowledgeable of what it takes to make the HSE culture.

Case studies have shown that companies can comply with the letter of an HSE management system without actually adapting the spirit of that system. This is primarily due to the fact that senior management has not understood the underlying goals of such a system that can bring sustainable performance for the company and their employees in the long term. Without such an understanding the senior management cannot lead by example nor will they promote the key elements of a strong HSE culture.

It has often been said that HSE and production are at different ends of the business process but results have proved that good HSE performance gives good business results. The theory and practice of risk management is equally essential in establishing, maintaining and enhancing the HSE culture. Again, this must start at the highest levels of management within a company.

HSE Organisation

HSE organisation has a broader meaning and is not limited to the HSE department. It generally means all the employees, line managers and leaders who have specific responsibilities pertaining to HSE management performance in their domain affecting their own health and safety as well as the same of others directly or indirectly linked to it and the environment. This clearly indicates that each and every employee in the organisation has a certain responsibility and role to play towards HSE matters.

For a HSE culture to flourish and strengthen, all levels of the organisation must: (a) be aware of lessons learned from past incidents and close calls, (b) hold HSE as a value - and not just a priority, and (c) benefit from an atmosphere that encourages a questioning attitude towards work processes.

HSE Processes

HSE processes for the organisation need to be developed by the employees involved in the processes and activities. This is the key for ensuring that these processes and activities will be implemented in a better way. To ensure this process related to communication, participation and consultation, and management review need to be designed, approved and monitored for effectiveness and continual improvement.

Improving an organisation's HSE culture takes time. An organisation passes through different maturity levels during its journey towards HSE cultural transformation.

1. **Supreme Petrochem Limited (SPL):** The Supreme Petrochem Limited has a state-of-the art manufacturing facility located at Amdoshi - Wangani - Roha Road, Roha District, Raigad. The company is engaged in the business of manufacturing and sale of polystyrene. SPL was formed and incorporated in 1989 at Mumbai. The company is committed to ensuring safety in all its operations, protection of health of every employee and protection of the environment. Continuous efforts for improvement in quality, environment, health and safety systems are made through:

- Setting objectives to continually improve performance in HSE.
- Compliance with HSE legislation.
- Adopting suitable safety and environmental practices and proactive means to minimise hazards and risk due to industrial activities.
- Designing process and plant operations to minimise risk and wastage.
- Enhancing the skill and competence of the employees through training.
- Reducing consumption of energy and natural resources.
- Maintaining safety, health and environment management systems in accordance with ISO 45001:2018.

SPL top management is fully committed to the cause of HSE. The factory has an IMS policy integrating OHS, environment, and quality systems. Occupational health and safety policy, environmental policy and quality policy are integrated into a management system policy (MSE). Deliberate efforts are made to improve employee participation and involvement through various schemes, activities, and systems related to HSE matters. Safety committee meetings are held every month and HSE matters are discussed and actions are initiated.

Implementation of active and reactive monitoring techniques such as inspections, audits, work environment monitoring, health monitoring, equipment monitoring for mechanical integrity is carried out effectively. Internal audit is carried out for Occupational Health and Safety Management Systems (OHSMS) and Environmental Management Systems (EMS). External audit is carried out once in a year for Integrated Management Systems (IMS). The company has a full-fledged HSE department with qualified and experienced HSE professionals. Also, the company is actively involved in CSR activities for improvement of the nearby community.

Appendix C

ILO Conventions Concerning Safety and Health in Construction

The General Conference of the International Labour Organisation, having been convened at Geneva by the Governing Body of the International Labour Office, and having met in its Seventy-fifth Session on June 1, 1988, and noting the relevant international labour Conventions and Recommendations and, in particular, the Safety Provisions (Building) Convention and Recommendation, 1937, the Co-operation in Accident Prevention (Building) Recommendation, 1937, the Radiation Protection Convention and Recommendation, 1960, the Guarding of Machinery Convention and Recommendation, 1963, the Maximum Weight Convention and Recommendation, 1967, the Occupational Cancer Convention and Recommendation, 1974, the Working Environment (Air Pollution, Noise and Vibration) Convention and Recommendation, 1977, the Occupational Safety and Health Convention and Recommendation, 1981, the Occupational Health Services Convention and Recommendation, 1985, the Asbestos Convention and Recommendation, 1986, and the list of occupational diseases as revised in 1980 appended to the Employment Injury Benefits Convention, 1964, and having decided upon the adoption of certain proposals with regard to safety and health in construction, which is the fourth item on the agenda of the session and having determined that these proposals shall take the form of an international Convention revising the Safety Provision (Building) Convention, 1937, adopts this twentieth day of June of the year one thousand nine hundred and eighty-eight the following Convention, which may be cited as the Safety and Health in Construction Convention, 1988.

Scope and Definitions

Article 1

1. This Convention applies to all construction activities, namely building, civil engineering, and erection and dismantling work, including any process, operation or transport on a construction site, from the preparation of the site to the completion of the project.
2. A Member country ratifying this Convention may, after consultation with the most representative organisations of employers and workers concerned, where they exist, exclude from the application of the Convention, or certain provisions thereof, particular branches of economic activity or particular undertakings in respect of which special problems of a substantial nature arise, on condition that a safe and healthy working environment is maintained.
3. This Convention also applies to such self-employed persons as may be specified by national laws or regulations.

Article 2

For the purpose of this Convention:

(a) The term "construction" covers:

  (i) building, including excavation and the construction, structural alteration, renovation, repair, maintenance (including cleaning and painting) and demolition of all types of buildings or structures;

  (ii) civil engineering, including excavation and the construction, structural alteration, repair, maintenance and demolition of, for example, airports, docks, harbours, inland waterways, dams, river and avalanche and sea defence works, roads and highways, railways, bridges, tunnels, viaducts and works related to the provision of services such as communications, drainage, sewerage, water and energy supplies;

  (iii) the erection and dismantling of prefabricated buildings and structures, as well as the manufacturing of pre-fabricated elements on the construction site;

(b) the term "construction site" means any site at which any of the processes or operations described in subparagraph (a) above are carried on:

(c) the term "workplace" means all places where workers need to be or to go by reason of their work and which are under the control of an employer as defined in subparagraph (e) below;

(d) the term "worker" means any person engaged in construction;

(e) the term "employer" means:

  (i) any physical or legal person who employs one or more workers on a construction site; and

(ii) as the context requires, the principal contractor, the contractor or the subcontractor;

(f) the term "competent person" means a person possessing adequate qualifications, such as suitable training and sufficient knowledge, experience and skill for the safe performance of the specific work. The competent authorities may define appropriate criteria for the designation of such persons and may determine the duties to be assigned to them;

(g) the term "scaffold" means any temporary structure, fixed, suspended, or mobiled, and its supporting components which is used for supporting workers and materials or to gain access to any such structure, and which is not a "likling appliance" as defined in subparagraph (h) below;

(h) the term "lifting appliance" means any stationary or mobile appliance used for raising or lowering persons or loads;

(i) the term "lifting gear" means any gear or tackle by means of which a load can be attached to a lifting appliance but which does not form an integral part of the appliance or load.

General Provisions

Article 3

The most representative organisations of employers and workers concerned shall be consulted on the measures to be taken to give effect to the provisions of this Convention.

Article 4

Each Member country which ratifies this Convention undertakes that it will, on the basis of an assessment of the safety and health hazards involved, adopt and maintain in force laws or regulations which ensure the application of the provisions of the Convention.

Article 5

1. The laws and regulations adopted in pursuance of Article 4 above may provide for their practical application through technical standards or codes of practice, or by other appropriate methods consistent with national conditions and practice.
2. In giving effect to Article 4 above and to paragraph 1 of this Article, each Member shall have due regard to the relevant standards adopted by recognised international organisations in the field of standardisation.

Article 6

Measures shall be taken to ensure that there is co-operation between employers and workers, in accordance with arrangements to be defined by national laws or regulations, in order to promote safety and health at construction sites.

Article 7

National laws or regulations shall require that employers and self-employed persons have a duty to comply with the prescribed safety and health measures at the workplace.

Article 8

1. Whenever two or more employers undertake activities simultaneously at one construction site -
   (a) the principal contractor, or other person or body with actual control over or primary responsibility for overall construction site activities shall be responsible for coordinating the prescribed safety and health measures and, in so far as is compatible with national laws and regulations, for ensuring compliance with such measures :
   (b) in so far as is compatible with national laws and regulations, where the principal contractor, or other person or body with actual control over or primary responsibility for overall construction site activities, is not present at the site, he shall nominate a competent person or body at the site with the authority and means necessary to ensure on his behalf co-ordination and compliance with the measures, as foreseen in paragraph (a) above;
   (c) each employer shall remain responsible for the application of the prescribed measures in respect of the workers placed under his authority.

Article 9

Those concerned with the design and planning of a construction project shall take into account the safety and health of the construction workers in accordance with national laws, regulations and practice.

Article 10

National laws or regulations shall provide that workers shall have the right and the duty at any workplace to participate in ensuring safe working conditions to the extent of their control over the equipment and methods of work and to express views on the working procedures adopted as they may affect safety and health.

Article 11

National laws or regulations shall provide that workers shall have the duty to-

(a) co-operate as closely as possible with their employer in the application of the prescribed safety and health measures;
(b) take reasonable care for their own safety and health and take of other persons who may be affected by their acts or omissions at work;
(c) use facilities placed at their disposal and not misuse anything provide for their own protection or the protection of others;
(d) report forthwith to their immediate supervisor, and to the workers' safety representative where one exists, any situation which they believe could present a risk, and which they cannot properly deal with themselves.
(e) comply with the prescribed safety and health measures.

Article 12

1. National laws or regulations shall provide that a worker shall have the right to remove himself from danger when he has good reason to believe that there is an imminent and serious danger to this safety or health, and the duty so to inform his supervisor immediately.
2. Where there is an imminent danger to the safety of workers the employer shall take immediate steps to stop the operation and evacuate workers an appropriate.

Preventive and Protective Measures

Article 13

Safety of Workplaces

1. All appropriate precautions shall be taken to ensure that all workplaces are safe and without risk of injury to the safety and health of workers.
2. Safe means of access to and egress from all workplaces shall be provided and maintained, and indicated where appropriate.
3. All appropriate precautions shall be taken to protect persons present at or in the vicinity of a construction site from all risks which may arise from such site.

Article 14

Scaffolds and Ladders

1. Where work cannot safely be done on or from the ground or from part of building or other permanent structure, a safe and suitable scaffold shall be provided and maintained, or other equally safe and suitable provision shall be made.

2. In the absence of alternative safe means of access to elevated working places, suitable and sound ladders shall be provided. They shall be properly secured against inadvertent movement.
3. All scaffolds and ladders shall be constructed and used in accordance with national laws and regulations.
4. Scaffolds shall be inspected by a competent person in such cases and at such times as shall be prescribed by national laws or regulations.

Article 15

Lifting Appliances and Gear

1. Every lifting appliance and item of lifting gear, including their constituent elements, attachments, anchorages and supports, shall -
   (a) be of good design and construction, sound material and adequate strength for the purpose for which they are used;
   (b) be properly installed and used;
   (c) be maintained in good working order;
   (d) be examined and tested by a competent person at such times and in such cases as shall be prescribed by national laws or regulations; the result of these examinations and tests shall be recorded;
   (e) be operated by workers who have received appropriate training in accordance with national laws and regulations.
2. No person shall be raised, lowered or carried by a lifting appliance unless it is constructed, installed and used for that purpose in accordance with national laws and regulations, except in an emergency situation in which serious personal injury or fatality may occur, and for which the lifting appliance can be safely used.

Article 16

Transport, Earth-Moving and Materials-Handling Equipment

1. All vehicles and earth-moving or materials-handling equipment shall-
   (a) be of good design and construction taking into account as far as possible economic principles;
   (b) be maintained in good working order;
   (c) be properly used;
   (d) be operated by workers who have received appropriate training in accordance with national laws and regulations.
2. On all construction sites on which vehicles, earth-moving or materials-handling equipment are used -
   (a) safe and suitable access ways shall be provided for them;and
   (b) traffic shall be so organised and controlled as to secure their safe operation.

Article 17

Plant, Machinery, Equipment and Hand Tools

1. Plant, machinery and equipment, including hand tools, both manual and power driven, shall -
   (a) be of good design and construction, taking into account as far as possible ergonomic principles;
   (b) be maintained in good working order;
   (c) be used only for work for which they have been designed unless a use outside the initial design purposes has been assessed by a competent person who has concluded that such use is safe;
   (d) be operated by workers who have received appropriate training.
2. Adequate instructions for safe use shall be provided where appropriate by the manufacturer or the employer, in a form understood by the users.
3. Pressure plant and equipment shall be examined and tested by a competent person in cases and at all times prescribed by national laws or regulations.

Article 18

Work at Heights including Roofwork

1. Where necessary to guard against danger, or where the height of a structure or its slope exceeds that prescribed by national laws or regulations, preventive measures shall be taken against the fall of workers and tools or other objects or materials.
2. Where workers are required to work on or near roofs or other places covered with fragile material, through which they are liable to fall, preventive measures shall be taken against their inadvertently stepping on or falling through the fragile material.

Article 19

Excavations, Shafts, Earthworks, Underground Works and Tunnels

Adequate precautions shall be taken in any excavation, shaft, earthworks, underground work or tunnel -

(a) by suitable shoring or otherwise to guard against danger to workers from a fall or dislodgement of earth, rock or other material;
(b) to guard against dangers arising from the fall of persons, materials or objects or the inrush of water into the excavation, shaft, earthworks, underground works or tunnel;
(c) to secure adequate ventilation at every workplace so as to maintain an atmosphere fit for respiration and to limit any

fume, gases, vapours, dust or other impurities to levels which are not dangerous or injurious to health and are within limits laid down by national laws or regulations;

(d) to enable the workers to reach safety in the event of fire, or an inrush of water or material;

(e) to avoid risk of workers arising from possible underground dangers such as the circulation of fluids or the presence of pockets of gas, by undertaking appropriate investigations to locate them.

Article 20

Cofferdams and Caissons

1. Every cofferdam and caisson shall be -
   (a) of good construction and suitable and sound material and of adequate strength;
   (b) provided with adequate means for workers to reach safety in the event of an inrush of water or material.
2. The construction, positioning, modification or dismantling of a cofferdam or caisson shall take place only under the immediate supervision of a competent person.
3. Every cofferdam and caisson shall be inspected by a competent person at prescribed intervals.

Article 21

Work in Compressed Air

1. Work in compressed air shall be carried out only in accordance with measures prescribed by national laws or regulations.
2. Work in compressed air shall be carried out only by workers whose physical aptitude for such work has been established by a medical examination and when a competent person is present to supervise the conduct of the operations.

Article 22

Structural Frames and Formwork

1. The erection of structural frames and components, formwork, falsework and shoring shall be carried out only under the supervision of a competent person.
2. Adequate precautions shall be taken to guard against danger to workers arising from any temporary state of weakness or instability of a structure.
3. Formwork, falsework and shoring shall be so designed, constructed and maintained that it will safely support all loads that may be imposed on it.

Article 23

Work Over Water

Where work is done over or in close proximity to water there shall be adequate provision for -

(a) Preventing workers from falling into water;
(b) the rescue of workers in danger of drowning;
(c) safe and sufficient transport.

Article 24

Demolition

When the demolition of any building or structure might present danger to workers or to the public -

(a) appropriate precautions, methods and procedures shall be adopted, including those for the disposal of waste or residues, in accordance with national laws or regulations;
(b) the work shall be planned and undertaken only under the supervision of a competent person.

Article 25

Lighting

Adequate and suitable lighting, including portable lighting where appropriate, shall be provided at every workplace and any other place on the construction site where a worker may have to pass.

Article 26

Electricity

1. All electrical equipment and installations shall be constructed, installed and maintained by a competent person, and so used as to guard against danger.
2. Before construction is commenced and during the progress thereof adequate steps shall be taken to ascertain the presence of and to guard against danger to workers from any live electrical cable or apparatus which is under, over or on the site.
3. The laying and maintenance of electrical cables and apparatus on construction sites shall be governed by the technical rules and standards applied at the national level.

Article 27

Explosives

Explosives shall not be stored, transported, handled or used except -

(a) under conditions prescribed by national laws or regulations; and
(b) by a competent person, who shall take such steps as are necessary to ensure that workers and other persons are not exposed to risk of injury.

Article 28

Health Hazards

1. Where a worker is liable to be exposed to any chemical, physical or biological hazard to such an extent as is liable to be dangerous to health, appropriate preventive measures shall be taken against such exposure.
2. The preventive measures referred to in paragraph 1 above shall comprise -
   (a) the replacement of hazardous substances by harmless or less hazardous substances wherever possible; or
   (b) technical measures applied to the plant, machinery, equipment or process; or
   (c) where it is not possible to comply with subparagraphs (a) or (b) above, other effective measures, including the use of personal protective equipment and protective clothing.
3. Where workers are required to enter any area in which a toxic or harmful substance may be present, or in which there may be an oxygen deficiency, or a flammable atmosphere, adequate measures shall be taken to guard against danger.
4. Waste shall not be destroyed or otherwise disposed of on a construction site in a manner which is liable to be injurious to health.

Article 29

Fire Precautions

1. The employer shall take all appropriate measures to -
   (a) avoid the risk of fire;
   (b) combat quickly and efficiently any outbreak of fire;
   (c) bring about a quick and safe evacuation of persons.
2. Sufficient and suitable storage shall be provided for flammable liquids, solids and gases.

Article 30

Personal Protective Equipment and Protective Clothing

1. Where adequate protection against risk of accident or injury to health, including exposure to adverse conditions, cannot be ensured by other means, suitable personal protective equipment and protective clothing, having regard to the type of work and risks, shall be provided and maintained by the employer, without cost to the workers, as may be prescribed by national laws or regulations.
2. The employer shall provide the workers with the appropriate means to enable them to use the individual protective equipment, and shall ensure its proper use.

3. Protective equipment and protective clothing shall comply with standards set by the competent authority taking into account as far as possible ergonomic principles.
4. Workers shall be required to make proper use of and to take good care of the personal protective equipment and protective clothing provided for their use.

Article 31

First Aid

The employer shall be responsible for ensuring that first aid, including trained personnel, is available at all times. Arrangements shall be made for ensuring the removal for medical attention of workers who have suffered an accident or sudden illness.

Article 32

Welfare

1. At or within reasonable access of every construction site an adequate supply of wholesome drinking water shall be provided.
2. At or within reasonable access of every construction site, the following facilities depending on the number of workers and the duration of work, be provided and maintained -
    (a) sanitary and washing facilities;
    (b) facilities for changing and for the storage and drying of clothing.
    (c) accommodation for taking meals and for taking shelter during interruption of work due to adverse weather conditions.
3. Men and women workers should be provided with separate sanitary and washing facilities.

Article 33

Information and Training Workers shall be adequately and suitably -

(a) informed of potential safety and health hazards to which they may be exposed at their workplace;

(b) instructed and trained in the measures available for the prevention and control of, and protection against, those hazards.

Article 34

Reporting of Accidents and Diseases National laws and regulations shall provide for the reporting to the competent authority within a prescribed time of occupational accidents and diseases.

Article 35

Implementation

Each Member shall -

(a) take all necessary measures, including the provision of appropriate penalties and corrective measures, to ensure the effective enforcement of the provisions of the Convention;

(b) provide appropriate inspection services to supervise the application of the measures to be taken in pursuance of the Convention and provide these services with the resources necessary for the accomplishment of their task, or satisfy itself that appropriate inspection is carried out.

Article 36

Final Provisions

This Convention revises the Safety Provision (Building) Convention, 1937.

Article 37

The formal ratification of this Convention shall be communicated to the Director-General of the International Labour Office for registration.

Article 38

1. This Convention shall be binding only upon those Members of the International Labour Organisation whose ratifications have been registered with the Director-General.
2. It shall come into force twelve months after the date on which the ratifications of two Members have been registered with the Director-General.
3. Thereafter, this Convention shall come into force for any Member twelve months after the date on which its ratification has been registered.

Article 39

1. A Member which has ratified this Convention may denounce it after the expiration of ten years from the date on which the Convention first comes into force, by an act communicated to the Director-General of the International Labour Office for registration. Such denunciation shall not take effect until one year after the date on which it is registered.
2. Each Member which has ratified this Convention and which does not, within the year following the expiration of the period of ten years mentioned in the preceding paragraph, exercise the right of denunciation provided for in this Article, will be bound for another period of ten years and, thereafter, may denounce this Convention at the expiration of each period of ten years under the terms provided for in this Article.

Article 40

1. The Director-General of the International Labour Office shall notify all Members of the International Labour Organisation of the registration of all ratifications and denunciations communicated to him by the Members of the Organisation.
2. When notifying the members of the Organisation of the registration of the second ratification communicated to him, the Director-General shall draw the attention of the Members of the Organisation to the date upon which the Convention will come into force.

Article 41

The Director-General of the International Labour Office shall communicate to the Secretary-General of the United Nations for registration in accordance with Article 102 of the Charter of the United Nations full particulars of all ratifications and acts of denunciation registered by him in accordance with the provisions of the preceding Articles.

Article 42

At such times as it may consider necessary the Governing Body of the International Labour Office shall present to the General Conference a report on the working of this Convention and shall examine the desirability of placing on the agenda of the Conference the question of its revision in whole or in part.

Article 43

1. Should the Conference adopt a new Convention revising this Convention in whole or in part, then, unless the new Convention otherwise provides-
   (a) the ratification by a Member of the new revising Convention shall ipso jure involve the immediate denunciation of this Convention, notwithstanding the provisions of Article 39 above, if and when the new revising Convention shall have come into force;
   (b) as from the date when the new revising Convention comes into force this Convention shall cease to be open to ratification by the Members.
2. This Convention shall in any case remain in force in its actual form and content for those Members which have ratified it but have not ratified the revising Convention.

Article 44

The English and French versions of the text of this Convention are equally authoritative.

Appendix D

Model Guidelines on Construction Safety

1. Safety Management at Construction Worksites

1.1 Safety Organisation

1.1.1 Every construction project should have safety considered by the architects and construction engineers during the preparation of the specifications, drawings and operational plans.

1.1.2 Each construction project should establish a written safety policy and make it known to workers and a safety committee consisting of members from each section.

1.1.3 Regular meetings should be held and chaired by the site or project manager at the worksite and assisted by a safety officer with relevant section heads or subcontractors represented.

1.1.4 The committee's main tasks would be to :

(a) review the safety plans and to highlight any shortcomings in term of the plan's practicality and execution of it;

(b) carry out site inspections and review results of the inspection and to take remedial measures;

(c) evaluate accident statistics and reports of accidents;

(d) co-ordinate site works which are liable to give rise to safety problems.

1.2 Site safety planning

1.2.1 The safety plan should be developed before work on site begins and should form part of the documentation prepared by the architects and the construction engineers. It should contain the detailed safety measures for each stage of construction and the emergency evacuation plan.

1.2.2 The site layout plan should include workflow and workstations, storage areas and vehicular routes.

1.2.3 The persons responsible for each detailed task and the chain of command are also to be spelled out.

1.3 Notification and other statutory requirements

1.3.1 Various aspects of the law applicable at the site should be reviewed and persons be appointed to carry them out.

1.3.2 Notifications and other reporting duties particularly spelled out in the law must be noted and carried out.

1.3.3 Accident reporting procedures and a list of dangerous occurrence items should be made.

1.4 Safety training and promotional activities

1.4.1 Scheduled safety training programmes should be organised and published among the staff.

1.4.2 Detailed training programmes should be drawn up for each group of tradesmen.

1.4.3 Ongoing promotional activities should be scheduled

1.5 Accident investigation

1.5.1 These should be carried out promptly and thoroughly by the safety officer.

1.5.2 The report complete with recommended remedial measures should be tabled at the following safety committee meeting.

1.6 Statistics and record keeping

1.6.1 All injuries or fatalities occuring at the worksite should be recorded, investigated and the report presented to the site manager and safety committee.

1.6.2 These injuries regardless whether they fall within the statutory definition of accident will be the statistical basis for identifying where the dangers lie.

1.6.3 Details such as the number of days absent from work and cost of medical treatment should be kept for analysis at the end of the year.

1.6.4 Other details which should be captured are time of accident, trade of the injured, weather conditions, physical condition (based on medical report), unsafe condition, tool or machinery involved, type of safety equipment worn by injured and whether there was any co-ordination or work procedure factors, the work experience of the worker, his age, whether he has undergone a safety training course.

1.7 Duties and responsibilities of the workforce

1.7.1 Employers or main contractors :

(a) organise the safety committee and appoint a safety officer;

(b) provide equipment that is safe both at purchase as well as regular inspection;

(c) assume overall control and responsibility of the worksite.

1.7.2 Architects and engineers designing the construction project should take into account the safety factors of construction and ensure that the use of materials harmful to workers is avoided.

1.7.3 Subcontracts and nominated subcontractors must co-operate with the safety committee and each must appoint his own safety supervisor.

1.7.4 Supervisors at construction sites should plan, organise and inspect worksite safety.

1.7.5 All workers at the site must comply with safety rules and work safely.

1.7.6 All workers at the site must comply with safety rules and work safely.

1.8 Sign posting

1.8.1 Standard safety signs should be used in the worksite.

1.8.2 Where there are dangers not easily recognisable such as excavations or overhead activity.

1.8.3 Where blind corners exist in the site, which might affect vehicular safety, mirrors should be used.

2. Preventive Measures Against Specific Hazards

2.1 Fall prevention at the workplace

2.1.1 All open sides of buildings from which a worker might fall and openings into which a worker might fall should be adequately covered or barricaded.

2.1.2 Where barricades could not be installed, safety nets should be installed close to the level at which there is danger of falls.

2.1.3 Where secure foothold is impracticable, safety belts or harnesses with secure anchorage points should be provided.

2.1.4 At the elevated places such as encountered in formwork construction for beams, secure access and foothold should be provided.

2.2 Falling material hazards prevention and protection

2.2.1 Falling material hazards should be prevented by means of overhead screens and hoardings.

2.2.2 Discarding of waste materials by throwing down the side of the building should be prohibited. The provision of refuse chutes for discarding small items and the use of lifting equipment to remove large ones should be clearly communicated and enforced. The use of cleaning teams working after work hours should reduce potential hazards.

2.2.3 Where there is danger of collapse of structures, steps should be taken to barricade off a safety zone until repairs are made.

2.3 Structural failure prevention

2.3.1 All structures must be properly designed.

2.3.2 Temporary structures such as formwork and its supports must be properly braced in all directions to prevent displacement.

2.3.3 Vibrating machinery, piping and other parts must not be brought in contact with structures unless they are specially designed for it.

2.3.4 Formwork supports should not be removed until the concrete it supports has attained its required strength.

2.3.5 Structures should be designed by competent engineers and be put up under the supervision of experienced supervisors.

2.3.6 They should be inspected before use by a competent engineer and thereafter inspected on a regular basis or where parts are damanged and repaired.

2.3.7 Ensure only materials of adequate strengh be used.

2.4 Refuse disposal

2.4.1 Refuse accumulation should never be allowed as the resulting loading may be excessive.

2.4.2 Discharge of waste concrete on platforms or accumulation bricks on work platforms could result in a sudden collapse and should be avoided.

2.4.3 Refuse must be removed daily to prevent accumulation. Materials liable to cause persons to slip or trip and fall should be cleared immediately.

2.4.4 Properly designed refuse chutes should be constructed and used. Throwing refuse over sides of buildings should be prohibited.

2.4.5 Refuse removal teams working after work hours should be organised where normal cleaning cannot cope with the build-up of waste material.

2.4.6 Projecting nails should be removed or bent over.

2.5 Ventilation, ligthing and noise control

2.5.1 Enclosed workplaces in which there is a build-up of heat, dangerous gases or dust should be adequately ventilated.

2.5.2 Lighting should be provided in dark passageways and workplaces. Low voltage lighting is recommended.

2.5.3 Equipment and processes generating loud noises should be avoided as far as practicable. Air compressors and generators, for example, should be located away from the concentration of workers.

2.6 Fire protection

2.6.1 At every level of construction where combustible materials are liable to be used or stored, a 9 kilogrammes dry ABC powder extinguisher should be provided. Where the travel distance exceeds 30 m, additional ones should be provided. Two and a half kilogrammes BCF extinguishers should be placed near switch boards.

2.6.2 There must be trained personnel available on the site at all times. All workers at the site could be trained to operate the extinguisher.

2.6.3 The fire alarm procedure and the evacuation sequence should be clearly spelled out and the workforce notified. In workplaces where fire hazards are potentially more hazardous (e.g. in underground tunnelling work), drills should be carried out.

2.6.4 Combustible and flammable materials should be stored away from the source of heat such as generators, welding sets and electrical distribution boxes.

2.7 Gas cylinders

2.7.1 Storage

(a) When not in use store LPG cylinders in a well-ventilated place in the open air, at least 3 metres away from excavations, basements, etc. where leaking gas can collect.

(b) Acetylene cylinders may be kept with LPG but oxygen cylinders, flammable liquids and flammable materials should be stored separately, away from LPG cylinders.

(c) Keep the cylinders in a lockable wire cage. The cage should be clearly marked and situated in a well ventilated place in the open air, at least 1 metre away from site huts, boundaries, etc.

(d) Never store cylinders in ordinary lockable metal huts.

2.7.2 Prevent entry of children and trespassers

Serious accidents have happened when unattended LPG cylinders and equipment have been tampered with. At the end of the working day, isolate gas supplies to equipment and return cylinders to store.

2.7.3 Lighting up and shut down procedures

Make sure your workers know and follow the procedures recommended by the equipment manufacturer.

2.7.4 Bitumen boilers

LPG cylinders at bitumen boilers or cauldrons should be kept at least 3 metres from the burner. Keep the boiler clear of site traffic to prevent damage to the flexible hose, which should be of steel braid reinforced construction or similar. Never leave a boiler or cauldron unattended with the burner alight.

2.7.5 Site huts and similar areas

(a) Install all cylinders and regulators outside huts unless they are an integral part of the equipment. After use turn off the gas supply at the cylinders as well as at the appliance.

(b) Provide very good ventilation at high and low levels, and for individual gas appliances.

(c) Use wire mesh cages or other means to prevent tampering with the cylinder valves.

3. Personal Protective Equipment

3.1 Personal protective equipment should be supplied and maintained by the employer and should be of suitable quality to provide adequate protection.

3.2 Safety promotion and training on the use and maintenance of personal protective equipment should be carried out.

3.3 Helmets should be provided for all who are exposed to the dangers of falling material or structures they might strike against.

3.4 Suitable eye protection should be provided for all who are exposed to flying particles, harmful glare and dangerous substances.

3.5 In the handling of rough objects, gloves should be provided and used.

3.6 Safety footwear which should be provided to all who are exposed to foot injury, should be good fitting and comfortable to wear.

3.7 Safety belts should be provided where other means are not practicable. Both the anchorage points and lifeliness provided for attaching safety belts should be of adequate strength. The umbilical line should be fixed in such a way that user's freefall will not exceed 1 metre.

3.8 Catch nets should be used where persons are liable to fall and these should be securely supported at a level as near as possible to the working level.

3.9 Noise defenders should be provided for work areas where noise level exceeds 85 db.

3.10 Respiratory protection should be provided by employers and used by workers where the dust level remains high and where control at source is not practicable.

4. Scaffolds

4.1 General provisions

4.1.1 Scaffolds should be provided for workers working at a height unless a safe alternative could be found.

4.1.2 Sufficient, suitable and sound material should be used and the scaffold be constructed in accordance with good engineering practice.

4.1.3 Tall scaffolds should be designed by a competent person.

4.1.4 Adequate bracings must be provided.

4.1.5 Inspection of the scaffold should be carried out by a competent person before it is used, and thereafter weekly, after inclement weather or after repairs were carried out on damaged parts.

4.1.6 Lifting appliance should not be attached to the scaffold as a rule. Should it be necessary to do so, the scaffold's load-bearing capacity should be considered and scaffold reinforcement should be added.

4.1.7 Avoid attaching to scaffolds dynamically induced loads such as concrete pumping conduits.

4.1.8 Putlog scaffolds should be avoided. Instead independent scaffolds should be used.

4.2 Wooden pole and bamboo scaffolds

4.2.1 Wooden poles should be of a long fibre straight grained variety.

4.2.2 Material with defects such as knots, splits, rotting, pinholes, boreholes and bents should be avoided.

4.2.3 The minimum diameter of the pole depends on the species. For medium hardwood timbers, the diameter should not be less than 50 millimetres at the tip.

4.2.4 New species of timber used for scaffolds should be tested for their load-bearing strength and resilience before being introduced.

4.2.5 Putlog scaffolds should be avoided. Instead independent scaffolds should be used.

4.3 Steel tubular scaffolds

4.3.1 Steel tubular materials should be rust free, straight and free from defects.

4.3.2 Tubing sizes should not be less than 48 millimetres and a wall thickness of about 2.3 millimetres.

4.3.3 The correct type of fittings or couplings should be used.

4.3.4 Access to the scaffolds should be provided by means of ladders or stairways.

4.3.5 Anchorage points for the scaffold must be adequate for the purpose.

4.4 Prefabricated scaffolds

4.4.1 Scaffold frames should have locking studs to allow for fitting cross bracings and where liable to be dislodged vertically, vertical locking braces.

4.4.2 Frames with serious defects such as bent legs, badly dented pipings, deformed braces, oversized tubing, cracked welding points and expanded tube diameters should not be used.

4.4.3 Besides interframe cross braces, the external faces of the whole scaffold should be cross-braced with steel tubes. Alternatively every four bays should be cross-braced.

4.4.4 Frame scaffold base plates must be used to avoid damage to the base frame and also to prevent sinking into soft ground.

4.5 Suspended scaffolds

4.5.1 Suspended scaffolds should be designed, constructed and installed in such a way that every component from the anchoring points to the platform are safe or of fail-safe design.

4.5.2 Anchorage points utilising expanded bolting or resin embedded studs should be tested before use.

4.5.3 In systems using counterweights, steps should be taken to verify that counterweights are secured and that the countermoment is three times that produced by the greatest load imposed by both the self-load and the live loads.

4.5.4 To ensure safety, four independent wire ropes is capable of taking the full load, should be used.

4.5.5 Suspension ropes should reach the ground level and with about 3 metres length to spare.

4.5.6 Motorised winches should have a means for changing over to the manual mode so as to allow workers to lower themselves down to ground level in case of power failure.

4.5.7 Motorised winches should have the means to test the emergency brake under load.

4.5.8 Regular inspection and testing should be carried by competent persons.

4.6 Inspection and record keeping

4.6.1 A record book should be kept for all inspections and testings carried out on scaffolds. Defects should be clearly listed and a memo should be handed over to the facilities section with a specific deadline for completion.

4.6.2 Defects rectified should be noted in the record book.

4.6.3 Where defects render the scaffold unsafe, they should be rectified immediately. Where this is not practicable, a sign should be put warning against using it.

5. Lifting Equipment

5.1 General

5.1.1 All lifting equipment should be load tested by competent engineers before use and the designated safe working load should be stamped on the equipment.

5.1.2 All lifting machines such as hoists, lifts, cranes, etc. should be provided with safety devices to prevent overloading.

5.1.3 Regular inspection and maintenance should be scheduled.

5.1.4 Standard signals should be implemented so that the operator and the user would be able to synchronise their communication.

5.1.5 Workers should not be allowed under suspended loads and operators should avoid swinging loads overhead the workers.

5.1.6 Where the material to be lifted is liable to disintegrate such as palletised bricks, proper containers or bins with proper lifting points should be used.

5.2 Hoists and lifts

5.2.1 Material hoists should be provided with interlocking systems so that they cannot be moved when the hoist gates are open for unloading or loading.

5.2.2 A safety catch or secondary braking system should be incorporated so that in the event of rope failure or in the case of rack and pinion, should the primary brakes fail, the hoist cage will not drop to the ground.

5.2.3 Hoists and lifts should be incorporated with the following safety devices:

(a) overhoist limit so that the cabin or cage will not go over the top of the tower;

(b) overload warning device or cut-out device;

(c) interlocks at each floor level.

5.2.4 Mast sections must be straight pieces and when installed, the mast must be plumb. Adequate tie-backs must be provided.

5.2.5 Mast footings must be adequately designed and constructed to prevent movement.

5.2.6 Overhead shelters should be provided at the base of hoists and lifts to prevent persons being struck by falling objects.

5.2.7 The base of hoists and lifts must be enclosed to prevent any person being struck by descending cages or cabins.

5.2.8 Wire ropes should be frequently inspected and replaced if excessive number of strands are broken or worn out.

5.2.9 Lifts should be the type using counterweights.

5.2.10 The safe working load and the number of passengers for lifts and for hoists should be clearly painted at approaches to them.

5.3 Cranes

5.3.1 All cranes should have the following safety devices :

(a) overhoist limit;
(b) overluff limits (for luffing cranes only);
(c) limits for trolley (for tower cranes only);
(d) overload warning or cut-off;
(e) automatic safe load indicators or moment limiters are preferred if possible;
(f) angle-radius indicator coupled with various loads at each radius;
(g) load charts.

5.3.2 Cranes should be tested before use by a competent engineer after major repairs, and thereafter at regular intervals not longer than one year.

5.3.3 For tall tower cranes that are subject to excessive wind loads or to buckling danger, they should be adequately braced e.g. to the building it is serving.

5.3.4 Adequate ground support must be maintained before any lifting is done. For mobile cranes, packing to spread load at the footing should be carried out. For tower cranes the pile cluster below the tower must be adequately designed.

5.4 Winches

5.4.1 Adequate foundations should be provided for winches.

5.4.2 Winch operators should be with overhead protection but his field of vision should not be obstructed.

5.4.3 Winches should have a "dead man's" control so that the brake automatically comes on whenever pressure is released from the control lever.

5.5 Lifting gear

5.5.1 All lifting gear and slinging should be tested before use and thereafter inspected regularly by competent engineers. Workers should also check the lifting gear visually before using them.

5.5.2 Each piece of lifting gear should bear its safe working load, its identification number and its last inspection date. It could in addition be colour coded according to due date of inspection.

5.5.3 Wire ropes should be preserved against rusting, kinking, fraying, bird-caging and heat damage. Defective wires should be destroyed to prevent recycling.

5.6 Operational safety

5.6.1 Crane operators should be trained before being engaged.

5.6.2 Site personnel co-ordinating lifting operations should be instructed in safe operating procedures. Dangerous practices such as slewing loads above workers, overloading, lifting without a signaller, unsafe slinging, carrying workers on loads or slings must be prohibited.

5.6.3 Load slinging should only be carried out by persons who have been properly instructed in the slinging methods for different materials.

5.6.4 Mobile cranes should not be operated without fully extending outriggers or where the ground is too soft.

6. Construction Machinery

6.1 General provision

6.1.1 Dangerous moving parts of moving machinery should be effectively guarded or made safe by position.

6.1.2 Cleaning, lubrication and maintenance of machines should only be done after they have come to a complete stop. In order to ensure that machines are not re-started while work is being carried out on them, a lock-out system should be implemented.

6.1.3 Operators should be instructed in the operation of the machine before starting work on it.

6.1.4 Machinery work areas should be fenced off to keep out those whose proximity to the machine might endanger them.

6.1.5 Scheduled inspection and maintenance of all machinery should be carried out.

6.2 Woodworking machines

6.2.1 Safety devices which should be incorporated onto the machine are:

(a) adjustable guard that should cover the cutting tool;

(b) anti-kickback device where the workpiece is liable to be ejected. In the case of circular saws a riving knife would be adequate.

(c) the use of push sticks should be standard practice.

6.2.2 Cleaning of woodwaste around the machines should only be carried out when the machine has stopped and then only with the use of brushes and not bare hands.

6.2.3 Portable woodwork machines should also incorporate guards.

6.3 Concrete mixers

6.3.1 Moving parts which are liable to become nip points, such as gears, chains and rollers should be guarded.

6.3.2 Hoppers into which a person could fall should be guarded.

6.3.3 Fencing should be provided to prevent entry around the zones where hoppers or mixing drums may be tilted or turned.

6.3.4 Operators cleaning concrete mixers should do so outside the rotating drums. Lock-out devices should be provided where workers need to enter.

6.3.5 Where concrete mixers are driven by internal combustion engines, exhaust points should be located away from the workers' workstation so as to eliminate their exposure to obnoxious fumes.

6.4 Concreting pumping system

6.4.1 Concrete pumping workstations should be provided with overhead covers.

6.4.2 Pipings should be secured to prevent movement.

6.4.3 Piplelines should not be attached to temporary structures such as scaffolds and form-work support as the forces and movements may affect their integrity.

6.4.4 Pipe connectors, particularly those installed at heights should be secured against dislodgement, thereby becoming a falling object.

6.5 Earth-moving equipment

6.5.1 Earth-moving vehicles should be provided with standard safety features such as lights, rear mirrors, and for large vehicles, an audible reversing signal.

6.5.2 No earth-moving equipment should be started up until all persons are cleared away.

6.5.3 Supervisors and operators should survey the route of the earth-moving equipment for:

(a) electrical lines which may be an obstruction;
(b) rating of bridges and overhead obstructions;
(c) underground conduits containing service lines;
(d) slope-gradients.

6.5.4 Where the equipment has to slew around as part of the operation, such as a back-hoe excavator, and where there are other workers in its proximity, a signaller should be present to direct the operator.

6.6 Pressure vessels

6.6.1 Steam boilers should incorporate the following safety devices:

(a) safety valves;
(b) gauge glass and water level control;
(c) firing controls linked with pressure and water level sensor;
(d) pressure gauge;
(e) alarm system activated by over pressure, high temperatures, low water levels or air blower malfunction.

6.6.2 Steam boilers should be blown down and cleaned regularly to remove scale and sludge.

6.6.3 Air receivers should be equipped with the following:

(a) safety valve;
(b) pressure gauge;
(c) inspection cover;
(d) drain cocks.

6.6.4 Steam boilers and air receivers should be inspected by competent engineers before being brought into use and thereafter annually or bi-annually. The inspection certificate issued should specify the safety working pressure.

6.6.5 Operators working in close proximity to air compressors should be provided with and be instructed to use noise defenders.

7. Electrical Safety

7.1 General

7.1.1 The electrical system in the worksite must be properly designed and regularly inspected by a competent engineer.

7.1.2 All components and installation practice must be in compliance with the designed standards.

7.1.3 Low voltage systems should be used as far as possible.

7.1.4 All electrical system components should be protected against damage. Electrical lines should be hung overhead and not run along the floor.

7.1.5 Moisture and all solvents must be kept away from wirings and components.

7.1.6 Regular testing of all electrical installations and tools must be carried out by competent electricians to detect defects before use.

7.2 Components

7.2.1 All components and conductors used must be in good condition.

7.2.2 Proper junction boxes and distribution boards from which electric power could be tapped should be provided at every floor level.

7.2.3 Proper and adequate extension cables must be made available to all tradesmen.

7.2.4 Where there is a requirement for special grade components such as that used in explosion proof environments or where they are exposed to the weather, suitable equipment and components must be used.

7.3 Safety devices

7.3.1 Components for each voltage system must be different to eliminate the possibility of wrong connections.

7.3.2 ELCB (Earth Leakage Circuit Breakers) of the appropriate rating should be installed for each electrical outlet to prevent electric shock hazards.

7.3.3 Means such as a pilot lamp should be provided on each distribution board to indicate whether it is energised.

7.3.4 Electric powered tools must be made safe :

(a) by effective earthing or by choosing double insulation tools;

(b) there must be an isolator within 20 metres of each worker using the tool and on every distribution board and junction box;

(c) regular inspection and maintenance of the tools should be carried out by a competent electrician;

(d) a 2.5 kilogrammes BCF fire extinguisher should be provided at every main electric switch-board.

8. Pile Driving

8.1 General provisions

8.1.1 Crawler mounted piling machines should be operated only on firm ground.

8.1.2 The pile leader should be kept plumb but where inclined piling is to be carried out, adequate counter-balance must be provided.

8.1.3 The rungs for accessing the leader should be provided and kept in good repair.

8.1.4 Two pile drivers used in the same site must be separated from each other by the height of the tallest leader.

8.1.5 Inspection of piling frames should be carried out before use and regularly thereafter.

8.1.6 When hoisting a pile into position in the leader, care should be taken to prevent the pile swinging out of control.

8.1.7 The hammer should be lowered to rest on the ground when not in use.

9. Special Building Operations

9.1 Prefabricated building erection

9.1.1 Prefabricated parts must be designed to allow for lifting and transportation.

9.1.2 Supports, struts and braces and anchorage points must be of adequate strength to hold the panels in position as temporary support.

9.1.3 Anchorage points should be pre-cast into the floor slabs for attaching safety railings.

9.1.4 The design of the embedded lifting loops as an integral part of the reinforcement bars should have adequate strength to prevent detachment.

9.1.5 Storage racks should be stable so that panels and other prefabricated parts would not fall over.

9.1.6 During transportation, prefabricated parts should be secured in position against overturning or dislodgement.

9.1.7 When lifting prefabricated parts into position, workers carrying out the emplacing of parts should keep away from the building edge to avoid being thrown out over the side by the swinging load.

9.1.8 Where workers are exposed to falling hazards such as work on the external walls or in the rubbish chutes, secure working platforms must be used.

9.1.9 Cranes should only be used where the ground is made firm e.g. with steel decking.

9.2 Steel structure erection

9.2.1 Structural members on which access may pose difficulties should be equipped with attachments for suspended scaffolds, lifelines, anchorage points, safety nets, ladders or rungs.

9.2.2 Structural members should not be heated or cut without the approval of the engineer in charge.

9.2.3 Wherever practical steel parts should be assembled on the ground level rather than at a height.

9.2.4 When lifting structural members into position, all loose components or waste materials should be removed from it.

9.2.5 Structural members whose load-bearing capacity should be significantly changed when its vertical orientation changes, should be prevented from doing so by bracing, shoring and other means until permanently secured.

9.2.6 Safety nets should be installed where workers are exposed to falling hazards. Hand-rails should be provided where persons are required to traverse beams.

9.3 Work on tall chimneys

9.3.1 Safe access should be provided at every stage of construction.

9.3.2 Scaffolds forming part of the slip-form framework should be closely boarded.

9.3.3 Slip-form tower frames must be adequately designed to take into account loads imposed. A larger safety factor should be used if an elevator is attached on it.

9.3.4 Anchoring points for suspended scaffolds should be embedded in the chimney shell as the chimney is built.

9.3.5 Overhead protection should be provided at the internals and external of the base of chimneys during construction or repair.

9.3.6 Suspended scaffoldings used on the external of the chimney should be anchored against wind forces.

10. Demolition

10.1 Uncontrolled collapse of walls or other structures under demolition should be prevented.

10.1.1 The throwing of materials over the sides of the buildings should not be permitted.

10.1.2 The dust level should be controlled and where dangerous materials such as asbestos are present, special procedures should be followed.

10.1.3 Walls and structures liable to collapse on workers, should be broken up gradually from the top downwards. Scaffolds should be provided to allow for this procedure.

10.2 Preparatory work

The following actions should be carried out before demolition is started.

10.2.1 Disconnect all service lines e.g. gas, electricity and water.

10.2.2 Fence off the building under demolition, providing a safety zone between the building and the fencing as much as possible. Where required, hoardings should be provided.

10.2.3 Remove glass pieces and sharp materials.

10.2.4 Support structures that might collapse dangerously.

10.3 Waste handling

10.3.1 Waste materials should be cleared and removed as demolition proceeds so as to prevent overloading which might lead to a cascade of floor collapse.

10.3.2 Where materials are removed manually, rubbish chutes should be installed.

10.3.3 Where demolition is carried out near public areas:

(a) hoardings sloping inwards should be erected around the building;

(b) protective nettings should be hung around the building to prevent materials falling outside the periphery shelter. Where asbestos materials are present, appropriate dust control and respiratory protection approved by the local authority must be used.

11. Excavation

11.1 General provisions

11.1.1 Prior to commencement of excavation, a survey of the following should be carried out :

(a) ground water and soil condition;

(b) underground service lines, particularly electrical, sewerage, water and gas;

(c) where necessary a detailed soil investigation, where soft soils e.g. marine clay is suspected.

11.1.2 Means for rapid access and egress should be provided.

11.1.3 Workers should not be exposed to dangers of being buried by excavated material or collapse of shoring.

11.1.4 Tests for toxic gases should be carried out where their presence is suspected.

11.1.5 Exposure of shorings to vibration such as that produced by engines or vehicular traffic should be kept to a minimum.

11.1.6 Provisions should be made to ensure that water is removed gradually from the excavation.

11.1.7 Exhaust from internal combustion engines should not be allowed to be discharged near the excavated trench.

11.1.8 Thorough and regular inspections must be carried out by competent engineers.

11.1.9 Warning lights must be provided to prevent persons falling into the trench.

11.2 Shoring design and installation

11.2.1 Shoring design should be carried out by a competent engineer taking into account the condition of the soil.

11.2.2 Installation of the shoring materials should follow the excavation closely and should preferably be carried out in such a way as not to expose the worker to danger in case of earth collapse.

11.2.3 Adequate quantities of suitable shoring material should be made available.

11.2.4 Wherever practicable, trenching boxes should be used.

12. Tunnelling

12.1 General - ventilation, fire protection, lighting

12.1.1 A detailed plan for shuttering, safety measures and evacuation should be drawn up.

12.1.2 Adequate ventilation and where necessary a back-up system should be provided.

12.1.3 An effective communication system between the tunnel-face team and the control centre above ground should be maintained.

12.1.4 Gas checking should be carried out at critical points (such as soil change) and at regular intervals. Continuous monitoring may be required if harmful gases are evolved.

12.1.5 Where flammable gas concentration could reach explosive levels, it may be necessary to provide intrinsically safe electrical equipment.

12.1.6 Adequate lighting and emergency lighting should be provided.

12.1.7 Fire protection equipment should be provided. Carbon dioxide extinguishing agents should not be used, neither should water-based extinguishers be used. First-air fire extinguishers should be BCF (e.g. 2.5 kilogrammes). 10 kilogrammes ABC dry powder extinguishers should be provided for larger fires.

12.1.8 Adequate evacuation stairways should be provided for rapid evacuation in case of an emergency.

12.1.9 Means should be provided to remove any accumulation of underground water.

12.1.10 Lifting equipment used should comply with the safety requirements listed earlier. But where electrical components are exposed to moisture, a waterproof system should be used.

12.1.11 The following should be prohibited in the tunnel :

(a) combustible materials (unless authorised);
(b) accumulation of waste materials;
(c) smoking;
(d) welding unless authorised and precautions taken;
(e) petrol engines.

12.1.12 In mechanical haulage systems, rolling stock or wagons should be coupled in such a way that the wagons will not be disconnected when negotiating bends or slopes.

12.1.13 Adequate clearance between the rolling stock and the tunnel sides should be provided for passageway.

12.1.14 Diesel locomotives should be shut off whenever stationary and standing cars should be checked.

12.2 Supports

12.2.1 Supports to prevent cave-ins should be emplaced as closely as possible to the tunnel face.

12.2.2 Adequate and suitable materials should be provided for supports.

12.2.3 The support system should be designed and inspected by competent engineers.

12.3 Dust control

12.3.1 Dust generated by the tunnelling process should be suppressed and controlled to safe levels.

12.3.2 Dust sampling should be carried out where its presence is recognised. Where siliceous dust is present control measures including the use of appropriate respirators for exposed workers should be instituted

12.3.3 Medical supervision of workers exposed to siliceous dust should be carried out following recommendations of the national authority or labour Department.

12.4 Work in compressed air

12.4.1 Precautionary measures should be taken to ensure that the air lock system will not be accidently decompressed.

12.4.2 Safety devices such as safety valves, pressure regulators, pressure guages, should be provided and maintained.

12.4.3 All workers employed to work in a compressed air atmosphere should first undergo medical examination and clearance and thereafter periodically.

12.4.4 Compression and decompression procedures should be followed strictly and a qualified look attendant should be employed for this purpose.

12.4.5 De-matching exercise should be carried out as no smoking should be allowed in compressed air.

12.4.6 Air intake points for air compressors should be pollution-free.

12.4.7 Bulkheads and airlock systems should be of sufficient strength and be designed by competent engineers.

12.4.8 Control of combustible materials entering the compressed air tunnel should be strictly carried out.

13. Health and Welfare

13.1 Shelter from adverse weather conditions should be provided.

13.2 Toilet facilities should be provided and maintained in a clean condition. Washing facilities with running water should be provided.

13.3 Potable drinking water should be provided.

13.4 Food supplies of adequate quality should be available or provided in canteens on site.

13.5 All rest areas, canteens, toilets and washing facilities should be adequate in number, convenient to the work areas and kept clean, well lit and ventilated. Toilet facilities should be easily accessible, specially to those working on very tall buildings.

13.6 Adequate facilities should be provided for the changing, drying and storing of clothing.

13.7 Waste disposal bins should be provided for the disposal of garbage.

13.8 All workers should be checked by the foreman before being hired to make sure they are physically capable of the work and are free from obvious signs of sickness.

13.9 Workers who have to work at heights should be checked by the foreman before starting work each day. Any worker showing signs of sickness should not be allowed to work at heights.

14. First Aid

14.1 All foremen on site should be trained in first aid and hold the relevant national certificate in first aid. An appropriate number of workers should be trained in first aid.

14.2 First-aid facilities should be available at convenient locations on each site. A first-aid centre should be established on each major site with appropriate equipment and should be clearly identified, kept clean, well painted, and be properly lit and ventilated. The first-aid centre should have clean running water. Adequate medical supplies of dressings, splints, stretchers, and medicines should be supplied.

14.3 Rescue equipment, stretchers and telephone should be close to the first-aid centre.

14.4 Sufficient first-aid boxes containing simple dressings and supplies should be provided on the site under the control of the foreman.

14.5 A procedure should be established and publicised for rescue of injured workers from the workplace, first-aid treatment, and where necessary, despatch off site for medical or hospital treatment. Local doctors and hospitals should be informed of this procedure.

15. Ergonomics and Workplace Climate

15.1 Job and workplace design should be such as to avoid too strenuous or monotonous work. The work area temperature should be kept close to the comfort zone 20° to 25°C by using sun shields or other protections.

15.2 Manual lifting and carrying of heavy loads should be avoided by using mechanical equipment or helpers. Workers should be instructed in the correct lifting and carrying methods and in the use of equipment.

15.3 The design of the cab of working machines must allow the operator to work in a comfortable position. Controls, access means, operator's seat, windows, etc., must be kept in good condition.

15.4 Hand tools should be of good design and material and without defects. They should be used for the purpose of design, well maintained and securely stored.

Source : A booklet entitled " Model Guidelines on Construction Safety" compiled and circulated by the National Institute of Construction Management and Research, Pune.

Appendix E

Workplace Health and Safety Survey

1. Health Survey

1. Name (optional) ______________________________
2. Place of work/department ______________________________
3. Job description/title ______________________________
4. No. of years worked ______________________________
5. How do you feel, generally? ______________________________
6. History of any of the following

Frequent colds	Yes	No
Frequent sneezing	Yes	No
Chronic cough	Yes	No
Frequent headaches	Yes	No
Shortness of breath	Yes	No
Chest pain or pressure	Yes	No
Allergy to drugs	Yes	No
Recent loss of weight	Yes	No
Impaired hearing	Yes	No
Ear discharge	Yes	No
Back ache	Yes	No
Joint pain	Yes	No
Intestinal trouble	Yes	No
Epilepsy	Yes	No
Any other	Yes	No

7. Do you have any problems with your skin (infections, rashes, painful itching, redness, other)?

   Did you have these problems before you begain the job?

   Yes No

   Do these problems disappear on weakends or during vacations?

   Yes No

8. Do you have any problems with your respiratory tract (coughing, running nose, coughing up mucous or blood, dry or sore throat, frequent colds, chest pain)? Did you have these problems before you began the job? Yes No

   When do these complaints occur?

   Morning Afternoon All day Specific days of the week Daily Other No noticeable trend

   Do they disappear on weekends or during vacations?

   Yes No

9. Do you have trouble with your eyes (itching, redness, watering, swelling, pain, vision changes)?

   Did you have these problems before you begain the job?

   Yes No

   When do these complaints occur?

   Morning Afternoon All day Specific days of the week Daily Other No noticeable trend

   Do they disappear on weekends or during vacations?

   Yes No

10. Do you have trouble with your ears (hearing, ringing in the years, ear infections, can't hear after leaving the workplace)?

    Did you have these problems before you begain the job?

    Yes No

    When do these complaints occur?

    Morning Afternoon All day Specific days of the week Daily Other No noticeable trend

    Do they disappear on weekends or during vacations?

    Yes No

11. Do you have allergies? Did you have them before you began the job?

    Yes No

12. Do you regularly feel ill at work?

    Do you have headaches, dizziness, drowsiness, stomach aches, loss of appetite, nausea, vomiting, weakness, irritability, nervousness, rapid heart beat, muscle cramps, back aches, pain or stiffness in arms, legs, joints, swelling of arms, legs, joints, other? (Please circle and explain)

Yes No

13. Please describe any problem or complaints from other people in your area that you may consider important
14. Does the organisation give any regular tests or examinations to any special group of workers in an area?

    Yes No

    If yes what tests? Do you see the results?
15. Do you go for physical examinations regularly (either to your own physician or the organisation's) Yes No

    If so, what tests are done?
16. Do you know of any medical problems that you now have? Have these been confirmed by a doctor? Yes No
17. Have you ever been hospitalised? If yes, please record when and for what reason?
18. Is there anything else you think is important to say about health on your job?

2. Safety Survey

Indicate "Yes" or "No" to the following questions.

Emergency exits

1. Are adequate safe exits and entrances provided?
2. Are all emergency exits free from obstacles?

Passageways

3. Are they clear of obstacles?

Fire prevention

4. Are no smoking areas clearly marked?
5. Are fire doors marked?
6. Are fire alarms adequate?
7. Are workers clear what to do in the event of a fire?
8. Do they know the evacuation drill?
9. Are the drills practised regularly?
10. Are fire extinguishers or similar measures provided?
11. When were they last checked?

Housekeeping

12. Are the walls and ceiling in good condition?
13. Are floors in good condition?
14. Are they cleaned regularly?
15. Are all moving parts of a machinery securely fenced?
16. Are all fixed guards in good condition and securely fastened?
17. Are all automatic guards properly adjusted?
18. Are any safety switches in working order?
19. Are they checked regularly?
20. Are all emergency stop buttons labelled and working?
21. Is all electrical equipment regularly checked?

Storage and handling

22. Are storage facilities adequate?
23. Is all racking and shelving in a safe condition?
24. Have workers been trained in the handling of materials?
25. Is there adequate equipment for handling materials?
26. Have safe procedures for the storage of materials been established?

Dangerous substances

27. Are there any dangerous substances used?
28. Have hazard data sheets been provided?
29. Are workers trained in handling and using them?
30. Is adequate information displayed on their presence?
31. Are containers holding dangerous substances clearly marked?
32. Are dangerous substances safely stored?
33. Are there adequate extraction facilities for fumes or dust?

Noise

34. Are noise risks assessed and danger areas identified?
35. Is there a programme of noise reduction/control?

Protective equipment

36. Is protective clothing necessary?
37. Is it suitable?
38. Is ear or eye protection necessary?
39. Is it suitable?
40. Are protective gloves necessary?
41. Are they suitable?
42. Are there any other hazards?

Environmental conditions

43. Is the lighting adequate?
44. Is the heating adequate?
45. Is the ventilation adequate?
46. Is the furniture (desks, chairs, etc.) suitable?
47. Does anything else make the workplace uncomfortable?

Training

48. Have all workers been trained in the safety aspects of their job?
49. Have workers facing particular hazards received specialised training?
50. Is there any first-aid training?

Welfare

51. Is there overcrowding?
52. Are the washing and toilet facilities adequate?
53. Are they kept clean?
54. Are there changing facilities and a rest room?
55. Is drinking water provided?
56. Are there adequate first-aid facilities

3. Occupational Hazards Survey

Rating

No problem at all (1)
A slight problem (3)
A sizable problem (5)
A great problem (7)

1. Air pollution from dust, smoke, gas, fumes, and fibres	1	3	5	7
2. Danger from fire or shock	1	3	5	7
3. Dangerous chemicals	1	3	5	7
4. Defective tools, machinery or equipment	1	3	5	7
5. Extremes of temperature or humidity indoors	1	3	5	7
6. Dirty or badly maintained areas at the workplace	1	3	5	7
7. Inadequate toilet facilities	1	3	5	7
8. Inadequate eating facilities	1	3	5	7
9. Inadequate lighting	1	3	5	7
10. Noise	1	3	5	7
11. Dangerous work methods or procedures	1	3	5	7

Appendix F

Cases of Industrial Accidents

Fire in a solvent extraction plant resulting in the death of 51 persons and injuries to 17 others.

In April 1966 in Maharashtra there was a big fire in the solvent extraction section of an oil mill. On the day of accident the plant was in operation and the process of extraction of oil from groundnut cake with the help of (N-Hexane) solvent was carried out. At about 6 a.m. there was a leakage of solvent vapours from the evaporation vessel of the plant. The operators on the plant tried to plug the leakage by putting gunny bags on the outlet but could not do so. The vapours of N-Hexane which are highly inflammable in nature spread all round the plant for about half an hour. At about 70 feet from the plant on one side, there was barbed wire fencing and immediate beyond the fencing, the open area was occupied by hutment dwellers. The vapours spread beyond fencing to the hutment.

One of the operators on the plant who was not properly trained, in order to stop leakage, in panic, instead of opening the valve leading hexane vapours to condenser, he closed it. This increased the pressure of vapours of the solvents in evaporator and vapours started gushing out vigourously through the leakage hole and engulfed the whole area with vapours of solvents. The vapours being heavier than air travelled about 100 feet from the plant in all directions, beyond the barbed wire fencing and settled on the ground.

Early morning somebody from the hut lighted the stove. The solvent vapour caught fire, and flashed back and the whole area in a radius of 100 feet from the plant was engulfed in fire. This resulted in the death of 19 persons on the plant and factory premises and 32 persons in the hutments and injuring 17 others. On enquiring it was found that leakage was through a hole in the body of the evaporator vessel which was blanketed by ordinary galvanised iron sheet found corroded with holes in it on examination. On the recommendations of the enquiry commission appointed by the State Government, the State Government later on made specific rules for approval of site, of solvent extraction plants (rule 3 of MFR 1963) and also declared it as dangerous operation under rule 114 read with Section 87 of Factories Act and put in under Schedule XXI making special safety precautions. This has resulted in no major accident in solvent extraction plants thereafter. Similar provision is made under Section 41 - Constitution of Site Appraisal Committees relating to hazardous process by amendment of 1987 in Factories Act.

- There was a collapse of the factory building of Shantinath Textile Mills in Surat in which about 90 workers died.

  The mill was working on the day of accident (in the year 1980). The mill is an old textile mill and the management constructed an overhead water tank on the top of the building. As soon as the tank was filled with water the old structure below could not sustain its weight and the building collapsed like a house of cards killing 90 workers who were working in the mill.

- There was collapse of the factory building of the textile process house in Taloja Industrial Estate near Mumbai in the year 1986.

  It was a new unit hardly 4 years old. The factory was working at the time. Due to the collapse, 17 workers died and 3 or 4 were injured. In the low lying area near the factory building, there was lot of water logging which might have resulted in sinking of the foundation on one side resulting in collapse of the building. The enquiry commission held the builder responsible for the collapse of the building.

- In an old textile mill in Mumbai there was a structural collapse of a portion of the building in the early hours of the morning.

  As the factory was not working at the time of collapse there were no casualties.

  On the top of the collapsed portion there was an overhead water tank. One of the guy ropes of the chimney of the factory was tied with the water tank structure. The wind pressure on the chimney was transmitted to the tank structure through the guy rope connected to it. That might have made the tank structure weak and it ultimately resulted in the collapse of the tank and the building structure below it.

  These mishaps could have been averted had certificate of stability been obtained from competent person as per Rule 3-A of MER 196 periodically and after addition, alteration or replacement of machinery, plant etc.

- There was a fatal accident in an engineering factory. The accident caused the death of a supervisor who was instructing the operation of dressing of weld seam by means of flexible shaft grinder using $^1/_8$" thick abrasive cut off wheel, which burst due to excessive speed. Enquiries revealed that dressing of weld seam at the depth of 132 mm. below the edge of the heat exchanger expansion joint was to be carried out for which a 16" diameter nylon reinforced abrasive wheel was used. After completing $^1/_3$ job, the supervisor was trying to turn the vessel which was placed on the mobile 4 wheeled fixture. This was done before stopping the machine. The wheel burst and one piece hit the supervisor in upper forehead causing death before any treatment could be given to him.
- A fatal accident occurred in a Paper and Board Mill

  In a paper and board mill, two workers died when they entered inside the pulper to clean the pulp waste, dirt etc., at the beginning of the first shift. Enquiries revealed that dead bodies of these two workers were taken out at about 9.45 a.m. While they had entered the pulper at about 8.30 a.m.
- A fatal accident took place in sugar factory

  In a cooperative sugar mill, a six year old son of a contractor's worker was drowned in the condensate tank.

  An additional turbine was being installed in the factory and excavation work for the foundation of the turbine was in progress. This work was being carried on by a contractor, who had employed female workers for the purpose. One of the female workers had brought her 6 year old son with her and left him on the R.C.C. slab cover of the condensate tank which was about 50 feet away from the worksite. The cover had three manholes, one of which was not provided with any cover at the time of accident and the boy fell through the manhole into the tank which was 10 feet in depth and had 4 feet of water at that time and was drowned.
- A fatal accident took place in a chemical factory due to explosion.

  There was an explosion in a process kettle of a petrochemical factory manufacturing liquid paraffin. As a result of the explosion two contractor's workers died and two got injured.

  In a liquid paraffin plant a work of fixing up of heat exchanger coils on a process kettle was in progress. To fix up the heat exchanger coils a structure was built on the processing kettle. The kettle in which the explosion occurred contained stripped oil and buffer solution emptied out few hours before the explosion took place. Buffer layer contains 5% isopropyl alcohol (Fl.Pt. 53°F). The buffer solution containing isopropyl alcohol and stripped oil was removed from the tank on the day of accident while the safety permit was issued to the contractor on the previous day. The vessel was not purged with steam or inert gas. As per the safety permit the contractor's workers started the welding unaware of the presence of alcohol vapours. There was a violent explosion in the vessel as it was containing alcohol vapours. As a result of the explosion one worker was thrown in air and fell on the roof of the godown at 50 feet distance and died on the spot. Another worker received burn injuries and died in the hospital.

- A serious accident due to fire in Carbon Dry Colour Works in Mumbai, in the year 1960-61 in Dongri area of Mumbai.

  A devastating fire took place in a unit under Factories Act, manufacturing dry colours. The factory was employing about 20 workers when the mishap took place. The premises was 30 feet in width in front and about 70 feet in length. The front had two fully openable doors and the door on the back side lending to godown of factory and outside from the back side was kept closed and locked.

  At the time of accident mineral turpentine was being heated in a vessel on a stove which was near the front door. As the mineral turpentine frothed and overflowed it came in contact with the open flame of the stove and there was a big fire. The fire spread so quick that both the front doors got blocked by the fire. As the back door was locked there was no second exit to go out and workers had to run through fire to the front door. 7 workers escaped with severe burn injuries and 13 workers who ran on back side could not escape and died.

- A serious accident due to leakage of Chlorine took place in a plant manufacturing Chlorine affecting 16 persons. The Chlorine from the cell-house is passed through drier where H_2SO_4 is sprayed. It is further filtered through glass wool filter and compressed. After compression it is passed through equipment named "Coke Filter" where traces of Sulphuric Acid are removed by passing chlorine through Coke Bed. In the Coke filter the pressure of Chlorine is about 3 kg/cm^2 and the Sulphuric acid which is removed is discharged periodically through a nipple joint provided at the bottom of the filter. This nipple joint is normally blinded up with dummy flange and fitted with a gasket. On enquiring it was found that Chlorine gas leaked through the flange of the Coke filter when the gasket of flange fitting failed. 9 persons were given first aid and 7 persons were treated at the hospital and were discharged.

- A fatal accident occurred in a co-operative sugar factory.

The deceased was engaged with two other workers in replacing the vapour pipe bend. The derric pole was fixed and the three workers were arranging to fix the pulley at roof level. They took rope and wire and came on the platform of hot and cold water channel which was at about 70 feet height from floor level. This platform level and the level of the A.C. sheet roof of the factory was parallel. One of the three workers came on A.C. sheet roof from the platform and went towards the centre of roof and sat there. The deceased tied the rope to the wire end and has given wire to the other co-worker and told him to wait on the platform. Then the deceased took the other rope end and came on the A.C. sheet roof and began walking towards the other worker sitting on the roof. He walked about 6 feet from the rear end of the A.C. sheet roof which was at the height of 70 feet from floor level. Suddenly his legs were slipped off and he fell on the roof and slided instantly from the roof and fell on the hot water channel which was empty.

The deceased received multiple injuries and his ears were bleeding. He was removed to civil hospital but unfortunately died.

- An emission of the SO_3 fumes affected about 445 dwellers residing about 500 feet from the chemical factory.

  The factory manufactures Zinc Hydrosulphite with the Oleum, Spent Acid, Sulphur and Zinc Dust. The Oleum about 3000 kg. was stored in a tank which is installed in an open shed 25 ft. × 25 ft. and about 11 feet in height with parapet around.

  As it rained on previous night water collected in the shed around the storage tank and the discharge valve of storage tank which was about 3 feet from floor level and partially immersed in collected water was leaking. SO_3 from the tank which came in contact with water formed dil. H_2SO_4 which reacted with the valve and the valve gave way. The SO_3 fumes gushed out in the form of a cloud at about 12:20 p.m. and were carried by the wind in the direction of dwelling houses which were more than 500 feet away. The factory workers were not affected as the wind direction carried the cloud away from the factory. 362 dwellers were treated on the spot and 83 had to be admitted to hospital from where they were discharged in a day or two. Nobody died in the accident.

- An explosion took place in a dyestuff factory due to bursting of pipeline. An explosion occurred in a dyestuff factory in one of their plants. A P.V.C. pipeline through which a mixture of sulphuric acid and disulphonic acid was being transferred, bursted and the acid splashed on HDPE bags containing sodium chlorate bags which had been brought earlier from an explosive godown and stored nearby.

This led to a violent explosion which blew off the roof of the plant, damaged the pipelines and one of the vessels. None was injured in the explosion.

- An explosion took place in a chemical factory resulting in the loss of life of two workers of the factory.

  On that fateful day, preparation of chromic acid solution was going on the second floor of the factory. The chemist of the factory left the place of work to see the general manager, leaving the operation of mixing to be done by a semi-skilled worker. Within five minutes the explosion occurred followed by small fire. Two workers received serious burn injuries and died in the hospital. Due to the impact of the explosion glass panes of the windows and the partition in the room were shattered into pieces.

- Three fatal accidents took place in a chemical factory due to the explosion of Ethylene Oxide Cylinder (200 kg.) in a chemical factory:

  The factory was manufacturing dimethyl ethanol amine by reacting dimenthyl amine with ethylene oxide. In the manufacturing process dimethyl amine was taken into the reactor and ethylene oxide was charged in and reacted with it at 35°C. and 15 PSI. The reaction requires 16-18 hours. When there was 10-20 kg. of ethylene oxide left in the cylinder, it was noticed that the cylinder was getting hot. It was decided to isolate the ethylene oxide cylinder and take out of the work room. While doing so the ethylene oxide cylinder exploded. As a result of the explosion, three persons died and twenty sheds in neighbourhood were affected.

- ***Bhopal Gas Tragedy:*** At midnight of December 2-3, 1984, destiny planted an industrial disaster. Around 40,000 kgs. of Methyl Isocynate gas leaked out of the factory of Union Carbide, a multinational company. The killer gas took 2,500 lives and affected about 25,000 persons. The mishap left everybody shocked and grief stricken for all time to come. The disaster shocked the entire world.

- Turkish worst mining accident happened in which 263 workers were killed in a gas explosion in Zonguldak (1992).

- Liquid ammonia bursts out from a high-pressure vaporising unit at National Fertilisers Limited, Panipat in India killing 11 people (1992).

- About 12 people were killed as a fire breaks out at Indian Oil Company's refinery near Jaipur in India after a leakage in a pipeline.

- About 203 miners died after an explosion deep in a coal shaft in South Western China (2005).

- About 90 workers were killed after a Methane blast rips through tunnels deep below ground in a coal mine in Ukraine's Donetsk (2007).
- About 53 people were killed in South Western Pakistan after a gas blast deep in a coal mine near Quetta (2011).
- About 60 people died after a landslide at a goldmine at Congo (2012).
- About 83 workers were buried by a massive landslide at a Gold mining site in Tibet (2013).
- An underground explosion and fire killed about 245 workers in one of Turkey's worst mining disaster. The deaths were caused by carbon monoxide poison (2014).
- About 29 people were killed in a boiler tube blast at the state-owned power firm National Thermal Power Corporation plant in Raebareliin India. Many workers were injured as a result of the boiler explosion (2017).
- Three people have been killed and fortynine injured (eight critically) in the explosion at the Reliance Industries owned Indian Petrochemical Corporation Limited (IPCL) plant at Nagothane in Raigad district, on June 6, 2008. Two blasts took place at polymer section of the plant where maintenance work was being carried out.

Appendix G

Safety Checklist and Safety Programme Rating Form

Safety Checklist

1. Are the premises generally uncluttered and free from obstructions?
2. Are there any structural defects in floors, stairways, walls and roofs?
3. Are floors and stairs uniform and slip-resistant?
4. Is the working space adequate for safe operation?
5. Are the benches, furniture and fittings in good condition?

Storage facilities

6. Are storage facilities, shelves, etc., arranged so that stores are secure against sliding, collapse, falls or spillage?
7. Are storage facilities kept free from accumulations of rubbish, unwanted materials and objects that present hazards from tripping, fire, explosion and harbouring of pests?

Sanitation and staff facilities

8. Are the premises maintained in a clean, orderly and hygienic condition?
9. Is drinking water available?
10. Are clean and adequate toilet (WC) and washing facilities provided for both male and female staff?
11. Are hot and cold water, soap and towels provided?
12. Is there a staff room for lunch, etc.?

Heating and ventilation

13. Is there a comfortable working temperature?

14. Is the ventilation adequate, e.g., at least six changes of air per hour, especially in rooms that have mechanical ventilation?

Lighting

15. Is the general illumination adequate (e.g., 300-400 lux)?
16. Are there dark or ill-lit corners in rooms and corridors?

Security

17. Is the whole building securely locked when unoccupied?
18. Are doors and windows vandal-proof?

Fire prevention

19. Is there a fire alarm system?
20. Are all exits unobstructed and unlocked when the building is occupied?
21. Is the fire detection system in good working order and regularly tested?
22. Are the fire doors in good order?
23. Do all exits lead to an open space?
24. Are all exits marked by proper, illuminated signs?
25. Is all fire-fighting equipment and apparatus easily identified by an appropriate colour code?
26. Are fire alarm stations accessible?
27. Are "No smoking" signs posted in areas where smoking is prohibited?

Flammable liquid storage

28. Is the storage facility for bulk flammable liquids separated from the main building?
29. Are flammable liquids stored in proper, ventilated containers that are made of non-combustible materials?
30. Are appropriate fire extinguishers placed outside but near to the flammable liquid store?
31. Are "No smoking" signs clearly displayed inside and outside the flammable liquid store?

Electrical hazards

32. Are all new electrical installations and all replacements, modifications or repairs made and maintained in accordance with an electrical safety code?
33. Does the interior wiring have a grounded (earthed) conductor (i.e., a three-wire system)?

Compressed and liquified gases

34. Is each portable gas container legibly marked with its contents and correctly colour-coded?
35. Are compressed gas cylinders and their high pressure and reduction valves regularly inspected for safety?

Personal protection

36. Is protective clothing for e.g., gowns, coveralls, aprons, gloves, of an approved design provided for all staff?
37. Is additional protective clothing provided for work with hazardous chemicals and radioactive substances e.g., rubber aprons and gloves for chemicals and for dealing with spillages?
38. Are safety spectacles, goggles and visors provided?
39. Is radiation protection in accordance with national and international standards, including provision of dosimeters?
40. Are respirators available, regularly cleaned, disinfected, inspected and stored in a clean and sanitary condition?

Health and safety of staff

41. Is there an occupational health service?
42. Are first-aid boxes provided at strategic places?
43. Are qualified first-aiders available?
44. Are notices prominently posted giving succinct information about the location of first-aiders, telephone numbers of emergency services, etc.
45. Are proper records maintained of illnesses and accidents?
46. Are warning and accident prevention signs used to minimise work hazards?

Chemicals and radioactive substances

47. Are all chemicals correctly labelled with names and warnings?
48. Are chemical hazard warning charts prominently displayed?
49. Are spillage clearance kits provided?
50. Are proper records maintained of stocks and use of radioactive substances?.

***Source*:** Laboratory Biosafety Manual, Second Edition, World Health Organisation, Geneva, 1993.

Safety Programme Rating Form

Poor = 0	Fair = 8
Good = 14	Excellent = 20 A.

A. Organisation and Administration Points

1. Statement of policy and assigned responsibilities ______
2. New and transferred employee selection, testing and placement ______
3. Direct management involvement and support ______
4. Emergency plans and security ______
5. Safety rules and standards ______
6. Safety organisation

   Total value

B. Hazard Control

7. Housekeeping ______
8. Guarding ______
9. Physical work area

   Protection and plant layout ______
10. Materials handling ______
11. Personal protective equipment ______
12. Fire Protection ______
13. Environmental health ______
14. General chemical hazards ______
15. Hazard identification and analysis ______
16. Safe work permits and confined space entry ______
17. Equipment safety devices ______
18. Maintenance ______
19. Solid waste, air and water pollution and spill control ______
20. Design safety Total value

C. Training and Motivation

21. Indoctrination of new and transferred employees ______
22. Employee training ______
23. Supervisor safety training ______
24. Safe operating procedures ______

25. Internal self-inspection or auditing ______
26. Safety meetings ______
27. Employee/supervisor safety contact and communication ______
28. Safety suggestions ______
29. Safety recognition and promotion

    Total value

D. Accident Investigation and Cause Analysis

30. Accident investigation by supervisor ______
31. Accident cause analysis ______
32. Claim investigation and follow-up ______
33. Reporting and record keeping

    Total value

E. Off-the-Job Safety

34. Organisation and administration ______
35. Investigation, reporting, and cause analysis ______

    Total value ______

Overall Programme Rating

Below 40	-	Poor or ineffective programme
40 - 70	-	Fair programme with inconsistent results
70 - 90	-	Good programme, results show improving safety performance
90 - 100	-	Outstanding safety programme with excellent results

Appendix H

Environmental Legislation at a Glance

Table 1 The Water (Prevention and Control of Pollution) Act 1974 and Rules.

An Act to provide for the prevention and control of water pollution and maintaining or restoring wholesomeness of water.

(i) Pollution Control Board (PCB) has the right:

- to obtain any information regarding the construction, installation or operation of an industrial establishment or treatment and disposal system
- to take samples of trade effluent for the purpose of analysis in the prescribed manner
- to enter and inspect any industrial establishment, record, register, document or any other material object
- to prohibit use of stream or sewer or land for disposal of polluting matter, not in accordance with the standards laid down by the PCB.

(ii) Restriction on establishment and the operation of an industry, process or any treatment and disposal system without prior consent of the PCB.

(iii) PCB's right to refuse or withdraw consent for discharge of effluents.

(iv) Industry to comply with the conditions stipulated in the consent.

(v) PCB's to grant consent within four months after the date of receipt of the application complete in all respects.

(vi) Industry to appeal to the Appellate Authority, in case of grievances against the order passed by the PCB regarding grant, refusal or withdrawal of the consent within the specified time in the prescribed manner.

(vii) Industry to furnish information to the PCB and other specified agencies in case of discharge of poisonous, noxious or polluting matter into a stream, sewer or land, occurred or likely to occur resulting in pollution due to an accident or any other unforcing event.

(viii) PCB's right to issue orders restraining or prohibiting an industry from discharging any poisonous, noxious or polluting matter in case of emergencies, warranting immediate action.

(ix) PCB's power to make an application to the court for restraining apprehended pollution of water due to likely 1disposal of polluting matter in a stream or on land.

(x) Bar of jurisdiction to civil court in respect of any matter under purview of the Appellate Authority constituted under the Act and no grant of injunction in respect of any action taken or proposed in pursuance of the Act.

(xi) Bar on filing of any suit or legal proceedings against the Government or Board Officials, for action taken in good faith in pursuance of the Act.

(xii) PCB's to maintain a consent register containing particulars of the consent issued and to provide access to industry at all reasonable hours.

Table 2 The Water (Prevention and Control of Pollution) Cess Act 1977 and Rules 1978.

An Act to provide for the levy and collection of a cess on water consumed by persons carrying on certain industries and by local authorities to augment resources for the Pollution Control Boards

(i) Only specified industry sectors to pay cess on the quantity of water consumed for specific purposes at prescribed rates.

(ii) Any specified industry liable to pay water cess:

- to affix meters of prescribed standards and at prescribed places by PCB's for measurement of quantity of water consumed
- to furnish water cess returns in the prescribed form at prescribed intervals

* to pay interest for delay in payment of cess not made within the specified time, as mentioned in the assessment order of the PCB
* to pay penalty not exceeding the amount of cess in arrears, but non payment of cess within the specified time as mentioned in the assessment order of the PCB.

(iii) Specified industries entitled to 25% rebate in water cess, if they comply with the prescribed consent provisions and consume a quantity of water which is not in excess of the prescribed quantity.

(iv) PCB's right to make enquiries for assessing water cess payable by any specified industry.

(v) PCB's right to recover any amount due under the Act as arrears of land revenue from industry.

(vi) PCB's right of entry and inspection to carry out provisions of the Act including the testing of the correctness of the meters affixed.

(vii) Industry to appeal to appellate authority in case of any grievance against the water cess assessment, within the specified time, in the prescribed manner.

Table 3 The Air (Prevention and Control of Pollution) Act 1981 andRules 1982 and 1983.

An Act to provide for the prevention, control and abatement of air pollution

(i) State Governments powers include:
  * to declare any area within the state as an Air Pollution Control Area
  * to prohibit use of any fuel or burning of any material which may cause air pollution in the Air Pollution Control Area.

(ii) Restriction on establishment and operation of any industrial plant in an air pollution control area likely to emit air pollutant(s) into the atmosphere without the prior consent of the Pollution Control Board.

(iii) PCB's to make inquiries in respect of grant of consent in prescribed manner.

(iv) PCB's to grant consent within four months after the date of the receipt of an application complete in all respects.

(v) Industry to comply with the conditions stipulated in the consent.

(vi) Restriction on emission of air pollutants in excess of the standard prescribed by PCB's.

(vii) PCB's right to make an application to the court for restraining an industrial plant, located in an air pollution control area likely to emit air pollutants in excess of the prescribed standards.

(viii) Industry to furnish information to the PCB's and other agencies in case of emission of air pollutants in excess of prescribed standards occurred or likely to occur, resulting in air pollution due to an accident or unforcing act or event.

(ix) PCB's rights include:
- * to enter and inspect any industrial plant, records, registers, or documents at all reasonable times
- * to obtain any information related with the implementation of the provisions of the Act
- * to take samples of air and emissions for analysis in the prescribed manner.

(x) Industry to appeal to the appellate authority in case of grievances against the order made by PCB's under the Act within a specified time and in the prescribed manner.

(xi) PCB's power to issue directions for:
- * the closure, prohibition or regulation of any industry, operation or process; or
- * the stoppage or regulation of supply of electricity, water or any other service to an industry in an prescribed manner.

(xii) Industry to comply with the directions of the PCB.

(xiii) Bar of jurisdiction to civil court in respect of any matter under purview of the appellate authority constituted under the Act and no grant of injunction in respect of any action taken or proposed in pursuance of the Act.

(xiv) Bar on filing of any suit or legal proceedings against the Government or board officials for action taken in good faith in pursuance of the Act.

(xv) PCB's to maintain consent register containing particulars of consent issued and to provide access to industry at all reasonable hours.

Table 4 The Environment (Protection) Act 1986 and Rules 1986.

An Act to provide for the protection and improvement of the environment.

(i) Central Government's power to take necessary measures for the purpose of protecting and improving the quality of the environment and prevention control and abatement of environmental pollution.

(ii) Central Government's powers include:

* lay down standards for the quality of the environment emission or discharge of environmental pollutants from various sources
* restrict or prohibit industries operations or processes in specified areas
* restrict or prohibit handling of hazardous substances in specified areas
* lay down procedures and safeguards for the prevention of accidents, which may cause environmental pollution
* enter and inspect any industrial establishment, records and documents to ensure effective implementation of the provisions of the Act.

(iii) Central Government has powers to issue directions for:
* the closure and prohibition or regulation of an industry operations or processes; or
* stopping or regulating the supply of electricity, water or any other service in the prescribed manner.

(iv) Industry to comply with such directions.

(v) Restrictions on discharge or emission of pollutants in excess of the prescribed standards.

(vi) Handling of hazardous substances in accordance with the prescribed procedures and safeguards.

(vii) Industry to furnish information to specified agencies in case of discharges, emission of pollutants, in excess of the prescribed standards, already occurred or likely to occur, resulting in environmental pollution, due to an accident or an unforeseen act or event.

(viii) Central Government has the power to recover expenses incurred by it on remedial measures to prevent or initiate environmental pollution from the defaulting industry, as arrears of land revenue or of public demand.

(ix) Central Government has the power to take samples of air, water, soil or other substances from any industrial plant for the purpose of analysis in the prescribed manner.

(x) Bar on filing of any suit or legal proceedings against the Government or officials empowered by it for action taken in good faith, in pursuance of the Act.

(xi) Bar of jurisdiction to civil court to entertain any suit or procedures in respect of anything done, action taken or directions issued by the Central Government or any other authority empowered by it in pursuance of the Act.

(xii) Industry, operations or processes requiring consent under the Water Act or Air Act or authorisation under the Hazardous Waste (Management and Handling) Rules, or both, to submit 'Environmental Statement' every year before September 30th for last financial year.

Table 5 The Hazardous Wastes (Management and Handling) Rules 1989.

(i) Occupier's responsibility to ensure proper handling and disposal of hazardous wastes, either by themselves, or through the operation of hazardous waste management facility.

(ii) Restrictions on handling of hazardous wastes without prior authorisation from the PCB.

(iii) PCB has the power to suspend or cancel an authorisation for handling hazardous wastes, after providing an opportunity to showcause and recording the reasons thereof.

(iv) PCB has the power to refuse grant of authorisation after providing reasonable opportunity of hearing to the occupier.

(v) Packaging, labelling and transportation of hazardous wastes to be done in the specified manner.

(vi) State Government to identify sites for disposal of hazardous wastes within the states, and publish inventory containing relevant information.

(vii) Occupier generating hazardous wastes, in the operating or handling the facility, to maintain records of such operations in the prescribed manner.

(viii) Occupier generating hazardous wastes or the operator handling the facility, to submit annual return in the prescribed firm.

(ix) Occupier or the operator handling facilities to report to the PCB in the prescribed forms, in case of accident occurred at the hazardous waste handling site or during transportation.

(x) Specified procedures to be followed for import of hazardous wastes to be used for processing or reuse as raw material.

(xi) Any person importing hazardous wastes to maintain record of the imports in the prescribed form for inspection purposes by regulatory agencies.

(xii) An occupier's right to appeal to the appellate authority in the prescribed manner in case of grievance(s) against any order of suspension, cancellation or refusal of authorisation by PCB.

Table 6 Manufacture, Storage and Import of Hazardous Chemicals Rules 1989.

(i) Occupier who controls an industrial activity involving hazardous chemicals is required to identify major accident hazards, and take adequate step to prevent and limit such major accidents, by providing on site information, training and equipment necessary to ensure the safety of workers.

(ii) Restrictions on industrial activity involving hazardous chemicals unless the occupier submits information in the prescribed form.

(iii) Occupier to send further information within the specified time mentioned in the notice, if so desired by the concerned authorities regarding the safety report.

(iv) Occupier to prepare and keep up-to-date on-site emergency plans, for preparation during major accidents, before commencing industrial activity, involving hazardous chemicals.

(v) The concerned authority to prepare and keep up-to-date off- site emergency plans for preparedness during major accidents, before commencing of the industrial activity involving hazardous chemicals.

(vi) Occupier to provide relevant information to the persons liable to be affected by the accident.

(vii) Occupier to develop information in the form of safety data sheets.

(viii) Occupier to label specified information on every container of hazardous chemicals.

(ix) Occupier to follow specified procedures for importing hazardous chemicals.

(x) The concerned authority to serve 'improvement notice' to the occupier, in case of contravention of the provisions of these rules, for remedying action.

Table 7 Rules for the Manufacture, Use, Import, Export & Storage of Hazardous Micro-Organism, Genetically Engineered Organisms or Cells 1989.

(i) Import, export, transport manufacture, process, use or selling of any hazardous micro-organisms or genetically engineered organisms/ substances is prohibited without prior approval of the Genetic Engineering Approval Committee (GEAC).

(ii) Deliberate or unintentional release of above organisms is not allowed.

(iii) Grant of approval shall be done by the State Biotechnology Coordination Committee (SBEC) or District Level Committee (DLC).

(iv) Provide immediate information to DLC/SBEC and the State Medical Officer in case of any accident or interruption of operations.

Table 8 The Public Liability Insurance Act and Rules 1991.

An Act to provide for liability insurance for the purpose of for providing immediate relief to the persons affected by accidents occurring while handling hazardous substances.

(i) Owner to give relief in case of death (other than workman) or damage to property, from an accident occurring while handling specified hazardous substances.

(ii) Owner to obtain insurance policies of amounts more than paid up capital of the undertaking but less than Rs. 50 crores, before starting the handling of specified hazardous substances.

(iii) Owner to pay, additionally to the insurer prescribed amounts, not exceeding the amount of premium for being credited to the Environmental Relief Fund to be established under the Act.

(iv) Verification, publication of accidents and awards of relief to be done by the District Collector.

(v) The Central Government has the right to obtain any information from the owner's, related to the requirements of the Act.

(vi) The Central Government has the right to entry and inspection at all reasonable times any premises where hazardous substances are handled.

(vii) The Central Government is empowered to search and seize if it has reason to believe that handling of hazardous substances is in contravention of the provisions of the Act.

(viii) The Central Government is empowered to issue directions in the prescribed manner for (a) prohibition or regulation of the handling of any hazardous substances; or (b) stoppage or regulation of the supply of electricity, water or any other service.

(ix) Industry to comply with such directions within a specified time.

(x) The Central Government is empowered to make an application in the court for restraining an owner from handling hazardous substances, if it has reason to believe that provisions of the Act are being contravened.

(xi) The Central Government is empowered to recover all expenses incurred by it in implementing the court's directions from the owner as arrears of land revenue or of public demand.

(xii) Bar on filing any suit or legal proceedings against the Government or empowered authorities for the action taken in good faith in pursuance of the Act.

Appendix I

Glossary

Accident: an unplanned and unexpected event or condition that causes or is likely to cause an injury, damage to property and environment.

Accident causes: a series of causes that leads to carelessness and bad habits; breaking the rules; not knowing the rules; using faulty equipment; not thinking safety all the time; and wrong perception that it happens to others but not to self.

Accident investigation: a method to discover hazardous conditions and practices so that further accidents from similar causes be prevented.

Accident proneness: an individual characteristic that leads to a greater number of accidents than would occur by chance.

Accident sequence: a sequence that may be stated as (a) a personal injury occurs only as the result of an accident; (b) an accident occurs only as a result of an unsafe action or exposure to an unsafe mechanical or physical condition, or both; and (c) unsafe actions or unsafe mechanical or physical conditions exist only because of human failure which are inherited or acquired.

Accident type: a manner of contact of the injured person with the object or substance, or the exposure or the movement of the injured person which resulted in the injury.

Agency: an object or substance that is most closely associated with the injury and which, in general, could have properly guarded or connected e.g. machine and prime movers.

Air borne contaminants: a group of contaminants those developed during the manufacturing process and are of six types – dust, fumes, mists, smoke, gases and vapours.

Air monitoring: a programme of location, type, duration and number of samples to be collected and steps to be followed to calculate the average weighted daily exposure; the samples are to be collected from the breathing zone and in the general room air.

Alarm system: an audible and visible indication to be installed at the control room, and at strategic locations of buildings and plants.

Alcoholism: a chronic and progressive illness manifested in repeated and uncontrolled drinking of alcoholic beverages in excess of dietary and social uses.

Apparent ignition temperature: a type of temperature required to begin or cause burning at a rate which is sufficiently rapid so that the burning will continue when the heat source is removed.

Assumption of risk: a defence which assumes that employer was not liable for injury caused by an unsafe work condition when the employee knew the facts and understood the risks of the employment.

Automatic sprinkler: a device that uses water to extinguish or control a fire before it becomes uncontrollable; it automatically operates and discharges water to control the fire and thus works both as a fire extinguisher and a fire alarm system.

Automation: an automatic system to move materials and parts between operations and also inspect products by automatically using control devices.

Behaviour-based safety (BBS): a set of techniques focusing on the behaviour of employees to prevent accidents and injuries.

Biological hazards: a hazard arising out of insects, moulds, fungi and bacterial contaminants.

Chemical hazards: a set of health hazards including fire and explosion hazards caused by different types of chemicals, either by inhalation, skin absorption or ingestion of chemicals in the form of toxic dust, fume, gas vapour, smoke, mist, aerosol and liquid.

Combustion: a chemical change accompanied by the evolution of heat and light.

Communicating safety: a means to develop safety consciousness, provide information on safety rules and procedures through posters, slogans, leaflets, and the like.

Construction: a reference to the building and civil engineering activities that includes work above ground, work in open excavation, underground work and underwater work.

Contributory negligence: a doctrine under which the employer may raise the defence that the accident occurred purely due to employee's negligence on his own part.

Damage: a consequence of an accident that occurs to a person (injury) or matter (property loss) in a complex system of work area.

Disease: a damage to the body which shows up only after a period of time from first exposure to a hazard.

Effluent: a type of material flowing out of a treatment unit or plant.

Electrical hazards: a number of occurrences arising out of portable electrical hand tools, static electricity, inadvertent activation of electrical equipment, electrical shock to persons, fire and explosion of power cables, and so on.

Emergency planning: an important formulation of information system to develop and communicate plans containing details of identification, description, and response mechanism to emergency situations.

Emergency preparedness: a programme of safeguarding the life, plant, and people around the vicinity of the plant; in order to minimise the loss of life and property in case of an emergent situation.

Environment: a sum total of all the external conditions and influences that affect the survival of organisations on earth.

Environmental agents: a group of chemical, physical, biological, mechanical/ergonomic or psychological factors present in the work environment.

Environmental factors: a series of factors giving rise to stresses that can cause sickness, impaired health or significant discomfort in workers and can be classified as chemical, physical, biological, or related to ergonomics.

Environmental monitoring: an evaluation of the industrial environment undertaken broadly (a) to determine levels of exposure among workmen to various atmospheric contaminants; (b) to examine the effectiveness of control measures; (c) to investigate complaints; and (d) to fulfil research purposes.

Epidemiological approach: a statement that injuries and damages are the measurable indices of the accident; but the accident itself is an unexpected, unavoidable, and unintentional act.

Ergonomics: a discipline which aims at the application of human biological sciences in conjunction with engineering sciences in order to achieve the optimum mutual adjustment of man and his work, the benefits being measured in terms of human efficiency and well-being.

Error agency: a factor facilitating an error which include human behaviour, environment, equipment, design and the work system.

Explosive range: a maximum and minimum concentration, expressed as percentage by volume of a flammable vapour or dust in air, which exists within definite upper and lower limits in order that the vapour or dust may burn when exposed to a source of ignition.

Exposure: a quantity of environmental agents impinging on a person.

Fault analysis: a technique which wishes to ensure that faults are reduced to a minimum and therefore wish to train operators to recognise, correct and prevent faults.

Fault-tree analysis: a system of making a detailed analysis of failure or potential failure characteristics in a system.

Faulty behaviour: an act of doing the wrong thing; failing to do the right thing; overdoing the right thing; not doing enough of the right thing; and adhering to wrong timing.

Fellow servant rule: a defence which states that an employer was not liable to pay an employee who got injured through the negligence of a fellow employee.

Ferrall theory: a theory that states that accidents are the result of a causal chain (as in the multiple causation theory), one or more of the causes being human error.

Fire classification: a grouping of A, B, C, D which provide different extinguishing agents for different types of materials involved.

Fire extinguishers: a number of extinguishers of different types such as water type, foam type, dry powder type, and carbon dioxide type, and are designed keeping in view the three classes of fire.

Fire prevention: a requirement to ensure that proper installation, operation procedures, structures, and safe and easy egress in the event of fire are designed and maintained as a fire safety measure.

Fire protection: a technique that necessitates the development and the use of design and methods for detecting and controlling fires so as to limit the probable damage from fire, if one does start.

Flash fire: an occurrence when there is a very rapid combustion.

Flash point: a lowest temperature of a liquid at which it will vaporise sufficiently to form a mixture with air capable of burning when ignited.

Flow chart: a usual method for analysing a plant's layout and operations indicating the flow of material as it passes from one stage to the next.

Frequency and severity ratios: a known standard measure by which a company appraises its work safety effort by expressing the incidence of major injuries and the severity of each in relation to the employee hours of work exposure during the period that is measured.

Harmful agents: a group of chemical substances in the form (direct contact with the skin), or ingestion (eating or drinking).

Hazard: an energy or environmental agent which could produce injury or disease.

Hazard analysis: a means whereby hazards or potential hazards are identified, evaluated and managed in order to eliminate or reduce to an acceptable level of risk of an incident occurrence.

Hazard and risk: a hazard is a danger; a risk is the chance of that danger turning into an accident or injury.

Hazard measurement/testing: a process of maintaining work environment regarding physical hazards (noise, heat, stress, illumination), chemical hazards (dust, fumes, gases, toxic chemicals), biological hazards, and physiological and psychological hazards.

Hazardous material: any substance which has been detrimental and be capable of imposing an unreasonable risk to a person or property.

Hazob: a hazob is a qualitative technique based on guide-words and is carried out by a multi-disciplinary team during a set of meetings.

Health: a state of complete physical, mental, and social well-being and not merely the absence of disease or infirmity.

Health and safety programme: an analysis of health hazards and factors that cause them; that are to be planned, organised, administered, stimulated and evaluated.

Healthcare: a health system that customarily described as consisting of a three-tier pattern, namely(a) primary, (b) secondary, and (c) tertiary.

Heat balance: an occurrence in human body mainly from two sources namely (a) metabolic, and (b) external or environmental heat.

Housekeeping: an important factor in accident prevention; it is more than cleanliness; it is cleanliness and order; and adherence to the principle of proper place for everything and everything in its proper place.

Incidence rate: an accepted means by which a company can appraise its occupational safety and health record and set goals for achievement.

Incident: an undesired event which under slightly different circumstances could have resulted in harm to people, damage to property or loss to process.

Industrial hygiene: a science and art primarily concerned with the appraisal and control of occupational health hazards that arise from various industrial processes or operations.

Influent: a type of material flowing into a treatment unit or plant.

Ingestion: a cause of serious health problem if people unknowingly eat, drink or inhale harmful chemicals and toxic dusts in the workplace.

Inhalation: a set of airborne contaminants that can be inhaled directly into the lungs and can be physically classified as gases, vapours and particulate matter including dust, fumes, smoke, aerosols and mists.

Injury: an immediate damage to the body caused by exposure to a hazard.

Injury report: a record of information which provides: (a) the time of injury; (b) the specific place where the injury occurred; (c) the name and address of the injured person; (d) nature of injury; (e) location/place of injury; (f) the severity of injury; and (g) name(s) of witness(s), if any.

Job safety analysis: a process of selecting the job; breaking the job into successive steps; identifying the hazards in each step; and developing the safety precautions to be adopted and ways to eliminate the hazards.

Lighting: a factor in accident prevention and assisting persons most effectively if it is adequate for the job requirements and properly installed.

Loss control: a system firmly based on the economic argument that accidents lead to a wastage of company assets and that an effective programme of loss control can lead to a reduction or elimination of the wastage.

Loss control activity: a programme which involves and coordinates the activities of management, supervisors and operatives in the prevention of all aspects of accidental loss, including those accidents that result in injuries to people.

Machine guarding: a device to prevent a worker from coming in contact with the dangerous and moving parts of the machines due to any mechanical or human failure.

Major accident: an occurrence resulting from uncontrolled development which causes a serious damage to persons whether immediate or delayed.

Major hazard: an industrial activity using, producing or storing dangerous substances in such a quantity that they possess the potential to cause extensive damage and to kill or injure persons within or outside the site or the boundary.

Manual lifting technique: a technique that consists of six steps from the view point of safety, namely: (a) correct position of feet; (b) straight back; (c) arms close to the body; (d) correct hold; (e) chin in; and (f) use of body weight.

Material handling: a systematic procedure which involves certain considerations like (a) the properties of the materials involved; (b) location from which the material is to be moved; (c) the quantities of material to be handled to maintain its constant flow; (d) the quantities of material to be stored to assure an adequate reserve supply; and (e) environmental factors that may influence the storage.

Multiple causation theory: a theory advocating that many contributing factors, causes and sub-causes combine together in a random fashion causing accidents; and the identification and removal of all undesirable conditions that would lead to prevention and re-occurrence of all similar accidents.

Near miss: an accident that does not produce an injury or disease.

Neutrons: a group of uncharged particles with a wide range of energy and power of penetration; they cause remote and severe damage to the tissue.

Noise exposure: an exposure that leads to various types of effects such as masking, annoyance, speech interference, hearing loss and even work efficiency.

Noise pollution: a wrong sound in the wrong place and at the wrong time.

Occupational disease: a type of disease arising out of and in the course of employment; and occur as a result of exposure to physical, chemical, biological and psychological factors at the workplace.

Occupational health: a concern with the effects of health on work and the effect of work on health.

Perception: an ability of the mind to transform sensory stimuli into conscious thoughts or sensations through five sense organs.

Personal protective equipment: a type of equipment that are used to avoid injuries to head, eyes, lungs, hands, legs, and other parts of the body.

Peterson model: a model that suggests that there are three broad categories of human error**:** overload, traps, and decision to err.

Physical hazards: a hazard that includes excessive levels of electromagnetic and ionising radiations, noise, vibration and extremes of temperature and pressure.

Plant illumination: an illumination customarily consists of four types namely, (a) general lighting; (b) localised general lighting; (c) supplementary lighting; and (d) emergency lighting.

Pollutant: a form of energy or matter that causes degradation and pollution in the existing natural balance of the ecosystem.

Preventive maintenance: a system mainly involving regular systematic investigation; a record of the inspection and findings; repair of equipment or parts as needed; and replacement before failure.

Product liability: a legal responsibility of a manufacturer or seller of a product to compensate a consumer who has been harmed by the product.

Programme on safety training: a safety training programme (a) for new employees; (b) when new plant and equipment or processes are introduced; (c) when safety procedures have been evolved on updated; (d) when employee safety performance need to be improved; (e) when there is increase in accidents and injuries year to year or as compared to other similar organisations.

Quality in safety: a new systems approach, integrating safety with quality as an organisation improvement tool deviating from the traditional safety management.

Radiation: can affect people according to whether it comes from outside or from inside the body; in industrial situation external radiation is common.

Risk: a chance of a hazard causing injury or a disease.

Risk assessment: a judgement as to the likelihood of an agent producing harm to persons under the circumstances of its use.

Risk avoidance: a strategy of conscious decision on the part of the company to completely avoid a particular risk by discontinuing the operation that is producing the risk.

Risk management: a minimisation of the adverse effects of pure – rather than speculative – risks within a business by involving the processes of identification, evaluation and economic control of the pure risks within a business.

Risk manager: a person with the responsibilities for designing, purchasing and administering the company's insurance policies as a measure of risk control.

Safety and health policy: a well-laid out statement setting down in clear and unambiguous terms the approach of the management to safety, its commitment to safety and health, and the responsibilities being clearly defined at all levels.

Safety audit: a process of auditing the management policy on safety training, PPE, plant operations, emergency plans, accident records, and so on.

Safety committee: a most frequently recognised means of employee participation in safety management.

Safety communication: a follow-up through (a) personal safety contacts; (b) safety talks; (c) safety postures and leaflets; and (d) employee participation.

Safety management: a system that drives safety planning process and purpose into action through long and short term objectives.

Safety organisation: a process of systematic safety programmes and activities which for their effectiveness must: (a) get top management approval and support; (b) be given important consideration in elimination of mechanical and personal hazard; (c) develop educational programmes for employees in safety and health and secure their active cooperation in accident prevention; and (d) be included in all phases of planning, purchasing, supervision and operation.

Safety perception: an internationalisation and integration of self and organisational safety beliefs, values and behavioural regulations.

Safety planning: a tool for implementing and improving safety by setting goals, monitoring progress, formulating decisions and listening to people at all levels.

Safety programming: a process of carrying out certain activities logically such as (a) securing management involvement; (b) organising for achievement of the goals; (c) laying down the operating plans; (d) inspecting operations; (e) considering engineering revisions;(f) using suitable guards and protective devices; and (g) providing education and training.

Safety rules: a set of specific rules made in order to protect employees from possible hazardous working conditions; for effectiveness the rules should be in a written form; should be practical; should reflect an unsafe act or condition; should be limited to safety matters only; should assist in implementing safety rules; and should be adopted and strictly enforced.

Safety sampling: a method of systematically observing workers in order to determine what unsafe acts are being committed and how often they are occurring.

Safety science: a mixture of arts, natural science, engineering, humanities, and social sciences.

Safety standards: a reference to those that are measurable, achievable, and realistic.

Safety suggestion scheme: a common participative management approach intended to provide some means of satisfaction to the employees.

Safety survey: a process of detailed study of plant operations, management policies and programmes, and so on.

Safety training: a detailed extension of the educational safety programme applied to specific occupations, processes, jobs, or activities.

Safe work: a means of preventing personal injuries, deaths, and all kinds of unwanted events through cost reduction.

Safe worker: a person who: (a) uses proper personal protective equipment; (b) understands the job and hazards; (c) obeys safety rules and regulations; (d) follows safe working practices; and (e) uses proper tools in proper way.

SCRAPE: a systematic method of measuring accident prevention effort actually by deciding what we want supervisors to do and then measuring to see that they do it.

Standardisation: a process of implementing and developing technical standards based on the consensus of different parties that include firms, users and interest groups.

Spontaneous combustion: an occurrence as a result of gradual development of heat generation by chemical changes.

Strategic safety planning: a course of action to achieve the desired objectives on safety and encompasses the operational details to translate the safety policies and goals into effective practice.

Systems analysis: an orderly examination of a system or subsystem designed to analyse systematically operations, processes, and procedures.

Systems safety: a complete orientation towards analysis and system improvement.

Systems safety analysis: an emerging concept and activity engaging contractors, governmental agencies, and a generalist safety specialist, as well as specialists in the application of systems analysis techniques.

Occupational Safety and Health Act (OSHA), 1970: an enactment intended to assure, in so far as it is possible that every employee in the United States has safe working conditions.

THERP: a means of calculating the probability of human operating errors causing other unwanted errors in manufacturing.

Total safety culture: a management system for implementing (a) awareness training on health, safety and quality; (b) involvement of the entire workforce in pursuit of the organisation's safety objectives; and (c) empowering the employees to develop a safety philosophy in the organisation.

Total safety management: a dynamic process involving all levels in an organisation to promote never-ending improvement in safety and health and to motivate the entire workforce in pursuit of organisation's safety objectives with the primary aim of better quality of working life.

Toxicology: a science that deals with the poisonous or toxic properties of substances.

Union: an organisation of workers acting collectively and seeking to protect and promote their mutual interest through collective bargaining.

Unsafe acts/practices: a series of unsafe acts and practices that include: (a) operating or working at unsafe speed; (b) taking unsafe position or posture; (c) making safety devices inoperative; (d) not using personal protective equipment; (e) using unsafe tools and equipment; and (f) adopting unsafe loading, placing, mixing, and combing in manufacturing process.

Unsafe conditions: a series of accidents that include: (a) bad housekeeping; (b) unguarded machines; (c) unsafe equipment; (d) hazardous arrangement/process; and (e) unsafe methods and procedures.

Unsafe personal act: a series of hehavioural acts that lead to injuries.

Unsafe personal factor: a type of human behavior that include mental or bodily characteristics such as unsafe attitude, lack of knowledge or skill, bodily defects, and mental state responsible for the performance of the unsafe act.

Unsafe physical conditions: a number of factors that lead to accidents such as unhealthy working conditions, errors in environment design, faulty planning, or omission of essential safety requirements for maintaining a relatively hazard free physical environment.

Vibration: a cause often associated with noise; its excessive exposure can adversely affect employee comfort, efficiency, safety, health and well-being.

Worker: a person employed, directly or through any agency, whether for remuneration or not, in any kind of work connected with the manufacturing process.

Workplace: a location where a worker performs certain trades for a relatively long period of time.

Workstation: a workplace or series of workplaces that may be occupied or used by the same workers sequentially when performing their jobs.

Appendix J

Software on Industrial Safety

Computerization and development of software have changed every activity around us, thus changing the face of every field of interest. The computerization has affected almost all subjects, including accounts, engineering, science and technology, medicine, etc. Industrial safety being purely an interdisciplinary subject, these changes brought in by computerization in relevant fields have changed the outlook of industrial safety apart from the changes that are taking place in industrial safety itself due to its computerization.

Areas of Industrial Safety

Different areas of industrial safety which are being influenced by computerization can be identified as follows:

- Accident data analysis
- Mechanical safety
- Electrical safety
- Structural safety
- Construction safety
- Chemical safety
- Process safety
- Risk analysis
- Consequence modeling
- Ergonomics
- Health surveillance and monitoring
- Toxicology
- Reliability study

Computerization in Industrial Safety

The computerization has led to the documentation and analysis of data in a systematic manner. The complex data analysis is simplified by the availability of computation tools. This is widely felt in the safety areas like accident data analysis, mechanical safety, electrical safety, structural safety, and so on. By making the computation easy, the development of simple programmes towards addressing the safety issues within the industry, has paved way for better understanding and appreciation of industrial safety.

The safety information system (SIS) also provides a base in sharing and making available safety related information by computerization, within the industry as well as towards the society.

Even though, there is an exhaustive software in the field of industrial safety, the most commonly used are related to chemical safety which requires lot of computations and such computations can never be done manually. Some of the widely used software in the field are listed as follows:

1. **ALOHA:** ALOHA (Area Locations of Hazardous Atmospheres) is a computer programme designed by U.S. Environmental Protection Agency for use by the people responding to chemical releases, as well as emergency planning and training. ALOHA models key hazards – toxicity, inflammability, thermal radiation and explosion blast force. It is a free download software available at www.epa.gov.us.
2. **CAMEO:** CAMEO is a software designed by U.S. Environmental Protection Agency to plan for and respond to chemical emergencies. It is also a free download software available at www. epa.gov.us.
3. **MARPLOT:** MARPLOT (Mapping Application for Response, Planning, and Local Operational Tasks) is a general purpose mapping application software developed by the U.S. Environmental Protection Agency with the following features:
   - Easy-to-use GIS interface.
   - Ability to add objects (such as schools or chemical facilities) to this map and mark them using MARPLOT's set of symbols or an inserted picture.
   - Allows to customize the maps by specifying which layer appears and whether objects in those layers (such as roads) are labeled.
   - Easily displays ALOHA threat zones.
4. **SAFETI:** SAFETI is a software developed by DNV to model key hazards like flammability, explosion, and toxicity.

5. **Fire and Explosion Index (F&EI):** The Dow Fire and Explosion Index calculation is a tool to determine the areas of greatest loss potential in a particular process. It also enables to predict the physical damage that would occur in the event of an accident. Software is available to calculate Dow's Fire and Explosion Index by feeding the required data.
6. **HAZOP:** HAZOP is a systematic technique to identify potential hazard and operable problems. It arrives at the deviations from the intended operating conditions that can lead to hazardous situations or operability problems. There are software available in market to conduct the HAZOP study of complex plants in a simplified manner.

The above are list of some of the software available in the field. But the list is exhaustive and it would be difficult to cover all these software in this write-up and such coverage is also outside the objective of the text. The fact is that the use of software has made the calculations very accurate so as to decide on strategies to prevent and control accidents apart from managing them as a last resort.

Bibliography

Anton, T.J., *Occupational Safety and Health Management*, McGraw-Hill Book Co., Singapore, 1989.

Blake, R.P., *Industrial Safety*, Prentice-Hall, Inc., Englewood, Cliffs, N.J., Maryland, 1963.

Chacko, G.K., *Systems Approach to Environmental Pollution* (ed.), Operations Research Society of America, Arlington, 1972.

Chissick, S.S., Dorricot, R., *Occupational Health and Safety Management*, John Wiley & Sons, New York, 1981.

DeReamer, R., *Modern Safety and Health Technology*, John Wiley & Sons, Inc., New York, 1980.

Desai, A.A., *Environmental Jurisprudence*, Vikas Publishing House Pvt. Ltd., New Delhi, 1998.

Garg, M.R., Bansal, V.K., Tiwana, N.S., *Environmental Pollution and Protection*, Deep and Deep Publications, New Delhi, 1995.

Grimaldi, J.V., Simonds, R.H., *Safety Management*, A.I.T.B.S. Publishers and Distributors, Delhi, 1996.

Heinrich, H.W., *Industrial Accident Prevention - A Scientific Approach*, McGraw-Hill Book Co., New York, 1959.

International Labour Office, *Accident Prevention, A Workers' Education Manual*, 1983.

International Labour Office, *Encyclopaedia of Occupational Health and Safety*, Vol. 1 and 2, Geneva.

International Labour Office, *Major Hazard Control - A Practical Manual*, Geneva, 1993.

International Labour Office, *Your Health and Safety at Work: A Collection of Modules, Instructors Guide to the Modules*, ILO, 1996.

Jain, A.K., *Descriptive Law on Pollution and Environment*, Akalank Publications, Delhi, 1998.

Jeyaratnam, J., Koh, D., *Occupational Medicine Practice*, World Scientific Publishing Co. Pvt. Ltd., London, 1996.

Krishnan, N.V., *An Introduction to Safety Engineering and Management*, OPS Publishers Pvt. Ltd., Calcutta, 1983.

Krishnan, N.V., *Safety Management in Industry*, Jaico Publishing House, Bombay, 1993.

Mansdorj, S.Z., *Industrial Safety*, Prentice Hall, New Jersey.

Mistry, K.V., Fundamentals of Industrial Safety and Health, Siddharth Prakasham, Ahmadabad, 2012.

Mohan I., *Environmental Pollution and Management*, Ashish Publishing House, New Delhi, 1989.

National Safety Council, *Accident Prevention Manual for Business and Industry*, USA, 1992.

National Safety Council Inc., *Accident Prevention Manual for Industrial Operations*, USA, 1951.

Nicholson, A.S., Ridd, J.E., *Health, Safety and Ergonomics*, Butterworths, London, 1988.

Petersen, D., *Analysing Safety Performance*, Garland STPM Press, New York, 1980.

Petersen, D., *Techniques of Safety Management*, McGraw-Hill Book Co. Ltd., New York, 1978.

Phelps, G.R., Safety for Managers, Gower Publishing Ltd., Hampshire, 1999.

Ridley, John, *Safety at Work*, Butterworths and Company Publishers Ltd., London, 1983.

Roche, J.M., *Safety and the Foreman*, National Foremen's Institute, Inc., Chicago, 1951.

Schilling, RSF., *Occupational Health Practice* (ed.), Butterworth, London, 1973.

Stranks, J., *Health and Safety at Work* (6th ed.), Kogan Page Ltd., London, 2001.

Stranks, J., *Health and Safety in Practice*, Pitman Publishing, London, 1994.

Stranks, J., *The Handbook of Health and Safety Practice* (5th ed.), Pearson Education Ltd., London, 2000.

Tiwari Committee Report, Department of Science and Technology, Government of India, 1981.

Trade and Technical Press Ltd., *Handbook of Industrial Safety and Health.*

Vaid, K.N., *Construction Safety Management* (ed.), NICMAR, Mumbai, 1988.

Waldron, H.A., *Occupational Health Practice*, Butterwort-Heinemann Ltd., London, 1989.

Index

Name

A

Adam 301

Adarkar, B.P 257

Agricola 184

Anderson, W 341

B

Bird and Loftus 41

Blake, R.P 108

Blum 305

C

Chernobyl 189

Cooper 28

D

Deathrage 215

Dhanwantri 301

Domino 28

Duffy, K.G 128

DuPont 39, 93

G

Gokhale, V.P 340

H

Harrell 305, 395

Heinrich, H.W 28, 34, 35, 38, 67

Hippocrates 184

Holum, J.R 303

K

Kirpartrick 114

N

Nightingale, F 125

P

Parkes 299

Pathak, R.S 342

Pliny 184

R

Ramazzini, B 72, 184

Robens, Lord 74

S

Selye, H 279

Singer, M.G 293

T

Tiffin, J 305

U

Ulrich, E 184

W

Waldheim, K 301

Subject

A

Accident, occupational 22

Advisor, safety 97

AFCONS 220, 221

AIDS 123

Analysis
- cost benefit 37
- event tree 117
- fault tree 116, 117, 190
- gross hazard 116
- job safety 190

Approach
- behavioural 30
- engineering 29
- ergonomics 52
- total safety 30

B

Bomb threat 337

Building and Construction Workers' Act 245

C

Causes
- behavioural 26
- environmental 26

Cell, disaster management 333

Code, Hammurabi 71

Commitment, safety 85

Committee, safety 10, 93, 267

Committee, Tiwari 323

Conditions, unsafe 5

Control, hazard 27, 40

Course, safety 111

Courses, induction 110

Critical, incidents 286

Culture, safety 5

D

Department, safety 100, 267

DGFASLI 77, 95

Director, safety 97, 98

Doctor, industry 124

Dust 31, 42, 43, 108, 151, 155, 193, 233, 311

E

Education, safety 11, 110, 269

Electrical plants 246

Elements, safety 82

Energy transfer 118

Enginear, safety 215

F

Factories Act 225, 227, 228

Failure, design 7

Fire
- causes 151
- checklist 151
- fighting 154, 155
- precautions 153, 154
- prevention 157
- risks 152
- safety 214

First aid 34, 58, 91, 111, 112, 124, 143, 158, 166, 170, 174, 175, 216, 235

Fumes 42

G

Gas tragedy, Bhopal 316, 320, 323, 340, 342

Guarding, machine 44

H

Haulage 233

Hauliers 248

Hazards, health, occupational

- biological 187
- chemical 8, 9, 186
- electrical 8, 9, 187
- ergonomic 188, 189
- fire, explosion 9
- mechanical 9, 186
- miscellaneous, hazards 8
- nuclear 189
- physical 9, 187
- psychosocial 188
- radiation 8, 189

Health, mental 125, 126

Heat 42

Housekeeping 9, 32, 45, 105, 152, 159, 174, 195, 207, 270

Human error 118

Hygiene 183, 184

Hygiene, industrial 161, 166, 184

Hygienist, industrial 166, 185, 186

I

ICMA 248

ILO 22, 32, 91, 162, 164, 165, 167, 169, 176, 199, 200, 238, 240

Inspection, safety 107, 268

Investigation, accident 49-51

J

Job safety 47, 48

L

Labour institutes, regional 78

Legislation

- environmental 316
- labour 173
- safety 209

Lighting 41

List, plant inspection 108

Loss control 128, 129

M

Management, process safety 11, 12

MARG 337, 338

Medicine, industrial

Mines Act 73, 184, 235

Model, behavioural 139

N

National safety council 79

Noise 43, 44, 188, 270, 304-306

Nurse, occupational 125

O

Occupational diseases 192

- anthrax 192
- asbestosis 193, 194, 200
- bagassosis 194
- byssinosis 194
- dermatitis 21, 192
- lead poisoning 194
- pneumoconiosis 193
- silicosis 193, 200

Officer, safety 95, 96, 242
Off-site 330
On-site 330
Organisation, safety 138, 139
OSH 177
OSHA 75-77, 184, 185

P

Physician, occupational 165, 166, 167
Plan, communication 141, 142
Plantations Labour Act 242
Policy, health 87
Policy, safety 218, 219, 267
PPE 53, 55, 56, 100, 108, 110, 121, 145, 165, 173, 187, 189, 195, 207, 241, 272
Preparedness, emergency 275
Principles, accident prevention 10, 38
Principle, SREDIM 47
Principles, safety 8
Process work 287
Professional, safety 4, 106
Programme, wellness 122
Proneness, accident 31
Protection
- eye and face
- hand
- head
- leg and foot
- lung
- skin

R

Rehabilitation 290, 291
Research, behavioural 140
Royal Commission on Labour 257
Rules
- health 109
- safety 110

S

Safeguarding, machinery 8
Safety approach, total 30
Safety
- activities 100
- audit 13, 14, 143
- design 214
- evaluation 119
- philosophy 17
- policy 13, 84-89, 218, 219, 267
- rating system 14
Safety, organisation 267
Smoking 152, 156, 296
Stressors, occupational 280
Stressors, social 281
Stress, sources 281, 285

T

Tata Steel 312
Techniques, critical incident 190
Theory
- accident prone 31
- biorhythm 30
- domino 28
- multiple causation 28

TQM 131

Trade unions 37, 67, 70, 138, 171, 174, 212, 268

Training, safety 110-112, 114, 142, 269

Tremcards 247

U

Union Carbide 338-342

W

WHO 15, 70, 121, 125, 167, 195

Work, safe system 39

Y

Yoga 292

www.ingramcontent.com/pod-product-compliance
Lightning Source LLC
Chambersburg PA
CBHW062059160726
48282CB00026B/600
* 9 7 8 9 3 9 5 0 3 8 7 9 9 *